THE RHYTHMS OF LIFE

THE
RHYTHMS
OF LIFE

Consultant Editors
Professor Edward S. Ayensu and
Dr Philip Whitfield

ME
MARSHALL EDITIONS
LONDON

Consultant editors : **Professor Edward S. Ayensu**
Director of the Department of
Biological Conservation
Smithsonian Institution, Washington DC

Dr Philip Whitfield Lecturer in Zoology
King's College, University of London

Text : **Dr Philip Whitfield**

Professor Paul Bohannan
Department of Anthropology
University of California, Santa Barbara

Dr John Brady Reader in Zoology
Imperial College of Science and Technology
University of London

Kendrick Frazier
Science journalist and
editor of *The Skeptical Inquirer*

Dr Martin Hetzel Consultant physician

Chris Morgan Writer and specialist in time,
science fiction and the future

Dr D.M. Stoddart Lecturer in Zoology
King's College, University of London

Dr Bryan Turner Lecturer in Zoology
King's College, University of London

Illustrations by : **Michael Woods**

Eugene Fleury
George Glaze
Tony Graham
Tom McArthur
Richard Orr
Jim Robbins

Editor: **Jinny Johnson**
Text editor: **Ruth Binney**
Art editor: **Mel Peterson**
Picture editor: **Zilda Tandy**
Assistant editors: **Rosanne Hooper**
                   Pip Morgan
Design assistant: **Linda Abraham**
Proof reader: **Gwen Rigby**
Production: **Hugh Stancliffe**

First published 1982

Marshall Editions Limited,
71 Eccleston Square,
London SW1V 1PJ

Printed and bound in The Netherlands by Smeets Offset B.V., Weert

© 1981 Marshall Editions Limited

ISBN 0 9507901 0 9

Contents

Introduction	6–7

Familiar rhythms — 8–9
The vital rhythms — 10–11
Rhythms of babyhood — 12–13

Cosmic rhythms — 14–15
The dawn of rhythms — 16–17
Energy from the sun — 18–19
The spinning Earth — 20–21
The orbiting Earth — 22–23
Winds and weather — 24–25
Rhythms of the Ice Ages — 26–27
Cycles of cold — 28–29
The pull of the moon — 30–31
Phases of the moon — 32–33

Rhythms of the seasons — 34–35
Temperate seasons — 36–37
Seasons of the tropics — 38–39
The changing hours of daylight — 40–41
The growing season — 42–43
A time to flower — 44–45
A time to reap — 46–47
Preparing for winter — 48–49
The sleeping season — 50–51

Rhythms within rhythms — 52–53
The vocabulary of rhythms — 54–55
Daily rhythms — 56–57
Tidal rhythms — 58–59
Yearly rhythms — 60–61
The biological clocks — 62–63

Rhythms of sex — 64–65
Rhythms of conception — 66–67
The annual display — 68–69
The breeding season — 70–71
Responses to weather — 72–73
Rhythms of courtship — 74–75
Rhythms of migration — 76–77
The long-distance travellers — 78–79
Spawning with the tides — 80–81

Population cycles — 82–83
Cycles of life — 84–85
Alternating generations — 86–87
Communities: the numbers game — 88–89
The rise and fall of populations — 90–91
Cycles of disease — 92–93

Rhythms of growth — 94–95
Cells: multiplication by division — 96–97
Patterns of growth — 98–99
The carbon cycle — 100–101

Rhythms of energy — 102–103
Fuelling body rhythms — 104–105
Rhythms of breathing — 106–107
The pulse of life — 108–109
Rhythms of eating — 110–111
A time for action — 112–113
Cycles of sleep — 114–115

Rhythms of motion — 116–117
The art of flying — 118–119
Hopping and jumping — 120–121
The stepping sequence — 122–123
On two legs — 124–125
The wave makers — 126–127
Rhythms through water — 128–129

Rhythms of health and disease — 130–131
The body clocks — 132–133
'Owls', 'larks' and jet lag — 134–135
Hormonal rhythms — 136–137
Rhythms and asthma — 138–139
Treatment and transplants — 140–141

Rhythms of fate — 142–143
Biorhythms — 144–145
Waves of body and mind — 146–147
Astrological cycles — 148–149
Sunspot cycles — 150–151

A sense of rhythm — 152–153
Social rhythms — 154–155
Contagious rhythms — 156–157
Learning the beat — 158–159
Music and movement — 160–161
Rhythms of worship — 162–163
Rhythms of work — 164–165
The family beat — 166–167
Communal rhythms — 168–169

Rhythms of time — 170–171
Concepts of time — 172–173
The first clocks — 174–175
Charting the years — 176–177
Clocking the minutes — 178–179
Ticking away the seconds — 180–181
The race against time — 182–183

Rhythm data — 184–193
Index — 194–198
Acknowledgements — 199

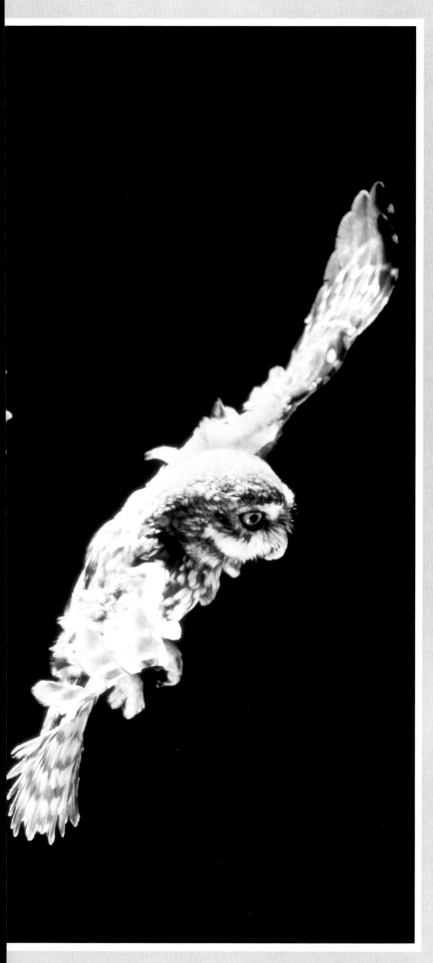

Introduction

While our ever-changing world appears more chaotic by the minute, the human mind yearns for the beauty of order and pattern. We turn up the hi-fi and the measured beats of music briefly smother the confusion. We stride out for a half hour's jogging, making paces to pre-empt the pace-maker. We re-form our family lives and links by seeking calm on an annual vacation.

What we probably fail to appreciate is that all these soothing influences are intrinsically rhythmic. Music has an obvious beat and rhythm, jogging is the rhythmic repetition of a cycle of movements and the calming beat of the ocean waves on the shore is a multi-dimensional rhythm in water.

Once you are aware of rhythms you find that they are everywhere, around you and also within. The earthly world the human race inhabits is orchestrated to create a rhythmical environment—albeit a complex one. Rhythms of day and night, the repetitive march of the seasons, the daily and monthly patterning of the ocean's tides and even the incredibly slow historical cycles of the earth's climate weave the rhythmical fabric of our surroundings.

Life takes on its own rhythms to match the rhythm of the environment. The sudden feel of your own heartbeat or awareness of your own breathing rhythm are just two everyday examples of the huge range of living rhythms. But the most staggering conclusion to emerge when these rhythms are analysed and examined in detail is that mankind, and nearly all the other creatures who share the planet, have an innate ability to measure time. With some of Superman's mystical powers, without reference to the hands of a clock, our bodies, and the bodies of animals and plants, beat in time with the environment. This is no magic trick, no special effect or sleight of hand or mind but the power that rules all life on earth.

If time is truly our planetary dictator, and rhythms the manifestations of that dictator's supremacy, then the study of rhythms becomes essential to the understanding of the way all life works. So the reason why this book was created—and why it should be read—becomes clear. Rhythms are the key to self knowledge and to knowledge of our surroundings. They put all life into a timely perspective.

Professor Edward S. Ayensu
Smithsonian Institution, Washington D.C.

Dr Philip Whitfield
King's College, University of London

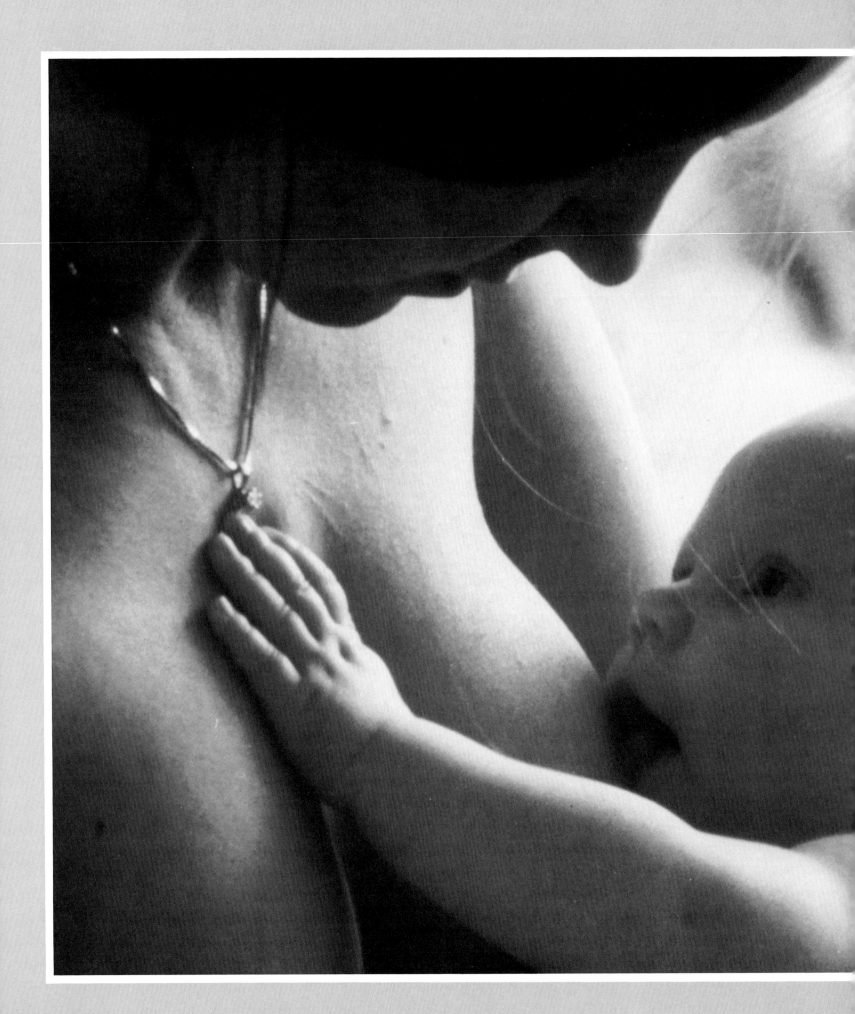

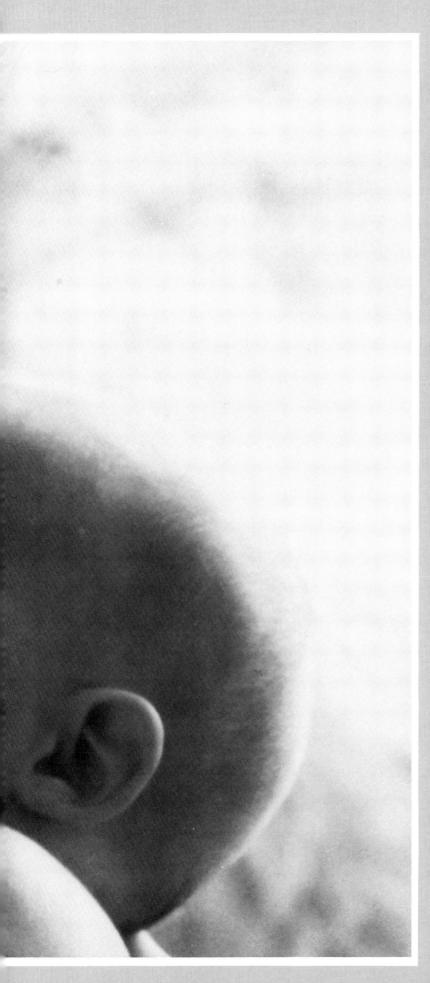

Familiar rhythms

In the ever-increasing complexity of human life we are surrounded by sights and sounds which are so commonplace that they pass unnoticed. The familiarity of everyday happenings can truly breed, if not contempt, then certainly a blunting of man's imagination. If our minds were not so hindered, then every sunrise, the birth of each new baby and the impact of every ocean wave on the shore would be recognized and greeted as the minor miracle that indeed it is. All these recurring miracles stem from one simple fact—life is rhythmic. This feature identifies the animals and plants of the planet Earth as effectively as a brand mark, but is so basic that even when it shapes our own lives we seldom honour it with a moment's careful attention.

Once the human mind has grasped the idea of rhythm, examples become obvious and all pervading. Our hearts beat rhythmically within us while our breathing follows a slower but equally repetitive pattern. We sleep, we wake. On a longer time-scale we can see that the lives of plants take the form of a yearly cycle. In a repeating and regular cycle a tree signals the seasons with its bare branches and inactive buds in winter, breaking buds in spring, a green profusion of growing leaves in summer and the fall of golden leaves in October and November.

A characteristic signature of every rhythm is its frequency, that is, the number of complete, even cycles that occur per second, minute, day, year, decade or even millenium. The frequency signature of a cycle provides a useful means of classifying biological rhythms and is a characteristic that can take on almost any value. So a baby's heartbeat might have a frequency of 100 beats a minute, a swallow's migratory rhythm a frequency of a single two-way journey per year. These two examples of frequency signatures reveal a startling diversity in the rapidity of life's rhythms—the baby's pulse is over 50 million times faster than the swallow's migratory rhythm. Such scales of difference make taxing demands on the imagination. At one end of the spectrum, for instance, insect wings vibrate at a frequency of hundreds of times per second, while at the other, it is possible that profound effects on animal evolution may have been brought about by rhythms in the earth's climate which have a frequency of one cycle every hundred thousand years or so.

Whatever their frequency, the cycles of life can be split into two distinct groups depending on whether or not the rhythms correspond to any external, rhythmical changes in the environment. So while the beat of the heart and the rhythm of breathing are not matched in frequency by any external rhythmical pattern, our cycle of sleep and wakefulness has a rhythm corresponding to the earth's external 24-hour rhythm of night and day. The breeding and migration of birds and the growth cycle of trees are examples of rhythms that occur in synchrony with the external cycle of the repeating seasons. Most scientists assume that by being 'tuned in' to the external rhythm of its surroundings an organism is gaining some benefit and, by inference, that the same plant or animal would be less efficient if it were operating out of step with its environment.

Man cannot isolate himself from the rhythmical imperatives of the world around him. He is part of the rhythms and the rhythms are part of him. Man has the same internal physiological rhythms as other animals, and similar daily and yearly rhythms, but he has created variations on the rhythmical theme which extend into every part of his life and society.

Seen and unseen, rhythms and cycles exist both around and within us. To gain a general insight into the ubiquity and importance of biological rhythms and cycles, it is illuminating to indulge in some self-analysis and to consider carefully the animals we each know most intimately and intuitively—women and men, babies and children—the 'naked apes' as Desmond Morris has called us. It tickles human pride to itemize the ways in which man is unique among the earth's animals. Man has been described as the only animal with a true spoken language, the most enquiring of animals and the supreme toolmaker. It might not be an exaggeration to say that man is also the animal that demonstrates more facets of rhythmical phenomena in his life-span and life-style than any other. What makes man an exception is not the nature of his bodily rhythms, but the fact that he has taken rhythms, ritualized and formalized them, and made them an integral part of his existence. By means of technology he can also distort rhythms.

Before investigating some of these aspects of man's nature in depth, it is essential to clarify a difficult issue of terminology. The words rhythm and cycle have so far been used more or less interchangeably, but the similarities and differences between these two related terms must now be scrutinized more closely. Stated baldly, a rhythm is regularly repetitive in time—drumbeats, heartbeats and sunrises are all rhythmical phenomena. Defined in a similar fashion, cycles are chains of events that occur in a set order, and the order itself is repeated. ABCDABCD is a cycle of letters, while spring, summer, autumn, winter, spring, summer . . . is the cycle of the seasons. But what is the relationship between a rhythm and a cycle? Rhythms relate primarily to the timing of events, cycles to the order in which they occur. To this extent they are different, yet if cycles are completed at regular intervals in time they can be highly rhythmical, and it is in such contexts that rhythms and cycles become almost—but not precisely—synonymous.

To make the distinction between a rhythm and a cycle more concrete, think of an ordinary mechanical clock. When properly adjusted, the hour hand on such a clock behaves in a cyclical way; it points in turn to the numbers 1 to 12 over and over again in the same unvarying order. The same hand is also behaving rhythmically because it passes any particular number at regular 12-hour intervals. If the same clock were placed in the middle of a sandstorm, however, the sand would make the hand move irregularly—first fast, then slow and sometimes stopping as it circled round the clock face. In such circumstances the hand no longer moves rhythmically and passes the number one, for instance, at varying intervals of time which may be more or less than 12 hours. Although the rhythm is disturbed, the cycle of events is still preserved. The hand continues to move from one to two, two to three and so on and then repeats itself. So cycles may or may not be rhythmic, while rhythms are often generated by underlying mechanisms that are cyclical.

Against this background of theory we can begin to discern the principal rhythms and cycles of human life. Some of these are easily perceptible through our ordinary, unaided senses. Others require the most sophisticated and detailed analysis of the internal workings of the nerves, brain and glands before they become obvious. Yet others only become apparent when careful records of human behaviour and activity are made over the course of many days, months

The concept of the breath of life in Judeo-Christian religions and similar descriptions in other religions testify to the ancient spiritual significance of the breathing cycle. To pre-scientific societies breathing was the essential attribute of life—when breathing stopped so did all other vital processes. A single breathing cycle is made up of one breath in, followed by one breath out. At rest, while sleeping or during intense activity we breathe rhythmically quite unconsciously. Driving this rhythm is a portion of the brain stem that sends rhythmic sets of nerve impulses to the muscles of the chest wall and diaphragm.

Even in repose, with shallow breathing and a quiet, regular heartbeat, this woman is seething with other internal rhythms. She may seem static but many of her tissues are in a dynamic equilibrium of growth and decay—new cell growth and old cell removal balancing one another. Skin, finger nails, hair, cells in the lining of the intestines and the millions of cells in the bloodstream are continually renewed. All of these renewal processes are modulated by rhythms which usually have a 24-hour period. The body is buzzing with cellular activity which is organized in long, slow pulses, each lasting a day.

The daily beat of the body is apparent in many other disparate functions: there is a rhythm of urine flow which normally peaks between 10.30 am and 2.30 pm; a daily blood pressure variation; a rise and fall in body temperature and distinct changes in the hormone content of the blood.

or even years and may even be a manifestation of the behaviour of whole human populations.

At the microscopic level the basic living units, the cells, from which the human body is constructed demonstrate cycles and rhythms which form the foundation of continued human existence. In order to work properly, many parts of the body need to renew themselves constantly, and to achieve this the cells multiply. Skin cells do so, and do so cyclically. Individual cells first double their genetic material which then splits into two identical halves. Then the cell itself divides so that each new daughter cell receives a complete set of genetic instructions. The whole process repeats itself in a cell cycle. Superimposed on this cell cycle is a larger scale rhythm in time because, in most tissues, more of the cells divide in the late evening and during the night than during the daylight hours.

Some body cells, such as those in the brain and nervous system, stop reproducing themselves once they have been formed. In specializing for their roles as the working components of a living computer, these cells seem to have forfeited their ability to divide and make replicas of themselves. Although nerve cells have given up their powers of division they still demonstrate their own high frequency rhythms. These are the rapid rhythmical changes in the surface electrical charges of nerve cells which create nerve impulses. Such impulses are the cell language by which nerves shuttle information around the body and the spikes of electrical activity, whose frequency is an information code, can occur at rates of hundreds per second.

Unlike cell divisions and nerve impulses, human breathing and heartbeat rates are directly perceptible. At rest our breathing and heartbeats are relatively constant rhythmical activities, the period or time gap between events being about five seconds in the case of breathing and about 0.85 seconds between heartbeats. These periods can be changed by a number of factors. Everyday experience tells us that exercise pushes up both rates—our hearts beat faster and we breathe more quickly and deeply. As with the pattern of cell division, there is also a daily rhythm of heart rate because, even when the body is resting, the heart generally beats faster during waking hours than it does during periods of sleep.

There are few physiological functions that man or his organs perform that do not conform to this general pattern of approximately 24-hour periodicity. Body temperature, urine flow, a huge diversity of hormone levels, the pattern of sleep and wakefulness and the timing of the beginning of labour are just some of the intricate body attributes that fluctuate with a daily rhythm. It is as if every organ in the body obeys the baton of an internal conductor who beats time very slowly, once every 24 hours. More remarkable still is the fact that many of these circadian or approximately daylength rhythms persist even when all external clues to the passage of time are removed.

Over periods longer than a day men and women continue to display rhythmical and cyclical behaviour. Most of these long-term rhythms seem to be generated by hormones, although when cycles have a yearly periodicity it is often difficult to disentangle the physiological from the psychological. A woman's menstrual cycle, however, falls neatly into the hormone-driven category. The menstrual cycle of a woman who is not pregnant, which ▷

A thousand years ago civilizations of Central and South America grasped one central fact about heart function. During their bloody, sacrificial rites they plucked out their victims' hearts and saw them still beat, though severed from the bodies. They knew that hearts have an intrinsic ability to beat. A heart is made up of cardiac muscle cells that, even in vertebrate embryos, have an innate facility for rhythmic contraction and elongation.

After puberty there is a rhythm in every non-pregnant woman's life so that, unlike most men, she can never be unaware of biological cycles. This so-called menstrual cycle is one of wonderfully orchestrated physiological and structural changes in the ovaries, the uterus and the vagina. The breakdown of the uterine lining and the consequent blood loss are the obvious parts of the cycle to its producer, but mid way between periods of bleeding is a key, often unperceived, event: an egg is released from one of the ovaries and, for a day or two, is available for fertilization.

Underlying the rhythmic changes of the menstrual cycle are complex cycles of hormone production. These hormones—estrogen and progesterone—have multiple effects so that the monthly changes coincide with cycles of mood, libido and other personality changes.

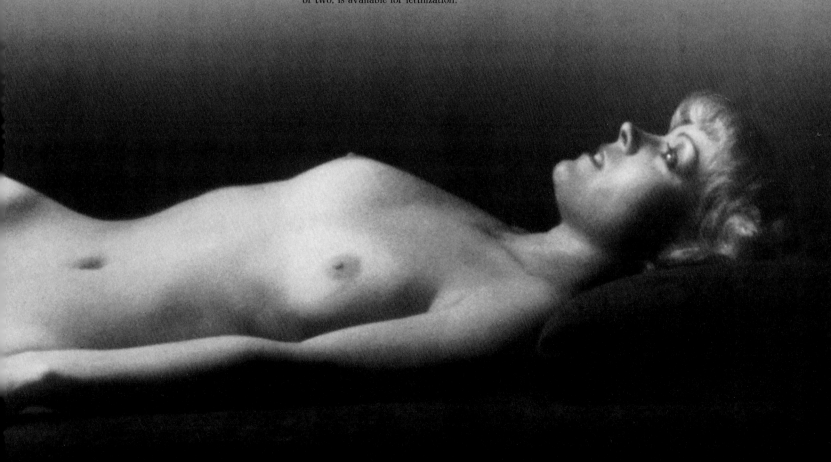

▷ normally averages between 28 and 29.5 days (it can vary between as few as 18 days and as many as 35), is often extremely regular, deviating only a day or so from its normal value over many years. Underlying the rhythm in time is a regular cycle of hormone-induced tissue changes in the ovaries, uterus and vaginal walls. A rhythmic cycle of mood and libido changes can often be linked with the hormone cycle and many female traits which are not immediately obvious, including sensitivity to painful skin stimuli, responsiveness to certain odours and the ability to distinguish changes in musical pitch, have all been shown to rise and fall in concert with the ups and downs of hormonal output during the menstrual cycle. We are rhythmic in more ways than we can know or understand.

While it is relatively easy to study many of the rhythms and cycles of particular cells, organs and systems of the body, it is impossible to comprehend the precisely coordinated, interlocking and interacting complex of human rhythms and cycles as a whole. It is not simply that each organ and activity has its own clocklike timekeeping ability, but there are obviously interdependent hierarchies of clocks within the body with unnervingly complicated patterns of dominance and subservience with respect to one another. We do not even know whether or not this hierarchy has an ultimate ruler—a single, central, all-determining time-giver.

In its changing activities and behaviour through time the human body is as intricate as an orchestra playing a symphony. Each organ, with its own internal rhythms, represents an instrument playing its own part in the symphony, but these rhythms only have a total significance in relation to the parts of all the other instruments. To discover more about the true nature of biological rhythms we must ask whether the full symphony of human activity is being played from the moment a baby is born. The answer to this crucial question seems to be 'no'. At birth a baby is playing only the rudiments of the final majestic opus and is performing on a few instruments alone. Through babyhood, childhood and puberty extra instruments—that is, extra organs and behaviour patterns—successively and accumulatively take up the overlapping rhythmical themes of the whole, so that it takes at least a decade of direct experience of day following night, the changing seasons and the succession of years before the human animal becomes fully attuned to the cyclically fluctuating world it inhabits.

The detailed story of our increasing understanding of the slow, stepwise development of human rhythms is a fascinating one. It shows how it has been possible for erroneous ideas about behaviour to arise, reveals how simple, but careful, observations can yield startlingly original results and demonstrates how even the youngest babies, with their simple responses to the world around them, already contain the embryo of adult rhythmicity. The aspect of a baby's development of circadian behaviour that is probably best understood is the cycle of sleeping and waking. This cycle is no more complicated than the alternating pattern in time of periods awake and periods spent sleeping. Most adults split each 24 hours into a reasonably regular array, with approximately 16 hours awake and eight hours asleep, each period of sleep or wakefulness starting at about the same time each day. And experiments on people with no access to watches or clocks, nor to external clues about time, such as the arrival

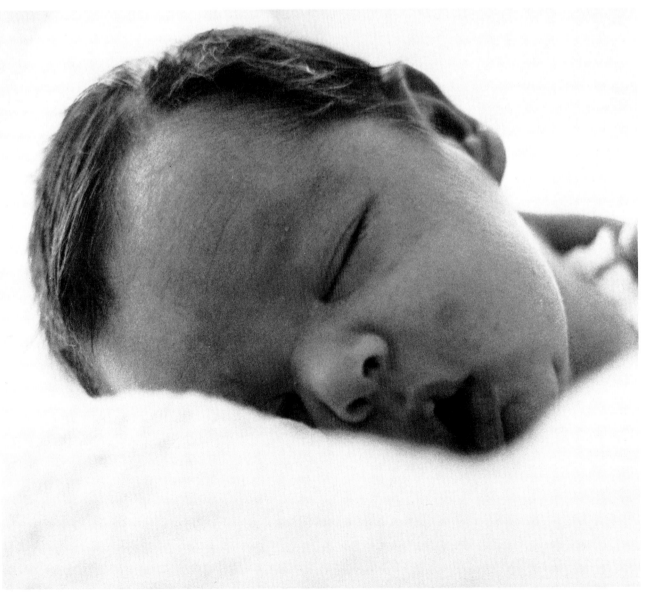

Coping with a baby's pattern of sleeping and waking is an aspect of parenthood that demands a considerable degree of compromise between adult and infant behaviour patterns. As the normal sleep rhythm develops, the growing child passes through a variable series of phases, from which the pattern of one period of sleep and one period of wakefulness in every 24 hours gradually emerges. Child-rearing gurus of the last two or three generations have argued as to how far this changing sleep pattern should be directly influenced by timed feeding regimes or left uninfluenced, as by the currently fashionable demand feeding method. Whichever stance is taken, a baby left to its own devices will not naturally wake at regular intervals during the first months of its life.

of dawn and the setting of the sun, show that such isolated human beings retain an internal personal ability to maintain a circadian rhythm of sleeping and waking. Expressed another way, man has an inbuilt or endogenous clock which commands that a period of sleep should begin once in about every 24 hours.

Parents the world over know that in their first weeks and months of life babies simply do not behave like this. Nights of disturbed parental sleep alone testify to the fact that a baby's pattern of sleeping and waking is very different from an adult one. A more detailed understanding of the pattern has, however, been hard to establish. In the first half of the century, for example, it was firmly believed that newborn babies, or neonates, needed to sleep almost continuously, that is for 22 hours out of every 24, and that during the first year of their lives that period gradually declined to more adult levels with the bulk of sleep taking place at night.

One piece of pioneering investigation, carried out in the United States in the 1950s by a pair of workers named Kleitman and Englemann, has done a great deal to change anecdote about infant sleep into hard data. Although their investigations were carried out on only 19 babies over the first six months of life, the central findings of Kleitman and Englemann's study are not only interesting but provide useful insights into specific aspects of child care, for example, whether babies should be fed every four hours by the clock or 'on demand'. The 19 babies were first monitored at three weeks and even at this early age were only sleeping for a total of approximately 15 hours out of 24, and there was already a slight excess of sleep in the nighttime hours. Between 8pm and 8am the babies slept for an average of $8\frac{1}{2}$ hours, while between 8am and 8pm they slept for only $6\frac{1}{2}$ hours. As the babies grew, this pattern gradually became accentuated and was accompanied by a gradual, but small-scale, decrease in the total amount of sleep. By six months the adult pattern was slowly emerging from the hint of the rhythm present at three weeks and the babies slept for an average of $13\frac{1}{2}$ hours in 24, but 10 of these were nighttime hours and only $3\frac{1}{2}$ were daytime sleeps.

In the developing child many of the high frequency physiological rhythms are obviously present before birth. Nerves produce rhythmic discharges and the heart is beating—albeit at a much higher rate than in adulthood—long before delivery. At the moment of birth, with the first breath, a true breathing rhythm is initiated with a marvellously articulated set of almost instantaneous changes in the heart, lungs and circulation.

Marked day-to-night or circadian rhythms arise in different body functions at characteristic periods of a baby's development. In the first days of life it is possible to show changes in the electrical resistance of a baby's skin between night and day. Body temperature fluctuations achieve a significant circadian pattern by two to three weeks, while similar changes in the rate of heartbeat and urine output arise between the fourth and twentieth weeks of life. Only with the onset of puberty and physical and sexual maturity in boys and girls, do the adult hormonal time patterns become established. With her first period, which in Western society occurs at an average age of 13, a girl is finally, and for the first time, demonstrating the full range of circadian and approximately month-long bodily rhythms and cycles.

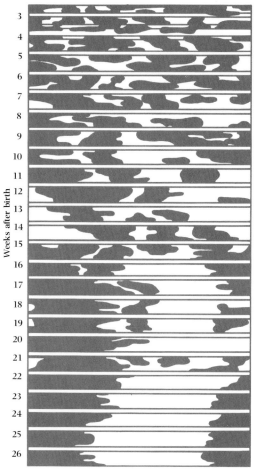

Weeks after birth

3
4
5
6
7
8
9
10
11
12
13
14
15
16
17
18
19
20
21
22
23
24
25
26

The **understanding of rhythmic behaviour** in babies was greatly increased by an experiment which monitored a child for the first six months of its life. In this simplified version of the data each horizontal bar represents the sleeping (dark area) and waking (light area) periods of one week and clearly shows the shifting pattern. After the first few chaotic weeks, a 25-hour rhythm emerges between weeks 5 and 25. This then switches to a 24-hour rhythm, with the bulk of sleep at night.

Age of child	Developing rhythmic functions
One week	Electrical skin resistance rhythm present
Two to three weeks	Body temperature rhythm begins to emerge
Four to twenty weeks	Sleep/wake rhythm nearly developed Body temperature rhythm present Heart rate rhythm present Excretion of urine, sodium, potassium and phosphate show daytime peaks
Five to nine months	All daily rhythms intensify and are present in nearly all observed functions
Sixteen to twenty-two months	Rhythm in excretion of creatine, creatinine and chloride develops

Significant day-night differences in bodily functions emerge through babyhood and childhood in a distinct and meaningful order. It is as though each process first operates in an unmodulated fashion and then, increasingly, comes under the influence and control of the body's inbuilt 'clock'. Once control is established there is a clear rhythm with a period of almost exactly 24 hours. Organs such as skin, which are well formed at birth, show a 24-hour rhythm in the first days of life. Others which do not fully mature until weeks or months after birth develop their rhythms correspondingly.

In his waking hours a baby gradually becomes aware of and attunes to the natural rhythms of his surroundings.

Cosmic rhythms

Imagine an alien creature, from a race infinitely more advanced than our own, whose life's task is to collect exotic organisms and then interpret their life-styles. She sits in a control room in her far-off galaxy, activates her gravity-sink-space-warp device and scoops up a random collection of animals and plants from planet Earth, hundreds of light years distant.

She has gathered the organisms, but how can she possibly deduce anything about the world from which they came? In fact, woven into the structure, physiology and behaviour of those animals and plants is a remarkably precise inventory containing not only information about the conditions that exist on Earth, but also several vital facts about the nature of our solar system. This information is not fortuitous but is the physical manifestation of 3.5 billion years of organic evolution. The inexorable effect of the selective process at the heart of the evolutionary mechanism is to nudge the adaptiveness of reproducing organisms farther and farther into intimate concert with the conditions in which they must live. With almost infinite slowness—but without pause—animals and plants increase their fitness for life on Earth. Each minute addition to the appropriateness of their lives and physical organization makes them reflect the nature of their environment more clearly.

So the interstellar 'butterfly collector' could come to know much about our planetary home. The physical construction and temperature sensitivity of her specimens would provide vital clues. They would show first the approximate gravitational field, and hence the size of our planet, and second that most life on Earth must live in areas with a temperature range between -10 and $+40$ degrees centigrade (14 and 104 degrees F). The chemistry and lack of extreme anti-desiccation devices in the organisms would suggest a planet where water was plentiful. Simple experiments on the effect of various gas mixtures would reveal the type of atmosphere we possess.

Amazingly, these creatures, from trees to cockroaches, could provide information about the nature of our solar system because the intrinsic organization of almost every living thing on Earth—cockroaches included—conforms to a pattern of design dictated by the motions of the Earth and moon in space. Our intergalactic investigator would have to perform only the most cursory of temporal examinations on her collection to find that rhythmical phenomena of remarkably few particular frequencies constantly recurred in her sample.

Almost all the plants and animals would have circadian or daily rhythms with a beat of about 24 hours. The longer-lived life forms would all show some activity with a rhythmical period of about 350 days, that is, circannual rhythms. Any marine organisms in her sample would probably show both twice-daily and 14 or 28-day periodicities, corresponding to the time patterns of the tides. Any astute alien would thus conclude that planet X rotates about its axis once every 24 hours, completes its solar orbit in about 350 days and probably has a single satellite moon with an orbital period about X of 28 days, producing tidal surges in the planet's large water masses.

At the heart of this whimsical tale of science fiction is a sobering, unflippant fact. Each of us on Earth, each of the 10 million species that has evolved on and now swarms over its surface, carries the time signature of the planetary motion of our world indelibly stamped within it. We can now examine the physical basis of these signatures—the mechanisms that generate our rhythmical environment, an environment that has made us rhythmical creatures.

From man's first cave paintings, documented human history stretches back a few tens of thousands of years. Yet as an insight into the life-span of the Earth, this popular view of history is almost useless. To understand the astronomical influences that have shaped, and continue to shape the patterns of regular change in our environment, the concept of history must be expanded many orders of magnitude in time to encompass the life story of our solar system—the sun with its attendant flock of gravitationally tethered planets, asteroids, moons and comets.

To put our personal star the sun—and with it ourselves—into an ultimate context we must first determine our position in the universe. This is conceptually difficult because the map of the universe has no bottom left-hand corner. Instead, all positional descriptions have to be relative to other objects. There is no centre of the universe, no definitive grid reference. We are part of the rapidly expanding debris of the primordial 'big bang', the unimaginably violent single event that initiated our universe. Within very hazy limits this most pregnant of events, the true zero, can be set in history. It probably happened between 10 and 15 billion years ago, and the background radiation generated by that fireball still reverberates round the universe.

In the simplest models of the big bang, evenly distributed atoms of low mass spread ever outward from the explosion into space. But if the matter was, and remained smeared out in this way there would be no galaxies, no stars, no planets and no human beings. By some combination of causes which is as yet unknowable, the expanding universe gained large-scale structure so that matter became congregated into millions of spaced-out galaxies. Each galaxy is a tightly packed archipelago of stars, and each star is a huge concentration of matter. Between the galaxies are vast, almost empty gaps of intergalactic space.

Galaxies come in a huge range of types, but all are island universes on a scale that the imagination cannot grasp. They are so big that astronomers have to use a daunting vocabulary of units to describe them, of which the most dramatic is the light year, that is, the distance travelled by light at the rate of 186,000 miles (299,000 km) a second in a year—about 6,000 billion miles (9,600 billion km). In the night sky our own galaxy, or at least a small part of it, can be seen as the Milky Way. Viewed from outside it is a spiral galaxy with a flattened, lenslike shape and is quite large as spiral galaxies go. From edge to edge it stretches 100,000 light years and contains some 100 billion stars. One of these myriads is our sun, which is about 30,000 light years from the mysterious galactic core. This puts us firmly in our place, for it not only positions us but also emphasizes the insignificance of our whole solar system—in galactic terms we are in the outer suburbs.

In the context of the universe our solar system is comparatively young, having condensed out of a cloud of atoms by some process, which is as yet unknown, about five billion years ago. The precise nature of the origin of the sun and planets is a matter of intense scientific speculation but there is no doubt that the sun ended up with a most interesting assortment of planets—Mercury, Venus, Earth, Mars, Jupiter, Saturn, Uranus, Neptune and Pluto—each circling the sun in an elliptical orbit and each a world in its own right, but with fascinating differences.

Mercury, Earth, Venus and Mars, the four small inner planets, are all dense

Our own Milky Way galaxy is like a multiple-armed spiral with a bulging centre. This has been deduced by examining the characteristics of far-distant galaxies and comparing them with our own. Since we are buried inside the Milky Way, along one of its arms, it is difficult to make direct observations of it. Optical and radio telescopes can discover the overall structure of typical configurations such as the whirlpool galaxy, *below*, which has a bright core and two prominent arms. The same instruments can provide information about the Milky Way from within. In a galaxy such as ours the stars are in spiral arms emanating from a dense centre, *left*. One arm of the Milky Way rotates about its centre every 200 million years.

Gas and dust cloud

Sun

Condensation of gas and dust in disc

Collisions cause randomization of orbits

Our solar system was formed from a huge cloud of dust and gas some five billion years ago. All theories agree that gravity played a large part in condensing this cloud to produce the sun and the planets. One theory suggests that a spherical cloud of matter collapsed to form a rotating disc with a bright radiating sun at its centre, *left*.

Huge energy releases from this sun created a wide temperature range across the disc—hot near the sun, cold at the fringe. Such a range of heat may have led to the present planetary types— hot and rocky ones near the sun, cold gaseous ones farther away. The three lower diagrams, *left*, show many embryo planets in orbit; an intermediate stage of collisions which disturb their orbits; and the present state where the mature planets have settled down again into the plane of the disc.

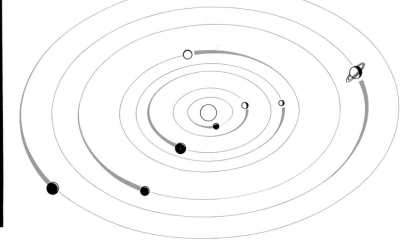

and rocky, with average densities between 4 and 5.5 times greater than water. The larger outer planets, excluding Pluto, are basically composed of gases. Their densities are much lower, being between 0.7 and 2.3 times that of water. To understand the conditions conducive to the existence of life, and the details of our global climatology, Earth is best compared with Venus and Mars. The initial surface temperature of Earth, dictated by the radiant energy reaching it from the sun 93 million miles (150 million km) away, was about 25 degrees centigrade (77 degrees F).

Carbon dioxide and water vapour, both gases arising largely from volcanoes, formed the original atmosphere. At the Earth's ambient temperature the carbon dioxide stayed as gas but most of the water condensed to form rivers, lakes and seas. Some of the carbon dioxide was removed from the atmosphere by being dissolved in the Earth's waters, while evaporation from the seas pushed water vapour into the atmosphere: clouds in the air and falling rain have been features of earthly weather for billions of years. The white clouds, however, are quite efficient at reflecting back into space some of the sun's short wavelength radiation, thus reducing the energy input into the planet below. As a result, the surface dropped to, and has remained at a mean value of about 15 degrees centigrade (59 degrees F).

To be provocatively simplistic, the presence of water, plus a physiologically mild temperature, led to the start of the evolution of life on Earth at least 3.5 billion years ago. Once cells evolved which, like those of modern green plants, could trap and use atmospheric carbon dioxide and release oxygen by the process of photosynthesis, the atmosphere altered dramatically. Oxygen was suddenly present in increasing quantities and today constitutes a fifth of the atmosphere's gases; a gigantic volume almost entirely biological in origin. The bulk of the atmosphere is now nitrogen, a gas which is only present in very tiny quantities in the gases emitted by volcanoes, but which has become accumulatively important because it is not appreciably consumed by any sort of biological activity.

Compared with Earth, Venus and Mars are worlds similar in size, rock composition and presumed early volcanic activity. So why is life absent on these planets? The answer seems to be simple: Venus is too near the sun, Mars, too distant. On Venus, only 67 million miles (107 million km) from the solar furnaces, the original surface temperature was about 87 degrees centigrade (189 degrees F). At this high temperature much of the water and all the carbon dioxide were in gaseous form. This thick gas shield produced an intense 'greenhouse effect', in which energy from the sun reached the planet's surface, but could not escape again because of the absorptive abilities of the gases. The resulting net energy gain has raised the Venusian surface temperature disastrously. It is now about 500 degrees centigrade (930 degrees F) with no possibility of life of an earthly type existing at all.

Mars is 142 million miles (225 million km) from the sun. Because of this remoteness its original surface was a chilling low of −30 degrees centigrade (−22 degrees F). At this temperature all the water vapour that reaches the atmosphere is immediately extracted again as ice. Even some of the carbon dioxide solidifies, resulting in a thin, denuded atmosphere in which no greenhouse effect of any significance can build up, so the temperature remains ▷

The Earth's atmosphere divides roughly into three zones of decreasing gas density—a dense troposphere, a thin stratosphere and a thinner upper atmosphere. Global weather occurs in the troposphere and is generated by the sun's radiation. Some radiation is reflected back into space by clouds, the atmosphere and by ice. The rest warms up the Earth, whose heat either fuels the weather machine or escapes into space. The poles receive less of this radiation and reflect more of it than the tropics, and this difference leads to the mass movements of air that are at the heart of the climate. The slice through the Earth, *right*, shows the extreme thinness of the atmosphere and the Earth's crust in relation to the size of the planet.

Atmosphere

Crust (5–25 miles) (8–40 km)

Mantle

1,800 miles (2,900 km)

Liquid core

1,380 miles (2,220 km)

Solid core

780 miles (1,255 km)

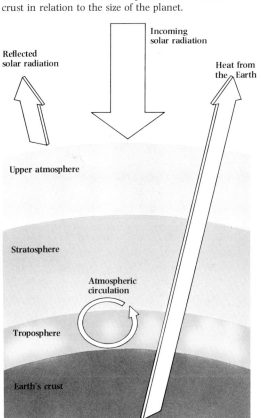

Reflected solar radiation

Incoming solar radiation

Heat from the Earth

Upper atmosphere

Stratosphere

Atmospheric circulation

Troposphere

Earth's crust

Spectacular images of Mars were taken as part of the NASA Viking expedition. The image, *right*, taken from 348,000 miles (560,000 km) away, shows three distinct volcanoes such as one would expect to find on a rocky planet. The image, *below*, taken from Viking Lander 1 shows the rocks and the drifts of dust on the Martian desert.

▷ very low. The surface of Mars is a freezing desert, just as that of Venus is a superheated one. Again life is apparently out of the question, but if earthly life is a consequence of being on a rocky planet at the right distance from a star, then it is perfectly logical to argue that our galaxy contains many living planets apart from our own.

Once life evolved on Earth the sun's central significance assumed another degree of importance. As the greatest energy source in the solar system, the sun became the ultimate and only noteworthy provider of energy for all life processes; and it still fulfils this role today, making worship of the sun perhaps the most rational of all religions. The sun is an energy source of gigantic power, a vast ball composed of little more than hydrogen and helium, the two simplest elements in the universe. For five billion years the sun has been pumping out this energy and it will probably continue to do so for another five billion. But how can 10 billion years' worth of power come from two gases, which on Earth are relatively unreactive, unexciting inhabitants of chemistry laboratories? Unfortunately—terribly unfortunately—these gases are found in our world in other states. Perched atop every intercontinental ballistic missile is at least one potential, tiny uncontrolled sun. An exploding hydrogen bomb—a fusion device in the bland jargon of physics—illustrates the awesome power locked in the atomic structure of hydrogen, or indeed any other element.

Essentially the sun is a vast, balanced and controlled hydrogen bomb. The mass of the sun is so immense that, at its core, gravity ensures that matter is hot enough, and hydrogen nuclei packed tightly enough, for hydrogen fusion to occur. At earthly temperatures hydrogen normally exists as neutral atoms, each consisting of a nucleus made up of one positively charged particle, a proton, around which circles a negatively charged particle, an electron. But at the sun's heart, as a result of the high temperatures there, hydrogen exists not as neutral atoms but as protons, rocketing about in a matrix of electrons, forming a 'soup' or plasma of particles.

In all ordinary circumstances the positive charges of the protons would mean that, despite their motion, proton–proton collisions would be impossible because, like two magnetic North Poles, they would push each other apart. But at the temperatures of over 10 million degrees centigrade which prevail at the core of the sun, the protons move fast enough to collide occasionally, and with enough energy for them to fuse together. At the centre of the sun a complex chain of fusion reactions takes place in which hydrogen nuclei fuse to make a type of helium nucleus, denoted as helium–4 because it contains two protons and also two neutral particles, or neutrons. The first step in this reaction is the linking of two protons to form a deuteron (the nucleus of deuterium or heavy hydrogen). The last step is the collision of two helium–3 nuclei, each consisting of two protons and one neutron. This is the crucial reaction because the total mass of the products of the collision is slightly less than the total mass of the particles that bumped in nuclear contact.

A tiny discrepancy in the mass budget of helium might seem insignificant, but, in fact, it is this lost mass that keeps us alive. The reason is that the lost mass is converted directly into energy, and the rate of exchange of energy for mass is spectacular. For every single gram (0.03 oz) of helium–4 produced in this reaction, 175,000 kilowatt hours of energy are produced: roughly the

equivalent of a single-bar electric fire burning constantly for 20 years. So the lost mass is not an accounting error. It is the energy that keeps the sun hot, which produced the conditions suitable for the emergence of life on Earth, which ultimately still drives all biological activity, and which generates our climate.

Unlike the hydrogen bomb, the sun has a means of controlling the fusion reaction, and this control mechanism is inherent in its structure. Fusion only occurs deep in the centre of the sun's massive bulk where extraordinary conditions of high pressure and temperature prevail. At the very heart of our star the temperature is about 15,000,000 degrees centigrade, and the density of the plasma is about 160 grams per cubic centimetre of material—a density 12 times greater than that of lead. This hideously energetic core does not 'blow up' the sun because upon it is piled a layer of nonfusing hydrogen and helium some 300,000 miles (500,000 km) thick. The bottom 250,000 miles (400,000 km) of this stack is a zone of radiation transfer, where the matter is transparent enough for the heat generated at the core to shine through it. Most of the outer 60,000 miles (100,000 km) is a convection zone, where the gases are opaque enough to retard the outward spread of heat by radiation alone. As a result, large local temperature gradients are set up, and vast convection movements of hot gas transfer most of the heat.

The brilliant white-hot sphere we see as the sun is a thin layer of photosphere at the top of the convection zone. The temperature of the photosphere is comparatively low at about 6,000 degrees centigrade, so it consists of neutral gas atoms rather than ionized plasma. It is also in continuous turbulent motion because of the violent convection currents of gas jetting up from beneath. This convection sets up a surface pattern, or supergranulation, on the sun which is visible in telescopic photographs. In reality it is nothing much more than a gigantic version of the pattern of rising bubbles in a simmering pan of syrup; except that the bubbles are tens of thousands of miles wide.

Between the relatively cool, neutral photosphere and the corona—the vast outer atmosphere of the sun—is a thin chromosphere which acts as a transitional zone between the two. The enormous corona, which stretches out beyond the earthly orbit as solar wind, is paradoxically much hotter than the photosphere, reaching temperatures of around 2,500,000 degrees centigrade. Compared with the photosphere the corona emits a much dimmer light and this means that it can be seen clearly only when the dazzling photosphere is blocked out by the moon in an eclipse.

Both the corona and convective supergranulation patterns are constant features of the sun's structure and behaviour. Other features occur spasmodically, but often with an overall underlying rhythmic pattern, linked in some way to the sun's magnetic properties. Huge flares and prominences are projected from the sun's surface, and dark, cool blemishes or sunspots move across it then disappear. These sunspots can easily be seen to wax and wane in number with a considerable degree of regularity. The cycle has a periodicity of approximately 11 years and is linked to a 22-year cycle in which the magnetic poles of the sun reverse and subsequently regain their original polarity.

The behaviour of planet Earth, and of the organisms that inhabit it, are ▷

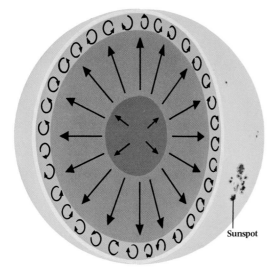

| 1750 | 1800 | 1850 | 1900 | 1950 | 1980 |

Increasing sunspots

Every 22 years the solar magnetic poles reverse twice, driving the 11-year rhythm of the sunspot cycle. The number of observable spots rises and falls during the cycle, directly indicating the amount of solar activity.

The sun is made up of four main layers:
Photosphere: the mottled surface is about 6,000°C, marked with sunspots.
Convection zone: 60,000 miles (100,000 km) thick, this is stirred by violent movements of hot matter.
Radiative zone: 250,000 miles (400,000 km) thick, up to 5 million°C, it increases in density toward the core. **Fusion core:** 12 times as dense as lead, 15 million°C it converts billions of tons of hydrogen into helium every second, creating solar energy.

Sunspot

A vast prominence of luminous gas streams out from the sun into space, *left.* Each surface granule is really a bubbling cell of heat turbulence 620 miles (1,000 km) wide.

The internal, unobservable structure of the sun can be understood only by using complex calculations. Its heart is the fusion core (red); a radiative zone (orange) transmits energy to the convection zone (dark yellow) where the energy is expelled outward to shine from the photosphere (light yellow), whose light and heat is what we see and feel on planet Earth.

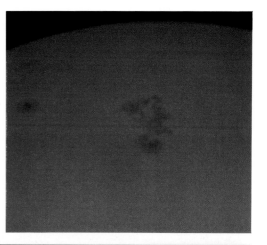

The discovery of sunspots spoiled man's idea of the sun as an unblemished sphere. They are relatively cool (4,000°C) areas of gas on the sun's surface. Usually they are 6,000 miles (10,000 km) across but can be 90,000 miles (150,000 km) wide. Appearing in pairs or clusters, they last a few weeks while moving with the sun as it rotates. Their strong magnetic field suggests that they are the visible effect of magnetic disturbances deep inside the sun.

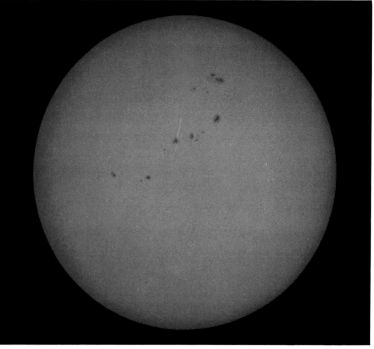

▷ intimately linked with the structure and working of the sun. The Earth is not an inert, unchanging rock ball, on which the story of the evolution of life is inscribed and played out by millions of species, but is in constant motion. Beneath the Earth's crust the great bulk of the planet's central mass is in a liquid or semi-liquid state. In this fluid, slow but massive convection currents, analogous to the immensely more powerful convection currents in the sun's substance, rise and fall.

Even the thin, solid crust that coats the outer surface of the Earth is more like a slow-moving sea of ice floes than a concrete platform. In a continual, slow jostling, which is clearly discernible only over millions of years, plates of the crust gradually creep past, over and under each other and then melt. The movements bring about the phenomenon of continental drift. They sometimes also cause earthquakes, in which the tensions set up by hundreds of years of relative plate motion are suddenly and violently released in rapid, spasmodic crustal movements—mere twitches on a global scale, but they can flatten cities. All the principal volcanic and seismic zones of the world, in which devastation is an awesome fact of life, are in the regions of the Earth where crustal plates make contact.

For all the living creatures on Earth, the most crucial pattern of planetary motion is the Earth's spin around its polar axis, the imaginary line linking the geographic North and South Poles. This spinning motion or, technically, rotation, generates day and night for any organism dwelling at a fixed point on the Earth's surface. The patterned division of time into a repeating cycle of dark and light periods is the strongest rhythmical environmental change to which

most earthly organisms are exposed. All the daily biological rhythms that life has evolved characteristically have a period of approximately 24 hours, and all such rhythms are a response to this dominant pattern of change.

The most obvious and dramatic differences between day and night is the discrepancy in light intensity. But the presence or absence of the energy-providing sun in the sky induces many other changes of great significance to living things. The most crucial pair are the ambient temperature and the relative humidity, that is, the degree to which the atmosphere is saturated with water vapour. Light, temperature and humidity are all enormously potent determinants of biological activity. The rapid changes in light intensity at dawn and dusk provide very clear-cut clues by which many organisms can monitor the procession of days. During the daylight hours, light itself is one of the forms in which the fusion power of the sun finally impinges on the Earth's life. The packets of light energy, or photons, from the sun, which strike green plants are used in the process of photosynthesis to create organic molecules.

Linked to the day-night cycle, temperature changes impose great constraints on all animals, plants and micro-organisms. All the biochemical processes that typify life are temperature dependent; they are speeded up by a temperature increase and slowed down as temperatures fall. Even a small, warm-blooded animal, such as any of the mammals or birds which can maintain a constant, high internal body temperature despite environmental temperature changes, must take note of the temperature alterations produced by night and day. The greater the temperature deviation up or down from the optimum, be it 37 degrees centigrade (98.6 degrees F) for man or 42 degrees

Solar radiation

Spin axis and geographic North Pole

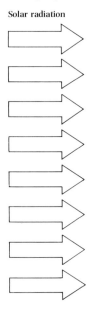
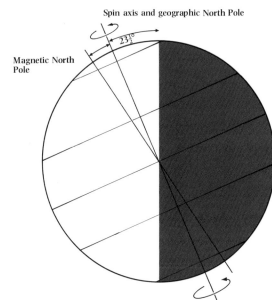

Magnetic North Pole

23½°

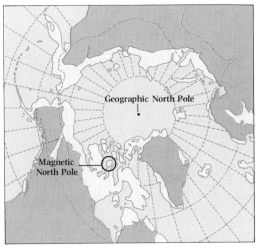

Geographic North Pole

Magnetic North Pole

Today's positions of the geographic and magnetic North Poles are shown, *above.* All compasses point to the magnetic North Pole whose location is known to wander gradually through geological time.

Every 24 hours the Earth rotates about its spinning axis, so causing the rhythm of day and night. At midsummer in the northern hemisphere the spin axis tilts 23½° toward the direction of the sun's rays, *above.* This axis lies between the geographic north and south poles. The magnetic axis is at a small angle to, and rotates around, the spin axis. Satellites and spacecraft carry cameras which, in the past 20 years, have enabled man to photograph the whole Earth. This view, *left*, was taken during the Apollo 11 mission.

The rotation of the Earth is gradually slowing down: 400 million years ago the solar day was about 22 hours long; today it is 24 hours 4 minutes long. Evidence for this comes from parts of fossil corals from the middle Devonian era, *left.* The growth of these corals varied with tidal and daily cycles forming bands and ridges that corresponded to the lunar and daily periods respectively. The pattern of these bands suggests that the lunar month in the Devonian era had 30.8 days, rather than the present 29.5 days.

centigrade (107.6 degrees F) for a duck, the more work has to be done by the animal's metabolism to produce an internal temperature that is thermostatically stable.

Humidity variations are of especial importance to a wide range of invertebrates (animals without backbones) and wet-skinned, terrestrial vertebrates such as frogs, newts and toads. Because of the increased evaporation of water from the body surface at high temperatures and in dry air—typical daytime conditions—many such animals are necessarily nocturnal in their habits, and are active at night, when cooler conditions and wetter air mean that they can operate efficiently without excess water loss.

The physical changes wrought by the cyclical succession of days and nights produce an environment with a time pattern which animals and plants ignore at their peril. So dramatic are the alterations in character of a single habitat between day and night that all organisms have to make vital strategic choices about their life tactics in relation to these fluctuations. One of the central features of any animal's life-style is thus its pattern of activity throughout 24 hours. If an animal can be described as nocturnal, diurnal (active in the daytime) or crepuscular (active only at dawn or dusk), then this description encompasses some basic definitions of the animal's way of life.

To order their lives according to the spinning period of their mother planet, organisms have developed clocks whose hands can be reset by external clues. The most powerful of these is the circadian clock, tied irrevocably to the Earth's period of rotation. For all practical purposes, this period can be assumed to be 24 hours but, like all large-scale natural phenomena, it is subject to both long- and short-term variations. Measured against the reference point of a star, the Earth's rotational period is in fact a fraction over 23 hours 56 minutes, a value described as the sidereal day. The solar day—the time between one noon and the next for a fixed, earthbound observer—averages just under 24 hours 4 minutes, but varies throughout the year by an amount never more than 16 minutes either way. Superimposed on this predictable, large-scale variation is a wide range of very minor perturbations caused by redistributions of mass within the Earth's solids, fluids and atmospheric gases.

Perhaps the most intriguing of such changes are the very long-term ones. Analysis of ancient records leads to the inescapable conclusion that the Earth's spin is slowing down. Irrespective of short-term changes our days increase in length by approximately one millisecond (a thousandth of a second) every century. If this finding is projected back in time, then in the Devonian geological age 400 million years ago, a day would have lasted only 22 hours and each year 400 days. Such seemingly wild speculation is in fact substantiated by patterns in the growth rings visible in some fossils.

In most parts of the Earth, conditions change annually with the succession of the seasons. The primary source of much of the seasonal, year-to-year variations in earthly existence is the complex interaction between the precise orientation of the spinning axis of the Earth and the plane of the Earth's orbit around the sun. It is also believed that, over long time-spans, changes in the orientation of the spin axis may induce slow climatic changes which affect the whole Earth.

If the spinning axis of the Earth and the plane of the Earth's orbit of ▷

The spinning axis of the Earth is commonly thought to point directly at Polaris, the pole star. This is not quite true, for really the North Pole is directed to a point in the star field close to Polaris. Proof of the Earth's rotation is shown, *left*, in a continuous eight-hour photographic exposure taken through a fixed telescope. Each star has made a streak consisting of a third of a circle. The brightest arc near the centre shows the motion of Polaris. Over the next few thousand years the spin axis will be directed at a point even further away from Polaris. In about 21,000 years it will return, to point toward Polaris once again.

Three important variables of the environment show clear dramatic rhythms that correspond to the day and night cycle. Changes in light intensity, humidity and temperature through a 24-hour period at one point on the Earth's surface are shown, *right*. The measurements were taken at a place in the temperate mid-latitudes where the length of day was about 15 hours. Light intensity obviously leaps to a much higher value in the day than at night. As the temperature slowly rises the humidity falls and vice versa.

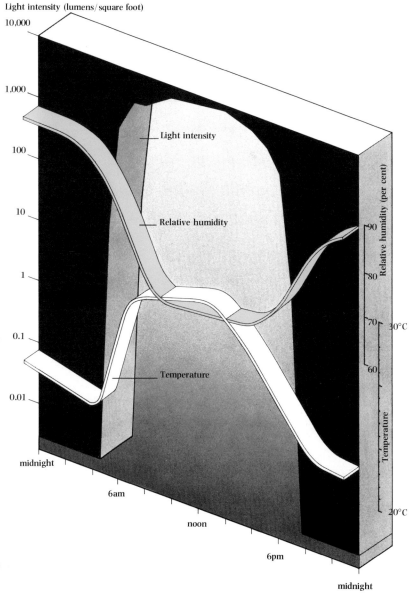

▷ revolution around the sun were orientated at right angles to one another, then the world would be an unrecognizable place. Everywhere on its surface, except for aberrant zones immediately around the North and South Poles, both days and nights would be unvaryingly 12 hours each throughout the year, and there would be no seasons as we know them. The vital role of the Earth's spinning axis in generating our seasons is difficult to grasp. If asked what makes summers hot and winters cold, most people, including those with some scientific training, would reply with a confused jumble of partially understood concepts. Perhaps the sun is farther away from the Earth in winter, perhaps the polar ice caps are involved in some way—or is it something to do with the height of the sun at noon?

The Earth's distance from the sun does alter with the seasons, although this has almost nothing to do with the differences between winter and summer, spring and autumn. But the answer to the question of seasonality has everything to do with the spin axis of the Earth, which is not at right angles to the planetary orbit, but inclined at about $23\frac{1}{2}$ degrees to it. And with gyroscopic stability, the axis points with great constancy—at least over many centuries— to one single point among the stars near the pole star, Polaris. This means that for half of each year the North Pole points obliquely at the sun, while for the other half the South Pole takes up this position.

The tilt of the Earth's axis of rotation divides the world into climatic zones, defined by the way in which the sun appears to move within them. At the top and bottom of the Earth are frigid zones, the outer boundary of each being $23\frac{1}{2}$ degrees of latitude from the Poles. In these areas there is constant night for at least one 24-hour day a year and, correspondingly, the sun is constantly in the sky for at least one other day. Here, within the context of cold, there are enormous extremes of temperature, and the maximum possible variation in day and night lengths. Round the centre of the Earth, bounded by the tropics of Cancer and Capricorn, each $23\frac{1}{2}$ degrees from the Equator, is the torrid or tropical zone. Here the sun is directly overhead at noon for at least one day a year, and throughout each 12 months daylengths alter very little from 12 hours. In all intermediate locations on Earth—the temperate regions—the sun never stays in the sky for 24 hours, nor is it ever directly overhead.

The interaction of the unchanging direction of tilt and the gravitationally driven orbit of the Earth produces the seasonally changing height of the sun in the sky at noon or, technically, its declination. At the summer solstice, 22 June, points $23\frac{1}{2}$ degrees of latitude north of the Equator experience a sun at the zenith, that is, overhead at noon. Anywhere in the northern hemisphere, this is the day on which the sun is at its highest in the sky and the daylight hours are longest. On 22 December, the winter solstice, the northern hemisphere experiences precisely the opposite set of conditions. The sun is at its lowest in the sky and the daylength is shortest. Within $23\frac{1}{2}$ degrees of the North Pole the winter solstice simply occurs in the middle of a long period of perpetual night. The southern hemisphere is the mirror image of the northern hemisphere in these respects, 22 June being the shortest and 22 December the longest day.

The $23\frac{1}{2}$ degree tilt of the Earth's axis of spin thus directly generates changing daylengths throughout the year and, during the same period, the alterations in height of the noonday sun. These fluctuations are both cyclical

The height of the sun at noon varies with the seasons. To an observer in the mid-latitudes of the northern hemisphere, *left*, the sun is always high or low in the southern sky. On 22 December the sun is directly above the Tropic of Capricorn: it is at its lowest point to the observer, whose shadow is long. On 22 June the sun is directly above the Tropic of Cancer: to the observer it is at its highest point and his shadow is short.

In winter the sun never climbs high in the sky, as this afternoon street scene in New York City shows. The Earth takes $365\frac{1}{4}$ days to orbit around the sun, *right*. The plane of this orbit is called the ecliptic, because only when Earth and moon are aligned in this plane can solar and lunar eclipses occur. The spin axis of the Earth is always tilted $23\frac{1}{2}°$ away from the perpendicular to the ecliptic. Because it tilts alternately toward and away from the sun the rhythm of the seasons is created. A tilt toward the sun means long days and more daily sunshine; a tilt away means short days and less sunshine. If the Earth were exactly perpendicular to the ecliptic the year would have only one season.

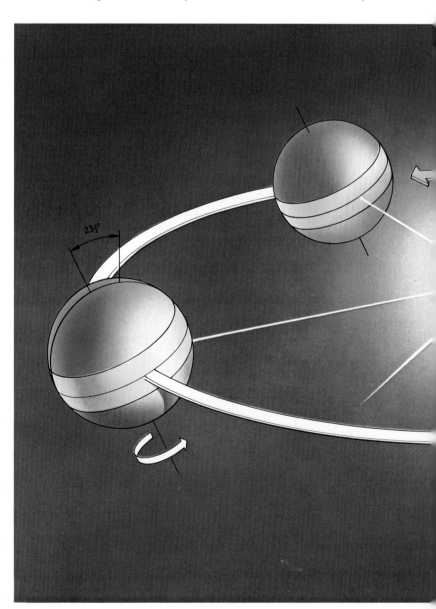

and rhythmic and, for the living creatures on Earth, provide a direct indication of the position of the Earth in its solar orbit. It is daylength, however, which communicates the march of the seasons most forcibly. On 22 December, at a position 60 degrees north, the sun is in the sky for close to eight hours. This means that the day must be 22 December. No other environmental clue such as temperature, rainfall, humidity or wind strength could give an organism that same information, although to receive it, the creature in question must have an accurate time sense in the form of some internal clock, against which external events can be measured.

As well as signalling the seasons, changing daylength and the altitude of the sun in the sky also generate the global pattern of climatic change that typify seasonality. All climate is sun-driven. Ultimately it amounts to no more than the heating effects of the sun and the movement of atmospheric gases—including water vapour—around the globe, although the movements are incredibly complex and the water vapour sometimes changes into tiny aerosol droplets (fog), large water drops (rain) or ice crystals (snow). Yet all the movements and heating are set into action and sustained by the interception of solar radiation by the Earth. Summer days are hot in the northern hemisphere because any point on the ground experiences longer days, and hence a longer period of heating, in each 24 hours. Also, the sun is at its highest so its radiation hits the ground at its least oblique angle, which means that there is a greater concentration of radiative energy on each unit area of the ground than at times when the rays are more oblique.

Despite the importance of the energy flux from the sun in determining the seasons, the distance of the Earth from the sun is not crucial in determining weather conditions. The Earth describes an elliptical orbit round the sun, but the sun does not lie at the centre of this orbit. Instead it is slightly displaced from it and is positioned at one of two foci. On average the Earth is some 93.5 million miles (149,589,000 km) from the sun, but the actual distance changes throughout the year in a cyclical way. And the Earth is not closest to the sun (in its perihelion position) in the middle of the northern hemisphere summer but in early January. At this time the Earth-sun distance is only 1.7 per cent shorter than the mean distance, so that any increase in the sun's heating power must be insignificant.

In the annual succession of the seasons, produced by the spin of the Earth around its tilted axis, animals and plants can use daylength to judge the time of year with considerable precision. Organisms need this information in order to organize their activities efficiently over long time-spans. So as well as dividing up their activities into appropriate patterns, with a periodicity of 24 hours to take account of night and day, they must also adjust their behaviour to be as successful as possible in the context of seasonal change.

Most of the seasonal changes in the environment are far more dramatic than the difference in length of day and night. Temperatures may vary by 50 degrees centigrade (122 degrees F) from one season to another, while rainfall can be completely absent in one season and more than an inch (2.5 cm) a day in the next. This sort of background change in the physical habitat generates secondary alterations in the environment during which both animals and plants respond to the climatic changes. Plants play a key role in determining ▷

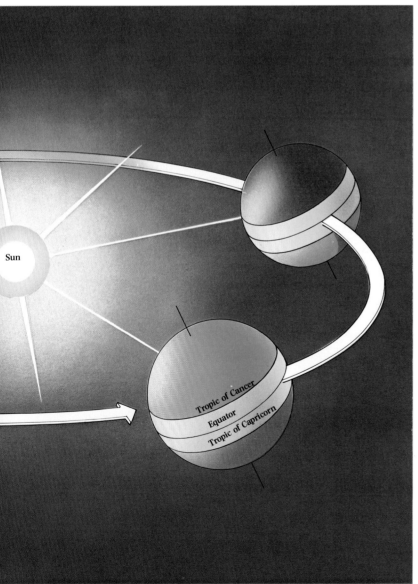

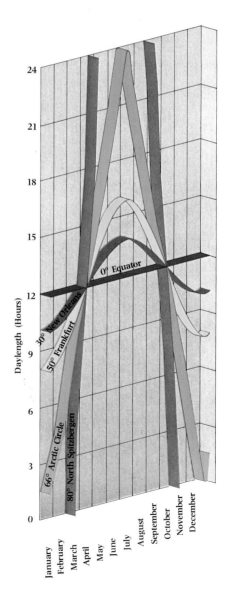

Spring

Summer

Autumn

Winter

Tropics

The length of day changes throughout the year in a pattern that is specific for any latitude north or south. For animals and plants this annual rhythm is the most predictable signal of the changing of the seasons. At the Equator the day is always 12 hours long. Everywhere else there is a seasonal variation: the same day of the same month will always be the same length in any year. The Earth's curvature, *below,* means that the density of solar radiation falling on it will be greater at the Equator than at higher latitudes.

23

▷ the structure of animal communities, and the way in which animals respond to the seasons, because in most ecosystems (the ecological units into which any environment is divided) they are the primary producers of nutrients on which all other living creatures ultimately depend. So in many tropical dry savanna regions, the annual changes in the environment can be reduced to a cycle of hot dry seasons alternating with hot wet ones. In such a habitat some of the seasonal responses of the animals can be triggered by internal clock mechanisms, or by direct responses to changing humidity, but a great deal of the annual pattern in their lives is tied to the powerful and immediate response of the plants in the environment to the onset of the rains. Rapid initiation of grass growth in the newly wet soil not only transforms the landscape scenically but, overnight, produces a mass of new food resource for herbivores and, after a short time, for the carnivores that prey on those invertebrate and vertebrate herbivores. Insect, mammal and bird breeding seasons and migrations are inextricably tied to the vegetational changes wrought by the rain.

Around the world there are scores of different types of landscape, each with a characteristic vegetation and climate. Only a few of these zones possess no clear seasons. Such zones include the almost rainless desert areas near the Equator and, at the other extreme, some portions of the tropical rain forest which are perpetually hot and wet. Everywhere else in the world, the cyclical nature of the year is perceptible to micro-organisms, animals and plants as a simple or complex pattern of seasonal and climatic changes.

Of the many influences that shape the Earth's climate, some are almost as old as the Earth itself: for example, the effect on our atmospheric constitution of

our precise distance from the sun. This distance, along with the gas-mix in the atmosphere, also plays a large part in determining the average temperature of the Earth's surface. Atmospheric conditions and mean global surface temperature are, in a sense, merely the most basic of ground rules in the game that global climate plays with the lives of all the organisms in our world. After over a century of accurate weather monitoring in many parts of the world, with weather ships dotted over the oceans, with the planet festooned with weather satellites and some of the world's largest computers at the disposal of meteorologists, we still have to admit that there are many detailed rules of the climate game that we do not yet understand.

The climate of our planet is often pictured as a huge machine—a heat engine whose working parts are the world's atmosphere, oceans, seas, lakes and rivers. For most of us the mental image of that engine would include wheels, cogs, pulleys and axles, all in concerted motion. Such an image is not inappropriate because, despite its enormous hierarchical complexity, the world's weather consists of an overlapping mass of cyclical phenomena.

The climatic heat engine of the Earth is driven by the sun. Radiated energy from the photosphere of our personal star continually reaches the Earth. This energy first interacts with the tenuous outer regions of the atmosphere, then with the denser lower layers of the Earth's gas shell, and finally strikes the Earth itself. Through the atmosphere, and on the Earth's surface, energy transfer takes place by radiation, conduction and convection. But because radiation is the dominant method of transfer, hardly any heating of the atmosphere has taken place by the time the incoming short wavelength energy from the sun

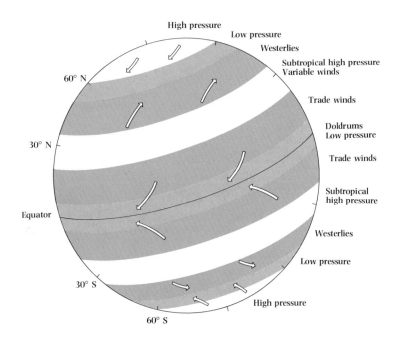

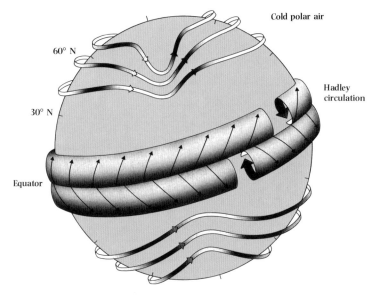

The Earth's atmosphere has several latitudinal bands, each with its own prevailing wind direction. The NASA mission to Jupiter and Saturn confirmed that these parallel bands were not peculiar to Earth. Pictures like the stunning view of Jupiter, *right*, show these patterns are the result of the interaction of the atmosphere of a spinning, spherical planet with solar energy. Features such as Jupiter's Great Red Spot are similar to some earthly atmospheric features.

Hot equatorial air rises, moves north or south and returns close to the land to produce the trade winds. This Hadley circulation is one general feature that underlies local transient weather variations. It interacts with cold polar air, creating the strong westerly winds blowing around the world. In the northern hemisphere the interaction also creates a high speed wind called the jet stream. The streak of cirrus cloud, *right*, marks the course of a high altitude jet stream.

reaches the Earth's surface. The Earth itself—or at least its outer layers—absorbs the radiation that has passed through the atmospheric gases, is itself warmed up and re-radiates much of the heat energy as long wavelength heat radiation. Although the atmosphere is relatively transparent to the incoming short wave rays, its constituent gases, especially water vapour and carbon dioxide in the lower atmosphere, are much more efficient at absorbing the re-radiated long wavelength heat. This means that the lower zones of the atmosphere, up to altitudes far above the summit of Mount Everest, 29,030 feet (8,848 m) above sea level, are heated from below upward rather than directly from the sun, which explains why the top of Everest is covered with perpetual ice, while the foothills of the Himalayas have a semi-tropical climate.

The tilt of the Earth's spin axis induces different heating levels at different parts of its surface. In a tropical belt around the Earth, stretching nearly 40 degrees north and south of the Equator, the atmosphere/ocean systems are in heat surplus, that is, they absorb more solar heat than they radiate back into space. North and south of this belt the same systems are in deficit and radiate more than they receive. If this set of relationships persisted for any length of time, temperatures would progressively climb in the tropics from year to year while it became colder and colder in the rest of the world. In fact this does not occur. The heat budget is made to balance by massive circulatory movements of air which transfer heat from the central belt to the cooler regions north and south of it. The weather machine may be 7,900 miles (12,720 km) across and, round the Equator, spin at 1,000 miles (1,600 km) an hour, but while the sun continues to emit radiation energy, that machine will keep running.

The climate of the globe may seem to us to be perfectly stable but, like the ground beneath our feet, it has undergone dramatic and rhythmic perturbations in the course of history. So while the 'solid' Earth is, in reality, a thin, moving skin of solidified rock on a slowly seething furnace of hot molten rock, the climate of the globe too is in a state of change. Indeed, man's evolutionary history has unfolded in a period of the planet's history that has probably witnessed some of the most violent changes of climate and environment that have ever occurred in Earth's five billion year life-span. The last two million years of this evolutionary period is usually termed the Quaternary era or period. Conventionally it is split into the Holocene, encompassing the most recent 10,000 years, and the very much longer Pleistocene period which immediately preceded it. In terms of the total history of the Earth the span of the Quaternary is remarkably brief, taking up only one twenty-fifth of one per cent of the total time that has elapsed since the Earth was formed. But during this mere flicker of geological time, man came into existence, and the form and climate of the planet were spasmodically and recurrently altered.

These Pleistocene changes in global climate are commonly called the Ice or Glacial Age, although it is only one of three such eras. About 20,000 years ago the temperate world of the northern hemisphere we know today was almost unimaginably different. The difference was ice. If all the ice in the world today were concentrated in one place, it would form an amazing cube measuring 200 miles (320 km) along each of its edges. In the coldest period of the most recent Glacial Age, the Earth carried a mantle of ice with a mass that could have produced a cube three times as large. That alteration in ice mass has brought ▷

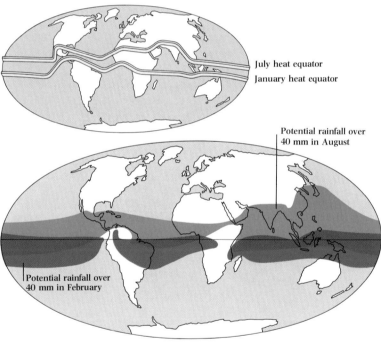

July heat equator
January heat equator

Potential rainfall over 40 mm in August

Potential rainfall over 40 mm in February

The patterns of winds and atmospheric circulation shown in the diagrams on page 24 are not constant throughout the year. The tilt of the Earth's spin axis forces these bands to shift up and down with a seasonal rhythm. The simple overall picture may be complicated locally by continents, mountains and oceans. The zone of the most direct solar heating shifts $23\frac{1}{2}°$ north to $23\frac{1}{2}°$ south and back again every year. This cycle of heating means the heat equator or the band of highest global temperatures itself moves cyclically. With a slight time delay after the summer and winter solstices, the heat equator, *above top*, reaches its northernmost and southernmost positions, respectively. Extreme solar heating over large oceans causes vast quantities of water to evaporate into the atmosphere and then precipitate as rain. So the zone of heavy and potentially constant rainfall generally coincides with the area of the heat equator. As a result, the band of heavy tropical rain, *above*, moves up and down the torrid zone, bringing the wet season.

The monsoon season in the Indian subcontinent and in Southeast Asia is one dramatic result of the seasonally shifting band of tropical rain. In winter huge evaporation occurs over the Indian Ocean and the west Pacific. In summer the water-laden air swings northward, dropping its cargo of rain on the land. For example, 95 per cent of Bombay's rain falls between June and September. Such rain is good for growing rice but can cause terrible flooding in low-lying areas.

▷ about, both directly and indirectly, incredible changes on the planet. Directly, much of the northern hemisphere was once buried under the ice. In the New World the ice sheets spread southwards to smother the eastern regions of Canada, much of the American Midwest and New England. More ice ground its way over Alaska, western Canada and the northwestern states of North America. On the other side of the Atlantic, from centres in Scandinavia and Scotland, the frigid juggernauts moved south to cover most of Britain and northern continental Europe. Smaller ice caps reached out from the Pyrenees and the Alps into the surrounding terrain. On a less extensive scale, ice caps grew in Argentina and Australasia, and tundra landscapes fringed all the ice sheets.

The indirect changes of the ice expansion were no less dramatic. With dropping temperatures and consequently changed habitats, animal and plant species were subjected to powerful new selection pressures. Some succumbed to extinction but others adjusted to the changing conditions by means of migrations or adaptations that made them more efficient in the new, cold world. The sheer weight of ice in some regions quite literally squashed the ground beneath it, sometimes producing rock deformations of many hundreds of feet.

The ice, locked up in solid form in the new ice sheets, had to come from somewhere and basically it came from the world's oceans. So great was the removal and conversion of water to ice, that worldwide ocean levels dropped as much as 350 feet (107 m). As this happened, extensive areas of the continental shelves around the world were exposed and became visible dry land. Continents expanded sideways and, as a result, some land masses which had previously been separate became continuous. Perhaps most dramatically, North America and Asia were linked by a land bridge in the area of the Bering Strait.

About 14,000 years ago the ice-spread reached its maximum extent. Net reconversion of ice into ocean water began, and the glaciers began to retreat as the levels of the oceans rose. About 7,000 years ago the ice cover of the Earth, its geographical climatic pattern and its vegetation systems were much as they are today. Thus, in broad terms, the Holocene period corresponds to this latter, most recent interval of raised global temperatures, shrunken ice sheets and high ocean levels.

This most recent spasm in the Earth's climate is the one that is best known, but it is reasonable—and also a wise precaution—to question its uniqueness. Has the world just passed through a completely unorthodox, single global twitch, or were these events just one of a succession of cyclical happenings? There seems little doubt that Glacial Ages themselves have an internal, periodic, rhythmic structure through time. It also appears very likely that the Pleistocene Ice Age was simply the last of three very widely spaced Ice Ages, which have gripped the Earth for relatively short periods during the last billion years or so of global history. Between these cold periods, most of the Earth's climatic history has been both warmer and wetter than it is during Ice Ages, and the ice sheets have been extremely restricted in extent or even non-existent.

The long-term climatic story of our planet is thus both cyclic and rhythmic. The Earth's weather pattern beats like a heart, but the beats are infinitely slow.

To be highly simplistic, the Earth's natural or typical climate is warm and wet, but superimposed upon this background has been a succession of colder, drier Ice Ages. The irregular alternations of the two climatic types constitute the cycles of the longest periodicity that are known about in the Earth's five billion years of development. Within the cold Ice Ages rhythmical patterns of climatic change are apparent, the planet cycling between relatively warm interglacial periods, such as the one we are experiencing now, and colder glacial periods in which the Earth's ice caps are much more extensive.

Our knowledge of the weather patterns that occurred thousands or even millions of years ago has not been acquired by chance. Rather it has only been made possible by a slow and painstaking accumulation and synthesis of many different types of evidence. To make informed speculations about climate during eras far back in geological history, an overlapping series of different techniques must be employed. With considerable precision and confidence we can back-track some 8,000 years in time. This is possible because of our ability to date organic material in the ground with radiocarbon (carbon–14) dating methods, cross-checked with tree-ring material of absolutely known age.

The radiocarbon technique measures the time that has elapsed since an organism died by measuring the proportion of different physical forms, or isotopes, of carbon, each having a slightly different atomic weight, in its preserved or fossilized remains. The technique is practical and accurate back to 50,000 years before the present day, but is particularly precise when calibrated against tree-ring material from bristlecone pine trees. These trees live above 9,000 feet (3,000 m) in mountains in the southwestern states of the United States of America. Some living trees are over 4,000 years old, which means that their seeds were germinating at the time when the first pyramids were being built. In the same area, dead wood with datable tree rings provides samples 8,000 years old. Using the radiocarbon technique it is possible to date fossil animals and plants over the past 50,000 years of the Quaternary era. And if the temperature requirements of these different types of organisms are known, it is possible to make detailed estimates about the climates in which they must have lived.

Another isotopic technique involves measuring the proportions of oxygen isotopes in ocean sediments laid down over the last million years of the Pleistocene period. This data provides a specific and subtle measure of the history of the global climate because, indirectly, it allows an assessment to be made of the volume of ice on Earth at any instant in the historical record. The scientific theory behind such a near-miraculous feat is as follows. When a volume of water freezes—leaving some water unfrozen—the oxygen isotope mixture in the ice and the residual water (both made of molecules consisting of one oxygen atom and two hydrogen atoms) is different. Compared with the original water the ice contains more light oxygen isotopes and the water fewer such isotopes. When the ice melts, the light isotope concentration in the water rises again.

During the expansion and contraction of ice sheets through a series of glacial and interglacial periods, these processes of freezing and thawing were going on not in a laboratory but throughout the world. Animals living in the oceans of the world were inhabiting an aqueous environment whose oxygen ▷

Each of the three Ice Ages of the last billion years had a rhythm of cold, dry glacials and warm, wet interglacials, *left*. We are probably now in an interglacial of the Pleistocene Ice Age.

Landscape changes generated by the expansion of ice and by glaciers and ice caps that persist in our interglacial world tell us about the last glacial period. The magnificent Bettmeralp glacier, *left*, in Switzerland shows the solid slow-moving river of ice, streaked with rocks, and dwarfing the town below. The gigantic isolated boulder, *below*, in California's Yosemite National Park, moved from a distant source by glacier power, is now a mute witness to former ice sheets.

Lakes in valley floors, U-shaped valley sides and parallel scratches on hard rock are signatures of a former glacier. The valley lake, *right*, is in the Black Cuillin mountains on the Isle of Skye off the coast of Scotland.

▷ isotope concentration was changing in step with the climate, and their preserved bodies still carry the chemical evidence of those changes in water composition. In ocean sediment samples it is possible to map these fluctuations through time and to reconstruct the climatic changes. Each time the polar glaciers expanded, light isotopes of oxygen were effectively extracted from the world's oceans. When melting ice sheets retreated, the locked-up stores of light atoms of oxygen were returned to the oceans, thus restoring its isotopic constitution.

Using these and other techniques, it has been discovered that apparently most of the two million years of the Pleistocene era has consisted of large-scale shifts, from glacial to interglacial conditions and back again, with a basic period of about 100,000 years. This extreme instability of climate imposed profound constraints upon all the animals and plants on Earth. In every continent it is possible to see patterns of species production, species distribution and types of adaptation which are all the evolutionary results of the selection pressures imposed by a pulsating Ice Age.

The record in the rocks tells us that there were at least two previous Ice Ages before the one in which we are embedded, albeit in an interglacial period. The first of these occurred during the late Precambrian era, approximately 700 million years ago. Due to rock movements and erosion since then, it is impossible to deduce a distinct pattern of glacials and interglacials in this Ice Age, but glacial deposits of the correct antiquity are certainly widespread in southern Africa, Brazil and Australia.

The second Ice Age has been termed the Permocarboniferous Glacial Age. It

began just over 300 million years ago and lasted for more than 50 million years. In it continent-wide glaciation affected parts of South America, southern Africa, India and Australia. At that time these areas, and Antarctica, were fused together as a single, huge continent. Later, continental drift moved segments or plates of this continent apart to produce the separate southern hemisphere land masses we know today.

Continental drift has been proposed as a general explanation for the timing of the three main, extended periods of massive glaciation that the Earth has experienced. This theory suggests that wherever substantial continental land masses are shifted into zones near the poles, an increasing accumulation of ice occurs in those high latitudes. This build-up may possibly take place by means of a positive feedback mechanism, whereby, once initiated, the accumulations of ice reflect more solar energy directly back into space, so allowing further accumulation to occur. This hypothesis certainly seems to fit both of the last two Glacial Ages. In the Permocarboniferous, for example, Earth's continents were fused into a solitary supercontinent, Pangaea, centred on the Equator but stretching to the position of the South Pole. It is precisely those portions of Pangaea nearest the Pole that were glaciated. The present Glacial Age was probably precipitated by the shifting of North America and Eurasia towards the North Pole. In the northern hemisphere glaciers first started appearing about 10 million years ago.

Because it occurs too slowly, and does not have regular rhythmic components, the theory of continental drift cannot explain the decidedly regular 100,000-year rhythm of climate within the present Ice Age, as

Long-term cycles describe changes in the tilt and direction of the Earth's spin axis, and also in the shape of its orbit around the sun. These changes gradually alter the pattern of solar radiation falling on the Earth and profoundly influence the extent of the ice cover on the planet. Three cycles, with periods of about 21,000 years, 41,000 years and 100,000 years probably played a causal role in generating the observed pattern of glacial and interglacial periods within the Ice Ages. While these cycles account for the rhythm within an Ice Age, they do not, however, explain how an Ice Age is triggered in the first place.

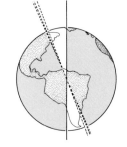

As the ocean waters froze during the last glacial of the Pleistocene Age, they expanded into huge glaciers. Spreading inexorably from the polar caps over great distances, the ice carved out the present northern landscape and altered the global climate patterns. The Arctic and Antarctic ice caps were much more extensive than they are today, and in the northern hemisphere what is now temperate land was then part of the polar region. The maps demonstrate the ice cover of the present day, *right*, and the extent of the Arctic ice sheet at the height of the last glacial of the Pleistocene Age, *below*, about 20,000 years ago.

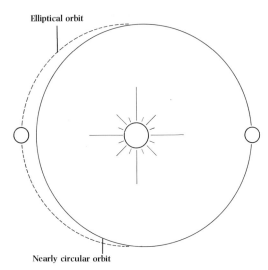

Elliptical orbit

Nearly circular orbit

The shape of the Earth's orbit changes every 90,000 to 100,000 years. The orbital path is basically elliptical with the sun at one focus of the ellipse. During one cycle the orbit oscillates between being more elliptical to being almost circular and back again, *above*. Together with the other two cycles, this produces complex changes in the amount of solar radiation reaching the ground at a particular latitude and at a particular season. These cycles and the changes they bring do not alter the total energy input from the sun. What they do alter is the way this energy is distributed over the Earth's surface. Global temperatures thus rise and fall enough to arrest or trigger glaciation.

The Earth nods up and down with respect to the sun. This is due to its spin axis tilting from 22 to 24½° and back to 22° every 41,000 years, *above*.

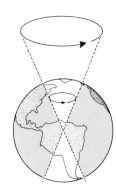

The Earth is wobbling in space like a top spinning very slowly. Its spin axis describes a huge conical path against the stars every 21,000 years, *above*. The southern hemisphere has hotter summers than the north since the wobble is superimposed upon the nodding of the axis. In 10,000 years this situation will reverse.

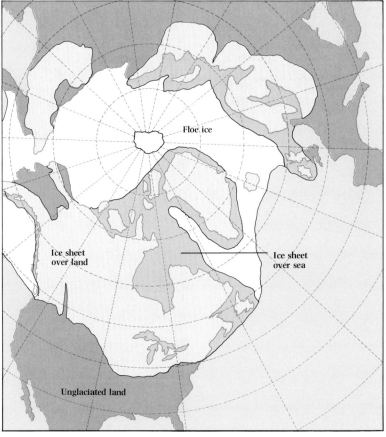

Floe ice

Ice sheet over land

Ice sheet over sea

Unglaciated land

demonstrated by isotopic oxygen and related studies. The best general theory so far propounded to explain the 100,000-year periodicity of climatic changes was first put forward by a Yugoslav mathematician Milutin Milankovitch, in 1920. In simple terms, Milankovitch suggested that three independent, rhythmic alterations in the Earth's celestial behaviour combine to produce temporal patterns in the amount of solar energy impinging on the Earth at different latitudes. The spin axis itself wobbles, with a complex dual periodicity of 23,000 and 19,000 years, in a movement known as precession. The tilt of the Earth's spin axis, relative to the orbit of the planet round the sun, can also vary cyclically between about 22 degrees and 24½ degrees, with a periodicity of some 41,000 years. Longest of all is a cycle, with a periodicity of between 90,000 and 100,000 years, in which the orbit of the Earth around the sun bounces between two configurations, one relatively elliptical, the other relatively circular. This is the eccentricity rhythm.

The eccentricity rhythm can obviously be tied to the dominant 100,000-year rhythm in the recent glacial/interglacial pattern. More startlingly, however, ocean sediment samples, analyzed by means of oxygen isotope estimations, also show minor periodicities of about 43,000, 24,000 and 19,000 years. These values are so close to the patterns predicted mathematically by Milankovitch, long before isotope studies were conceivable, that they represent a very powerful vindication of the basic truth of his theory.

It appears, then, that the Earth's cyclic and rhythmic climatic history is the consequence of a double interaction. First, the combined effects of planetary motion on the influx of radiant energy from the sun, and second, the

combination of these mechanisms with the painfully slow creeping of the continental masses over the Earth's surface. Every smooth-sided glaciated valley, whether it be in Scotland, Scandinavia or Alaska, is the final, physical evidence of the effects of the different motions of the whole planet—motion through space and the motion of the Earth's crust.

Apart from the partisan fact that we happen to be living in it there is nothing unique about the present moment of planetary history. Milankovitch cycles do not stand still and the cyclical history of the Earth's climate, which is elegantly explicable in terms of the rhythms of planetary spin and orbital motion, must continue on into our future. What does this future hold? Can the evidence that has been gathered about previous Ice Ages, glacials and interglacials enable mankind to predict what the global climate will do over the next fifty, hundred, thousand or million years?

The usable evidence in this sort of speculation comes in three forms. First, we can use the mathematical predictability of the Milankovitch cycles which should, in theory, provide us with a broadly outlined picture of the major climatic changes due in the future. Second, we can attempt to extrapolate the detailed weather recordings of the past few centuries into the near future. Finally we must take account of the evidence suggesting that, in fact, man's presence in this planetary ecosystem is not climatically irrelevant because we are physically changing the atmosphere.

Taking these three categories of clues in turn, it seems that the basic cyclical patterns within the last million years suggest that we are only at the very beginning of an interglacial period. This evidence suggests that we are only just ▷

The temperature record of the last million years is based on oxygen isotope measurements taken from deep-sea sediment cores, *right*. The 100,000-year cycle of changes in the shape of the Earth's orbit could account for the seven or eight cold glacial peaks in the last 700,000 years.

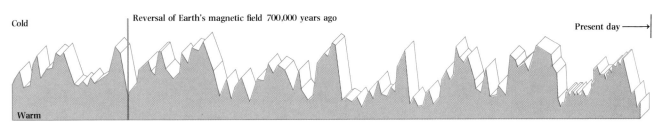

Cold

Reversal of Earth's magnetic field 700,000 years ago

Present day ⟶

Warm

Hooded crow

The hooded crow, *Corvus corone corvix*, has an unmistakable grey nape, back and chest.

Carrion crow

Carrion crows, *Corvus corone corone*, live in fairly open country and are all black. They form flocks, especially at roosting time, and nest in pairs on cliff ledges or in trees. Where they interbreed with the hooded crows, in narrow belts of land in Scotland and central Europe, the offspring all have darker feathers.

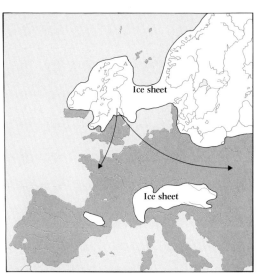

Ice sheet

Ice sheet

The ebb and flow of the ice sheets during the glacial and interglacial rhythm of the Pleistocene Ice Age has imposed profound evolutionary influences upon animals and plants. The ice changed their habitat so that they had to adapt, migrate or die. Carrion crows and hooded crows are both members of the same European crow species. The map, *right*, shows the distributions of these two subspecies which overlap only along a narrow zone of hybridization where they interbreed. An ancestral stock was possibly pushed south by the last glacial, see map *above*, and broken up into eastern and western populations. Over many years a segregation into two distinct forms has occurred.

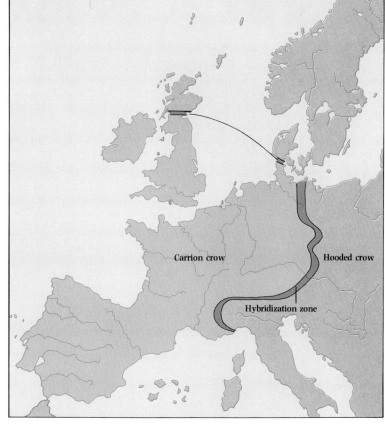

Carrion crow

Hooded crow

Hybridization zone

▷ entering many tens of thousands of years of relatively mild, moist weather. Looking at the past few hundred years of climate in the temperate mid-latitudes of the Earth we can distinguish many decades of unusually cold weather in the thirteen and fourteen hundreds and again in the seventeenth century. From around the year 1700 until the present day average temperatures seem to have been rising.

The possible influence that man may have on the future of his planet's climate is hard to judge, although it seems likely that his increasing utilization of forests might indeed have some impact. Since the beginning of the Industrial Revolution in the early nineteenth century the burning of coal, timber and then oil has led to an accumulative increase in atmospheric carbon dioxide levels. Tree destruction augments this change by reducing the extent to which carbon dioxide gas is removed from the air by the process of photosynthesis. It is impossible to make accurate estimates of this increasing carbon dioxide level but it is likely that today the natural equilibrium concentration of carbon dioxide in the atmosphere is 335 parts per million, compared with a figure of 270 parts per million in 1850, and is increasing by one or two parts per million each year.

These increments in carbon dioxide concentration, tiny though they are, could have profound effects on the Earth's weather. They could eventually increase the average temperature of the world a few degrees by means of an increasingly powerful greenhouse effect. Such a change could have dramatic consequences on the melting of the polar ice caps, sea levels, rainfall distribution and many other variables. Taken together, however, such changes tend to nudge the climate in the direction of typical interglacial weather so that man's activities over the foreseeable future are most likely to stabilize what is currently the Earth's natural state, namely that of interglacial mildness.

One of the cornerstones of the Milankovitch theory of variable planetary climate on the Earth is the effect of the precession of the Earth's spinning axis on solar energy input. The precessional movement of the Earth is exactly analogous to the rotating wobble that a spinning top develops as it begins to slow down. While the top is spinning fast, the orientation of its spin axis changes little, but as it slows down, so the rate of change of that orientation speeds up. In its precessional movement the spin axis of the Earth follows, in an approximate fashion, the surface of a cone whose tip-angle is 47 degrees, and it takes about 20,000 years for the orientation of the spin axis to describe a complete circle in the heavens.

The precession wobble of the Earth is essentially a gravitational effect, brought about by the influences of a small, close moon and a distant but massive sun on an Earth that is not a perfect sphere. Both the moon and the sun exert a complex gravitational pull on the Earth which has a bulge around the Equator—a sort of planetary paunch. Compared with the rest of the planet, this bulge experiences a slightly different pull from the moon and the sun and it is this that induces the precession wobble. Precession is an effect of lunar gravitation, not directly observable by any single ordinary person. Only over many generations is it possible to observe that geographical north points to slightly changing positions in the star field. At present the night sky in the northern hemisphere appears to revolve around a point near Polaris, the pole

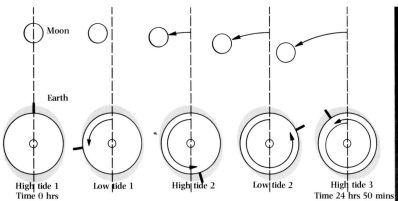

High tide 1   Low tide 1   High tide 2   Low tide 2   High tide 3
Time 0 hrs   Time 24 hrs 50 mins

The moon's gravitational pull combines with centrifugal forces on the oceans to produce the daily rhythm of the tides. Gravitation is an attraction between the huge masses of Earth and moon. Centrifugal forces are due to these masses moving around a common centre of rotation. Together they produce a double water bulge on the Earth's surface. How the Earth's 24-hour rotation combines with the moon's motion to produce a tidal sequence is shown *above*. One spot on the earth, indicated by a vertical marker, moves through the water's bulges and troughs in a cycle lasting just longer than a day since the Earth rotates and the moon orbits in the same direction. If this were a seashore a biological zone would arise, each height on the shore experiencing different immersion times per cycle.

The seashore at Puerto Vallarta, Mexico, *right*, is a colourful example of an intertidal zone. Over just a few yards relatively large changes in light, temperature and moisture influence the organisms' way of life. A typical beach profile, *below*, shows the four main tidal levels and their associated zones where some plants live according to their metabolic and reproductive needs.

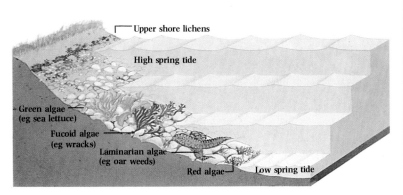

star, but because of the phenomenon of precession, this point is constantly—if slowly—on the move.

There is another effect of the moon's gravitational attraction that is simply and directly observable. Any child will tell you that at the seaside the tide goes in and out. It is common knowledge that in some way the tides—the cyclical advance and retreat of the ocean's fringe up and down the shoreline—are driven by the moon. But the gravitational dance of sun, moon and Earth that generates observed tidal patterns is, however, extremely complicated and, without recourse to rather abstruse mathematical formulations, only the major patterns and influences are comprehensible.

To begin at the beginning; all objects in the solar system attract one another gravitationally. Tidal systems are the particular result of one pattern of cyclically changing gravitational attraction on the waters of the Earth. In a more general sense, the passage of the moon over the Earth's surface in its approximately 28-day orbit raises tides other than oceanic ones. Atmospheric tides follow the circling moon and minute rock tides do the same. Ignoring the sun for a moment, and considering the moon poised over the Earth at a single instant in time, water particles on the side of the Earth closest to the moon will be attracted to it. If they are able to move these water particles will heap up to form a tidal bulge of deeper water beneath the moon's position, but that is not the only effect.

The moon does not really revolve around the planet Earth. The truth is that both the Earth and its solitary satellite revolve around a common centre of gravity placed on a line exactly joining their centres. The centrifugal acceleration that results from this movement varies with the distance from the common centre of gravity, and is, consequently, much larger on the portion of the Earth's surface directly opposite the moon's position than it is underneath the moon. The acceleration affects the Earth's waters in the same way as gravity. So, in effect, the moon's gravitational pull produces one bulge directly underneath it, while apparent centrifugal forces induce another on the opposite side of the Earth. If the Earth were a perfect sphere completely covered with water and only the moon affected that water, the tides would be produced by two symmetrically disposed bulges, through which any point on the rotating Earth would pass approximately twice every 24 hours.

That is the basic theoretical situation but, in reality, other factors complicate the real tidal phenomenon considerably. First, the sun cannot be ignored. It has a gravitational effect similar to that of the moon on the Earth's water, but the strength of its influence is only about half that of the moon's. When sun, moon and Earth are in a straight line the two tidal bulges are much larger than those produced by the moon alone. These are the spring tides which occur approximately every two weeks and which, inevitably, coincide with the appearance of a full or a new moon, because it is only at such times that the required linear configuration is achieved. Tides of this sort generate the highest high tide levels and also the lowest low tide levels. When, in contrast, sun, Earth, and moon are at right angles, the sun's gravitational influence will partially counteract that of the moon. The resulting less extreme, or neap, tides occur approximately one week after spring tides.

Real tides have other complex periodicities and irregularities. These result ▷

▷ from factors such as the friction between the ocean bottom and moving water masses, the physical blocking of water transfer by solid land masses, the resonance of water masses in relatively enclosed conditions and a number of other subtle physical effects. Taken together these perturbations make tidal prediction in the real world a highly sophisticated science. Specific predictions have to be made for every different geographical region of the Earth's ocean mass, simply because local conditions have an enormous influence on the basic tidal patterns. Some water masses, such as the Mediterranean, experience limited tidal effects while others, such as the coastlines on either side of the North Atlantic, demonstrate dramatically larger tidal changes. This particular difference is said to have caused the failure of one of Julius Caesar's attempts to invade England. Conditioned by the almost tideless Mediterranean, it appears that his captains underestimated the large tidal ranges along the English Channel coast, and this caused much of one invasion fleet to be wrecked on the shore.

The shoreline, the interface between two great ecosystem complexes, the sea and the land, is a life zone of staggering diversity. Even on a tideless planet the oceanic fringe of any land mass would be interestingly different from either the ocean or the land, in respect of the specialized organisms that lived there. All biological boundary regions have this property because, in them, a wide range of important physical parameters, such as light, temperature and the availability of water, undergo rapid change. On a tidal world such as planet Earth, the tides have the effect of spreading out the interface zone so that it extends over a considerable distance both horizontally and vertically. Tides also accentuate the way in which environmental factors fluctuate in the area of the shoreline.

The intertidal zone, that is, the region between the lowest tide marks and the highest, includes a complex community of animals and plants which often demonstrate very high productivity. The plants that live there, which are mostly algae (seaweeds) but include also lichens and a few flowering plants, have plentiful water, usually an unlimited supply of mineral salts from the sea, and sufficient light for the manufacture of foodstuffs by means of photosynthesis because, in this fringe zone, water levels are never very deep.

In all parts of the world the different nature of the intertidal zone depends on the physical form of the land abutting the sea, the exposure of the shoreline to strong wave action, the ambient climatic conditions and the physiological nature of the animal and plant life that inhabits it. Two extreme intertidal types would be a gently shelving mud or sandy shoreline, exposed only slightly to wave action, and a vertical cliff, facing the prevailing winds and pounded by waves with a long oceanic fetch. In between these two extremes are a wide range of intermediate types with moderate shore-profile gradients. But no matter how the intertidal band is stretched out, whether it is an almost flat expanse of mud several miles wide or a vertical cliff face, tide movements will provide extremely marked gradients of conditions for life throughout the zone.

The most basic of the gradients of the shoreline is that of immersion. At the base of the shoreline there will be a zone which is only exposed to the air during one or two low spring tides a year. At the top of the intertidal shore there will, likewise, be a band that is only covered by the sea on the same number of days.

The changing phases of the moon are the altering shapes an earthly observer sees of the sun-illuminated face of the moon. Probably the easiest of astronomical phenomena to observe, it must have been the earliest cosmic cycle to be recognized and revered by ancient man. Apart from sunspots, other easily detected features of the lunar disc are the only surface characteristics of any astronomical object that can be seen with the naked eye. Good unaided human vision can distinguish the broad dark plains called seas, the highland regions and a few of the largest craters. Through low-power telescopes and even binoculars much more detail can be seen, including the bright radiating rays that stretch out enormous distances from craters such as Tycho and Copernicus. These rays show the material ballistically thrown outward when the impact craters were formed by huge solid objects hitting the moon's surface. The lunar cycle, from crescent moon to full moon through to old moon, presents different opportunities to view the various surface features. Most detail is shown up when the moon is obliquely lit by the sun because then the topography casts large shadows. Thus the waxing or waning moon is the best time for lunar study.

Direction of solar illumination

New moon

Crescent moon

Old moon

Half moon first quarter

Half moon third quarter

Waning moon

Gibbous moon

Full moon

The moon is the only natural satellite of the Earth and has no mechanism for generating light of its own. It shines only by reflecting the light of the sun. From the moon the astronauts saw the Earth in precisely the same way—earthlight is also reflected sunlight. Apart from lunar eclipses, when the Earth comes exactly between the moon and the sun, half of the moon's nearly spherical surface is always illuminated by the sun: one half is always brightly lit, the other almost utter darkness. This is shown diagrammatically, *left*, in the inner ring of moon images. The appearance of the moon from the Earth varies, however, and the regular sequence of apparent changes of shape are termed phases. When the moon is between the Earth and the sun, its dark side is turned towards us and the moon cannot be seen. This is the new moon. When the Earth comes between the moon and the sun, the moon's face is turned toward us and we see a full moon. The intermediate phases of the moon, the crescent, the half moon, the gibbous, the waning, the last quarter and the old moon, are caused by a cycle of intermediate viewing conditions shown, *left*, in the outer ring of moon images. If the new moon is exactly between an observer and the sun a solar eclipse occurs.

In between there will be a more or less regular gradient with organisms being immersed for different proportions of their lives. This immersion gradient generates a wide range of secondary gradients in such features as temperature and desiccation. As a result of this superimposed series of gradients, all shorelines with marked tidal influences support a series of animal and plant communities, arranged in parallel from the top to the bottom of the intertidal area.

If you stand at any point on a graduated shoreline, it is possible to determine your position up or down the shore simply by identifying the animals and algae under your feet. Particular molluscs, worms and crustaceans inhabit—and are specifically adapted for—restricted zones of the shore. In the same way, the seaweed plant-cover of the intertidal zone consists of a series of specialized algal types, each operating most efficiently at a different degree of tidal immersion. On rocky temperate shores, for example, large brown laminarian kelps often dominate the low part of the shore, fucoid wracks, the centre zone and a mixture of green algae and rock-covering lichens, the upper shore.

For all these highly adapted organisms the cycle of high and low tides is the dominant environmental variable. This rhythm is more important in inducing adaptive behavioural patterns in the organism's life-style than any other physical cycle, including the changes that take place between day and night. So the animals of the intertidal zone organize their hunting and grazing, their meeting and reproduction according to tidal imperatives. The photosynthesis and reproduction of algae are likely, in a similar fashion, to be highly ordered by tidal sequences as well as day–night cycles.

The seashore may seem to be an unremarkable piece of seaside landscape, but it provides in microcosm an image of all rhythmic life processes. In a peculiarly direct way it is possible to perceive how an environmental rhythmic cycle—the pulsation of the tides—imposes both spatial patterning and behavioural patterning in time on all the creatures that are exposed to that elemental cycle. More than that, it is easy in this case to understand how the environmental rhythm is the direct consequence of a cosmic motion. All societies have realized that the moon governs the tides—no one with a boat and eyes lifted above the horizon could fail to grasp the inexorable linkage between the phases of the moon and the height of the tide.

The causal bonds between moon motion, tides and the behaviour of intertidal life are simply one set of ties, in a whole web of such connections, that link the motions of the solar system to the ebb and flow of life on our planetary home. The cyclical patterns can be so slow that we have no racial memory or human record of their passing. The glacial/interglacial pulsations, with their periodicities of many thousands of years, come into this category. The most rapid of the cycles, such as the twice-daily tides, occur at frequencies which demand that all but the most short-lived of organisms must perceive them and be influenced by them.

The astrologers of the past have been guilty of propounding some astonishing nonsense. Yet at the centre of their cosmic view they had grasped one jewel of truth, namely that all creatures on Earth, including man, have grown up reflecting in their daily lives the patterns and rhythms of the motions of the heavens.

The moon's pull is the main gravitational force on the Earth's oceans. Its nearness to Earth makes up for the moon's small size, and its pull is about twice that of the sun. The width of the broad arrows *left*, shows the relative size of the two forces. When the Earth, moon and sun are in a straight line, the combined pull of the sun and moon produce very high and low 'spring' tides. These occur about every two weeks, at the new and full moon. When the Earth, moon and sun form a right angle in space, *lower left*, in the days between the spring tides, the sun partly offsets the moon's pull. This results in the smaller high and low neap tides and occurs at a half moon. Such a rhythm, with its 14 and 28-day periods, is exemplified by the lunar month tide sequence, *below*, on the Atlantic seaboard of the United States near New York.

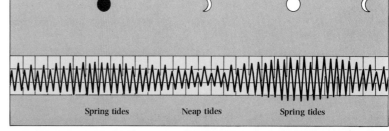

Spring tides Neap tides Spring tides

33

Rhythms of the seasons

When analyzing the attributes that have made *Homo sapiens* the most successful and widespread animal species on planet Earth, brain power is usually the skill that is emphasized first, followed, rather as an afterthought, by man's upright, two-legged stance. The blinkers of self-esteem stop us from seeing two other characteristics that must have been important to the evolutionary and social ascendancy of the human race—when considered alongside most of the other animals on earth we are comparatively huge and extremely long lived.

We human beings are not used to thinking of ourselves in such a comparative way, but more than 99.9 per cent of animal species are smaller and live shorter lives than we do. This combination of characteristics is in no way extraordinary because, for many years, biologists have recognized that there is a direct relationship between size and longevity or generation time. What is unusual·is the additional degree of insight the acceptance of these two human attributes provides.

The large size of the human frame probably meant that during the early, crucial stages of man's evolution in the Pleistocene era, which began about two million years ago, there were few animals that he could not hunt and kill when operating in a cooperative group. Man's longevity opens some evolutionary doors but shuts others. He takes so long—a dozen years or more—to reach sexual maturity that his natural rate of increase can never approach that of fruit flies or white mice. But man's long life-span—three score years and ten according to biblical rule of thumb—allows for developmental possibilities and perceptions of the world forever barred to such 'boom and bust' reproducers.

The intellectual reward of human longevity is that it endows us with the long view. Emotionally and instinctively we never measure out our lives in minutes or seconds. The pattern of days is only the most immediate and fleeting aspect of our world view of time. Our natural milieu is the grand progress of seasons and years so that, if asked to describe the pattern of our lives, most of us naturally fall into a discussion of periods of this duration. For man, such a perception of time is a mental creation, but it is also rooted in fundamental biological processes which are only partly conceived in cerebral terms. Physiologically we respond to the changing seasons much as other large, long-lived organisms do.

If they are to survive to old age all large animals and plants must respond to the alterations in their habitats produced by the recurrent cyclical sequence of the seasons. Experience of these changing patterns by untold past generations has imposed constraining selection forces on the evolution of all organisms with a life-span of 12 months or more. Those strains or species of animals and plants which did not have appropriate behavioural responses to the onset of seasonal weather built into their genetic codes perished more often than those that did. Perennial insect-pollinated plants without the genetic instructions telling them to flower in the season when the appropriate insects were active, died out. They were simply fertilized less often, so their offspring came to represent less and less of the plant population.

Inexorable evolutionary forces shape responses to seasonal change, just as they act on structure or outward appearance. The result, seen in animals and plants throughout the world, is that in every latitude living things are precisely suited to the pattern of seasonal change characteristic of their own particular global location.

For most people living in Europe or North America talk of the seasons almost inevitably conjures up a fourfold sequence of periods that make up a year. Our languages give these four annual segments—spring, summer, autumn and winter—names that evoke a vivid mental picture of life in which the weather is only one element. This classical progression of the seasons, the sequence beloved of poets, is not a pattern that is equally valid all over the world.

The particular seasonal cycle experienced by an area of the globe is crucially dependent on its precise position. More specifically, latitude is the cycle's main determinant. Considered in the broadest terms, the apparently orthodox four-season year is a phenomenon typical of the mid-latitudes, the temperate regions of the earth sandwiched between the equatorial tropical zone and the frigid polar zones at the top and bottom of our planet. Where the quartet of seasons does occur, all four are characterized by particular patterns of the three most vital physical influences that mould the environment for all living creatures, including ourselves—temperature, light and water.

Temperature is important because it is the essential physical regulator of the rate at which the chemical processes that determine life can occur. This regulation is a direct consequence of the thermodynamic fact that most chemical and physical processes that need energy in order to take place are speeded up as temperatures increase, and slowed down as they decrease. This is, provided that the temperatures are not so extreme as to kill the organism or place it in a state of suspended animation.

In terms of energy, it is light from the sun that drives the biological ecosystems made up of animals and plants. Ultimately sunlight is the basic energy source for all net conversions of inorganic molecules into organic ones, such as proteins, fats, sugars and all the other complex substances of life. This conversion is achieved in plant cells containing the green pigment chlorophyll by means of the process of photosynthesis, which is driven by sunlight. Pared to its bare essentials, photosynthesis involves the combination of carbon dioxide from the atmosphere with water drawn up from the soil to make sugar molecules. As a result of this process, oxygen is released into the air. The products of plant photosynthesis provide the food for herbivorous animals, and these, in turn, are the food for carnivores, occupying higher and higher positions on the pyramid of predatory hierarchy. When a lion kills and consumes an eland it is, in terms of biological conversion, only one step away from eating grass. Sunlight also provides the necessary milieu for all seeing creatures. Without this light, vision, the most precise and specific of the senses, would not only be irrelevant but unimaginable.

Water constitutes a surprisingly large proportion of all living things. All living cells consist of at least 90 per cent water and all the chemical processes that go on in cells take place in aqueous solution. The amount of water available in an environment can often prove to be the overriding factor, enabling a biological activity to occur or prohibiting it.

Spring, summer, autumn and winter each have a consistent signature of light, temperature and water availability, but there is little doubt that in the temperate parts of the world, say between the latitudes of 30 and 70 degrees (from New Orleans to Murmansk in the northern hemisphere and from Durban to Alexander Island on the Antarctic fringe in the southern hemisphere),

Spring has always sparked off the annual farming cycle in temperate areas, as temperatures then rise above the crucial 43°F (6°C) necessary to initiate new plant growth. Seeds are sown, and the breeding season begins, setting in train the essential processes of life.

The Shepheard's Kalender

Summer begins in June, in the northern hemisphere. Medieval man observed the varying positions of the stars during the year, and linked the signs of the zodiac to the seasons. As illustrated in *The Shepheard's Kalender* he noted that the signs changed in mid-month.

Winter months, with freezing temperatures and long nights, are a feature of the cool temperate zones (between latitudes 30° and 70°). Melting snow may fall at 38°F (3°C), if the air is saturated and, as the air grows colder, the flakes become finer and settle.

Autumn, to medieval man, signified the harvesting of the year's basic food reserves. *The Shepheard's Kalender*, shows how life in the early 1500s revolved much more closely around the cycle of the seasons than do our technologically 'cushioned' lives today.

temperature is the key factor. In the mid-latitudes it is almost always temperature that modulates, for instance, the seasonal pattern of plant growth. In fact it is in relation to the absolute levels of temperature that the meaning of the word temperate, as applied to geographical zones, can be understood. In this context temperate certainly does not mean unchanging or moderate. Because of the tilted spinning axis of the Earth, the temperate regions experience temperatures intermediate between the consistently high values characteristic of the tropics and the much lower temperatures that prevail all the year round in the Arctic and Antarctic zones.

As well as producing a mean intermediate temperature level in mid-latitudes the tilt also, of course, produces a profoundly fluctuating one. When the tilt nods one hemisphere toward the sun, producing longer daylengths and less oblique rays of sunlight, then the temperature is high. When the same tilt angles the hemisphere away from the sun, generating shorter daylengths and more oblique sunlight, then the diametrically opposite seasonal period occurs and the temperature is low.

In the temperate zone seasonality is thus inextricably tied to a cyclical change of average temperature, from a summer high through the intermediate values of autumn to a winter low and then, via intermediate spring temperatures, to summer again. The two wide temperate bands around the earth between 30 and 70 degrees north and south contain a relatively distinct series of subzones, each with its own particular variants of the seasonal temperature cycle. These regions have been given a variety of names but those that are most immediately and directly understood are the warm temperate,

cool temperate and cold temperate regions.

These regions lie in sequence from Equator to pole, north and south of the Equator. Each region experiences its own cyclical seasonal pattern of temperature variation, but the range of temperature in each zone is different. This variable amplitude of temperature change is particularly vital to plant growth because almost all plant growth, including that of leaves, stems and flowers, stops at temperatures below 6 degrees centigrade (42.8 degrees F). Thus the proportion of days in a year on which the temperature is above this threshold constitutes, in a direct sense, the potential growing period for plants in that zone. The length of this growing period must inevitably be, in turn, a prime determinant of the amount of new plant material—an area's primary product—in any given year. The longer the growing period, the greater is the mass of new plant substance, and this productive effort constitutes the only net input of new organic material into communities of animals and plants.

Warm temperate areas are those in which the average temperature of the coldest month of the year does not drop below the magic, growth-enabling figure of 6 degrees centigrade (42.8 degrees F). In the northern half of the globe in the Old World, the warm temperate zone includes the whole of North Africa north of the Sahara, most of the Mediterranean region, and some of the Middle East, as well as Afghanistan and southern China. In the New World it takes in the sunny southern states of the USA. In the southern hemisphere, Chile, Argentina, South Africa and the south of Australia are largely in this zone, which stretches, broadly, from latitudes 30 to 40 degrees.

Next in the poleward sequence are the cool temperate geographical strips. ▷

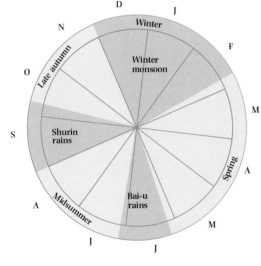

Japan, unlike other temperate countries, has six distinct seasons, three wet and three dry. The winter monsoon, the Bai-u rains in June and the late summer Shurin rains are each followed by a dry period.

The snowdrop, *Galanthus nivalis,* sometimes known in Britain as 'the fair maid of February', emerges through the melting snows of late winter. It heralds the transition into temperate spring, which brings longer days and fresh plant growth.

Temperatures fluctuate more dramatically inland than in coastal areas. In land-locked Winnipeg the annual temperatures range from −20°C to 20°C, while Vancouver, warmed by the heat stored in the oceans, rarely falls below 3°C.

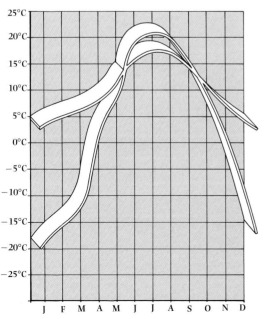

▷ These regions of the globe contain most of the rich and developed nations of the world, for in the northern hemisphere the cool temperate band includes most of the USA, the southern portion of Canada, most of Europe, southern Russia, northern China and Japan. There are hardly any cool temperate land masses in the southern hemisphere. It is a matter of some speculation whether the present coincidence of developed economic status and a particular climatic condition is fortuitous or causal. The cool temperate zone has at least one month in every year with a mean temperature below 6 degrees centigrade (42.8 degrees F) but not more than five such months. This frosty period is the true winter, which hardly exists in the warm temperate zone, and is characterized by a cessation of plant growth. It is in the cool temperate zone above all others that the four-season cycle is most characteristic.

Bordered on one side by a cool temperate zone and on the other by the treeless tundra of the polar regions is the cold temperate zone. Here it is normal to have temperatures below 6 degrees centigrade (42.8 degrees F) for at least six months out of every 12. The natural ecosystems in this zone have no deciduous trees of ordinary size, only miniature or creeping kinds, and plant life is dominated by vast tracts of coniferous, evergreen forests.

In the temperate parts of the world the dominant climatic theme is the cyclical seasonal variation in temperature. The tropical regions of the earth are, in contrast, typified by continous high temperatures but seasonal variations in rainfall. Everywhere between the Tropics of Cancer and Capricorn the sun is directly overhead on at least one day a year. Despite local variations, all this tropical zone is hot throughout the year, so that the sort of temperature-induced seasonal variation in plant growth, which is so vital in temperate regions, is of little importance near the Equator. Instead, where strong seasonal influences exist, they are centred around the availability of water, and the year is split into hot dry and hot wet seasons.

To synchronize the important activities of their lives, such as reproduction, both with each other and with the appropriate external climatic conditions, animals and plants that inhabit the temperate zones respond to the temperature changes or marked daylength alterations throughout the year. At first glance it might appear unlikely that animals living in the tropics in general, and those in regions on or near the Equator in particular, could behave in a seasonal fashion like their cousins in temperate zones. The reasons for this would be the absence of strong seasonal changes in either air temperature or daylength, implying an absence of clear-cut external clues, to which seasonal activity might be tied. In fact, even on the Equator itself, most animals still exhibit strong seasonality in their activities. This seasonality is present, despite the absence of any dramatic temperature variation to make an emphatic seasonal change in the appearance of the habitat.

In the tropics, the seasonal behaviour of animals can be the result of inbuilt annual activities or can be directly or indirectly tied to external seasonal clues, such as rainfall. In many of the tropical regions of the world there are broadly predictable changes in the amount of rainfall throughout the year, and there are also significant differences in this rainfall pattern. Nearest to the Equator there are many regions in which the seasonal variation in rainfall is low, but the absolute input of rain to the land masses is high. It is in these constantly wet

and hot zones that the true non-seasonal rain forest vegetation—the jungle— develops. These rain forests gradually merge into more seasonal forest belts where, for some periods of the year, drier conditions make a significant impact on plant success, restricting the amount of photosynthesis that occurs, and thus the amount of growth that can take place.

In many but not all of the zones within the tropics, the annual climatic cycle consists of two relatively wet and two relatively dry periods per year. It is these seasonal ups and downs in rainfall that trigger most of the clear-cut seasonal behaviour of tropical animals and plants. In any particular tropical location the total amount of rain that falls in the wet seasons depends on a number of factors, but of these distance from the sea and altitude above sea level are most important. As a rule, the wet seasons are wettest in sites closest to the sea because they receive air which has become water-laden as it has travelled landward over the ocean. Similarly, the higher an area is above sea level the fewer rainless months there are and the more rain there is in each rainy period. The reason for this is that water vapour condenses out into liquid raindrops more readily at the low temperatures of high altitudes. These variations within the tropics mean that the most dramatic transitions between wet and dry seasons occur in lowland zones, often those with savanna-type vegetation, in which the total annual rainfall is small. In such regions the seasonal changes are staggering, with the rains catapulting the landscape from arid, desiccated dormancy and inactivity to bursting plant growth and intense animal activity almost overnight.

The causes of the twice-yearly pattern of rains in the tropics can, like many other seasonal variations, be attributed to the Earth's tilted spinning axis. Because of the constant orientation of the Earth's axis, in the tropics, a zone of rainfall follows the apparent motion of the sun in a year—first northward, then southward then northward again. The rainfall belt is perhaps best described in relation to a specific location such as East Africa. Here the rain occurs in zones where water-laden air masses converge. Air rises in these zones, cools as it expands at lower pressure, and condenses into rain.

Of all these convergences, the one that is most important as far as East Africa is concerned is the intertropical convergence zone or equatorial trough. This zone is closely related to the heat equator of the globe and in it the massive tropical trade wind systems of the northern and southern halves of the world meet head on. If our planet were smooth and perfectly spherical like a billiard ball this convergence would create a girdle round the world that would follow the sun's apparent motion. On the earth's actual surface the mix of land masses and oceans means that the precise course marked out by the convergence zone is much more tortuous. None the less, it still sits astride Africa, and once each year moves up and down the central parts of that continent. The movement of the convergence zone alone can provide the essential clue to the puzzle of the double rainy period. At any spot in the tropics rain is likely to be associated with the twin passages of the convergence zone over that area—once as it moves north and once as it travels south again.

North of the Arctic Circle and south of its Antarctic equivalent, seasonality is tied to the third great ecological determinant—light. The polar year is divided, in different proportions at different latitudes, into periods of continuous ▷

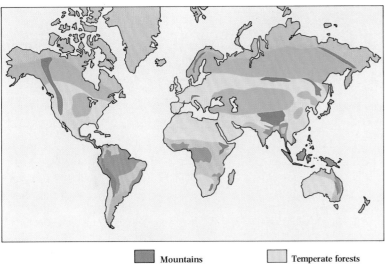

	Mountains	Temperate forests
Desert	Snowlands	Tropical forests
Savanna	Taiga	Steppe

Tropical seasons are controlled by rain rather than temperature. At the Equator, the humid rain forests experience only slight variations in the intensity of rainfall. The huge belts of tropical grassland and savanna on either side, however, have extreme wet and dry seasons. Dry trade winds sweep across the area from June to October, and in January and February, bringing drought, *left*. Rainfall is minimal, existing moisture evaporates rapidly and the sun blazes. Animals huddle round the few remaining water holes and scour the scorched brown grassland for nourishment. In November and December, and from March to May, bands of low pressure bring the rains and transform the landscape almost overnight, *right*. Parched river beds become torrents, the dead, cracked earth melts under the moisture and becomes fertile again. Within days the strawlike tufts of burnt grass give way to fresh green vegetation. With this new abundance of good grazing, animals disperse and begin breeding, so giving their young the maximum chance to grow and thrive before the next dry season devastates the land.

The eight main vegetation types, which mirror precisely the world's climatic zones, weave irregular latitudinal patterns over the earth's surface. Jungle, steppe and desert are all created by weather systems.

These two views of the Tsavo National Park in southeastern Kenya reveal the extreme seasonal swings in the tropical savanna. The same landscape, brown and sun-parched in the dry season, *left*, is dramatically transformed by the first rains, *right*.

▷daylight, continuous darkness and twilight periods. In the polar regions the truism that the sun is the ultimate source of all earthly life is dramatically illustrated as the sun appears and disappears for months on end. When darkness is the seasonal signature, animal and plant life plunges into the depths of dormancy. With the arrival of eternal day, frenetic activity ensues, so that organisms can rush through their essential reproductive processes, and collect and store enough food reserves to last them through the months of darkness.

Some of the most clear-cut and unambiguous clues that the weather machine can manufacture occur in various tropical regions of the world. The abrupt, almost instantaneous, transition of a landscape from arid, desiccated desert to a brilliant expanse of flowers, humming with insects, is not a signal that is easily ignored by other living creatures. But unambiguous though it is, such a stimulus cannot, by its very nature, be an accurate seasonal time check. Similarly, in the mid-latitude temperate zone of the world, the major climatic transitions of the seasons are roughly, but never exactly, at the same date each year. Autumn gales do, by definition, start in autumn, but in some years they are early, in others late—or they may be virtually non-existent. The last night frost of the year, a key event for any keen gardener with treasured tender specimens that need to be planted out, can never be pin-pointed in advance on the calendar, and may occur unexpectedly late when summer seems to be well on its way.

Meteorological signs can thus provide approximate signals about the time of year, but these messages can become garbled by the large part that chance plays in the pattern of weather at any specific spot on the earth's surface. The fact that weather signals, with a code of rain, temperature and winds, can give false information about seasonal time means that animals and plants can never place complete reliance upon such signals to order their own activities. But order their activities they must. Mates have to be found, breeding synchronized, nests built or flower buds made and opened at the correct time of year. Bad timing, a wrong extrapolation from freak warm weather in early spring, or a premature frost in late summer could make all the difference between disaster and a successful breeding season for an animal or abundant seed production for a plant.

Because of the inbuilt inaccuracy of forecasting by the weather, exact and consistent information about seasonal time is of life and death importance to living things. So how can organisms order their seasonal lives with apparently magical precision? Why do all the leaves of the trees in a beech wood spring out from their buds at very nearly the same time each year? And how do swallows, swifts and martins arrive back at their European nesting sites each breeding season with metronomic accuracy? In some ways it is surprising that the answer to these, and similar specific questions, was not arrived at until the 1920s and '30s. The answer is a single, technical word, photoperiodism. Dauntingly scientific it sounds, but all it means is changing daylength. Photoperiodism is nothing more esoteric or complicated than the fact that in temperate latitudes a day—that is, the number of sunlit hours—is longer in summer than it is in winter.

The transitions of daylength from, say, 16 hours in high summer to eight hours in midwinter, and the inverse change of the duration of nighttime, are

The number of daylight hours provides plants and animals with an accurate seasonal clock. As the tilted Earth orbits the sun

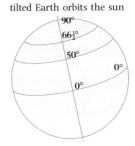

each year, only the Equator experiences a constant 12-hour day. The higher the latitude, the more extreme the

variations in daylength. At the poles, the days fluctuate between 24 hours of light in summer and none in winter.

Latitude	Maximum daylength	No of days of max length	Minimum daylength	No of days of min length
90°	24	189	0	176
66½°	24	1	0	1
50°	16¼	7	7¾	6
0°	12	183	12	182

In temperate areas deciduous trees must produce well protected buds which will eventually replace the leaves that fall in autumn. The rudimentary leaves and stems held dormant inside the bud are ready to emerge the following spring when it opens. Changing daylengths induce the production, opening and falling of the leaves.

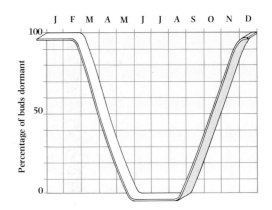

The annual activity of the colorado beetle, *Leptinotarsa decemlineata,* is ruled by the number of daylight hours. Only in the summer when there are more than 14 hours of light is this potato pest able to reproduce, grow and develop. When the days shorten in late summer and fall below this crucial level, the beetle becomes dormant.

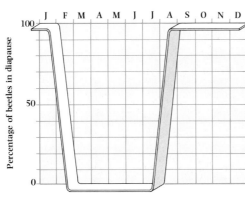

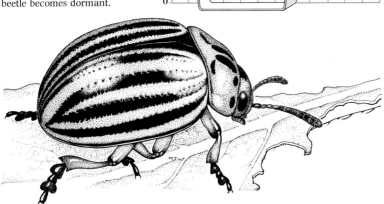

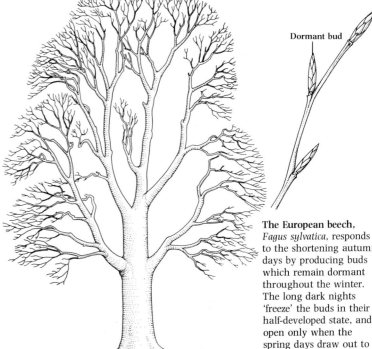

Dormant bud

The European beech, *Fagus sylvatica,* responds to the shortening autumn days by producing buds which remain dormant throughout the winter. The long dark nights 'freeze' the buds in their half-developed state, and open only when the spring days draw out to more than 12 hours.

not stepwise and abrupt. Daylength changes in a smooth, rhythmical cycle of alteration from a maximum at the summer solstice down to a minimum on the day of winter solstice, then back up to a summery maximum again. It needs chronometers and accurate record-taking to plot the smooth curve of change, but it takes no scientific sense to appreciate the reality. Everyone who lives in Europe, North America or Japan knows of long, balmy summer evenings and short, depressing winter days when the sun is low in the sky and the promise of spring far off.

Perhaps it was the very mundaneness of the phenomenon of the seasonal change in daylength that made zoologists, botanists and physiologists take so long to realize that this normal and all-pervading change in the environment is one of the most vital clues by which organisms impart temporal order to their lives. The fact that photoperiodism is normal and perceivable everywhere is central to an understanding of just why daylength is the external time signal by which living things run their seasonal lives efficiently. Unlike the weather, unlike wind speed or direction, unlike temperature, daylength is a message that simply cannot be jumbled or distorted. To change the signal you would have to stop the spinning, orbiting Earth in its celestial tracks. That the absolute length of a solar day on Earth has lengthened steadily over the billions of years of history of our planet is irrelevant. The changes are so small, amounting to perhaps a thousandth of a second a century, that every animal and plant has had more than enough evolutionary time in which to adjust to these minute alterations.

Except for a few underground, cave dwelling or deep-sea species, animals can see the enormous differences in light intensity between night and day. Day by day they can then integrate the total time-span of daylight (or night) and gather information about absolute daylength and whether it is becoming longer or shorter. The process might seem straightforward but, in reality, careful strategic safeguards and subtleties have to be built into the system. Gloomy overcast days must not be allowed to confuse the amount of daylight hours, and an accurate internal biological time sense is a necessity. With all this information computed, an animal located in a temperate climate can set its biological affairs in logical and efficient order. Most such animals are most active and reproduce during the warm months of late spring, summer and early autumn, timing their courtship and mating so that the young are born when conditions are most favourable for their survival. During the winter—if they do not avoid local climatic stresses by migrating to warmer areas—animals will usually reduce their activity so that they conserve as much body energy as possible, and will simply set about the business of survival on the minimum amount of food.

When firmly tied to external daylength, this overall pattern of change in the repertoire of an animal's activities offers other advantages that reach out farther than the individual life of a single animal. If all the members of a species use a common daylength as a trigger for the initiation of sexual development, the start of courtship and mating or nest building, then this will have the powerful and inevitable effect of synchronizing the breeding of all the animals in that species. In the mid-latitude climate of violently changing seasonal weather, tight synchronization of reproductive effort can be a necessity for the ▷

The male rock dove, *Columba livia*, like most birds, experiences a seasonal swelling and shrinking of its sexual organs. The lengthening spring days trigger hormonal changes which cause dramatic enlargement of the testes, equipping the bird for the breeding season. After breeding, their job done, the testes shrivel until the lengthening days of spring.

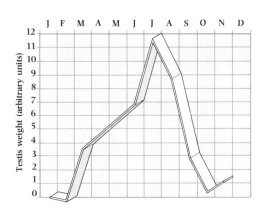

The lambing season coincides with the appearance of fresh spring grass, which will sustain the increasing sheep population. The sexual cycle begins in late summer when the shortening days induce ovary stimulation in the ewe and sperm production in the ram. Within four months the pair are ready to mate, and young are produced the next spring.

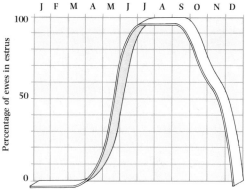

▷ continuation of that species. It is never so common a practice, nor as vital to survival, in more climatically stable tropical ecosystems.

Like man, other animals use their eyes to acquire information about daylength and its photoperiodic change. But it is perhaps not quite so obvious that a plant—be it a garden weed or a giant redwood—can also monitor the changing period of nights and days throughout the year. In fact a plant's existence is, in every respect, infinitely more light-centred than that of an animal. The vital process of photosynthesis, in which organic matter is miraculously constructed out of gas and water, is powered by sunlight and all the rest of a plant's existence circles round this core of its activities. This being the case, it would be strange indeed if plants were not minutely aware of changing light levels in the world around them, and all green plants can not only perceive and react to light level changes but also have the capacity to measure time.

Plants are static. Compared with animals, which can change their position in small or majestically large ways, a leafy plant must stay in one spot and cope with whatever seasonal weather variations the local climate throws at it. In the winter an animal can move into a cave, nest or roosting place to reduce the amount of heat its body loses; in a desert noon it will disappear down a burrow or hide in the shade of a rock to achieve the reverse advantage and augment heat loss from its body. These are small-scale adaptive movements compared with those at the other end of the distance scale. These are migrations, mass movements of whole species, sometimes half way round the globe, to avoid, either wholly or partially, the vicissitudes of the climate. A plant in a mid-

latitudinal location never has these options, so in one extra but crucial respect, it must be sensitively aware of its time position within the march of the seasons. On each day of the year it must be performing efficiently in the prevailing conditions, but it must also be ready for the next inevitable phase of environmental change. Whatever happens the plant or its offspring will have to live through those seasonal changes, since it cannot move away.

The growing season of a plant is the section, or sometimes the sections, of a year in which the variables of the environment—light, water and temperature—are at appropriate levels to allow new plant cells to be produced, thus enabling the plant to grow. In most plants growth will begin when temperature levels are capable of sustaining metabolic activity and when there is enough water available for this active metabolism. The beginning of growth in different types of green plant may mean quite different sorts of change. The initiation of growth may be the germination of a seed, and the rapid development of the plant embryo lodged inside it into a seedling and then a whole plant. Or the beginning of growth may be the time at which buds open, so that the active cells (the primordia) of leaves and shoots, which were laid down in miniature in the preceeding season, are suddenly pumped up to their full functional size and burst out of their protective bud scales. Or growth may begin in less obvious areas, for example in underground bulbs, tubers and rhizomes. Once formed, the new, leafy portions of the plant above the ground will grow, and ultimately produce structures for reproduction—most usually flowers and fruits, but less showy structures in 'primitive' plants, such as seaweeds, mosses and ferns. This continues until external conditions fall short

The life cycle of maize, *Zea mays*, or corn on the cob, is closely synchronized with the tropical rainy season. Although a native of the American tropics, the grain of this famous grass has become a vital annual crop in many climatic zones. The large grains, sown at the beginning of the rainy season, swell with moisture and germinate, sending out roots and a shoot. The plant grows to over 6 ft (1.8 m) tall, producing spiky, pollen-bearing male flowers at the top and smaller tassels of female flowers, emerging from a cob, at the base of the leaves. After cross-pollination, the cobs develop, ripen and are hung to dry in the sun. The grain 'after ripens' during the dry season and is ready for sowing when the rains come again.

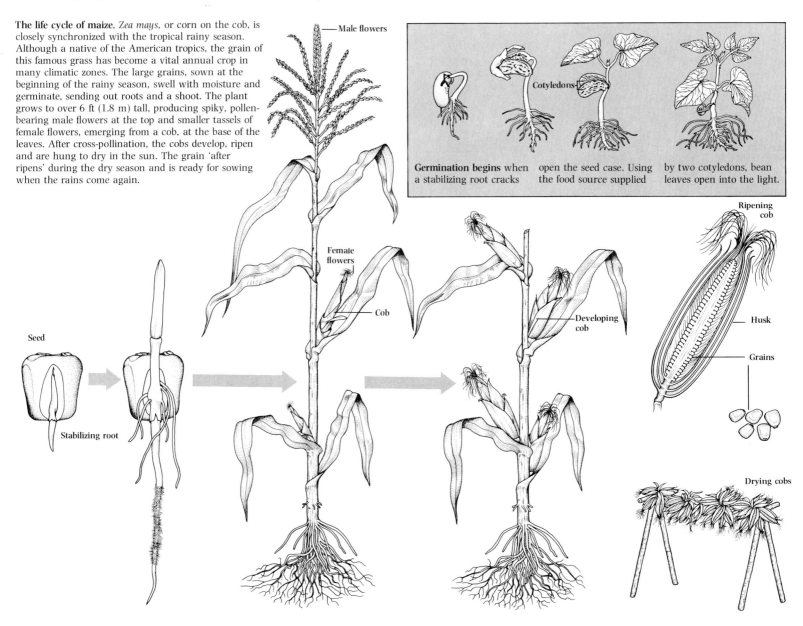

Germination begins when a stabilizing root cracks open the seed case. Using the food source supplied by two cotyledons, bean leaves open into the light.

Cotyledons

Male flowers

Female flowers

Cob

Seed

Stabilizing root

Developing cob

Ripening cob

Husk

Grains

Drying cobs

of the necessary threshold levels in some way.

The unfavourable season for plant growth is not the same the world over. In the northern and southern temperate zones, conditions hostile to plant activity are usually related to cold. With the onset of winter, average temperatures drop too low for growth to occur. In the subtropical and tropical portions of the globe (except on mountain peaks), temperature is never, by itself, the limiting factor. Water is the growth factor whose absence makes a habitat inhospitable, and plants cease their activities near to the onset of the dry season. So plants have to be adaptable to cope with a cyclically varying pattern of seasonal climatic change most efficiently, although these adaptations vary between temperate and tropical regions. Whatever the specifics of such adaptation, one phase of each yearly cycle encourages plant growth and development, while the other prohibits it.

To solve this problem of seasonal change, plants have adopted utterly different life strategies. Annuals, which include nearly all weed species, are the plants that can grow quickly from a germinating seed at the beginning of the growing season, reach reproductive maturity rapidly and set seed, all within a matter of a few months at most. As resistant seeds they can last out the next danger period, be it one of drying heat or intense cold. Such plants are usually small in stature, have no woody tissues and can often squeeze in many seed-plant-flower-seed cycles during a single growing season.

Woody perennials, the world's shrubs, bushes and trees and many of its climbers, have a permanent structural framework of woody trunks and branches above ground which persists through many winters or dry seasons.

During these periods of inactivity, however, many tree species in the temperate zone, such as the deciduous oaks, beeches and maples, shed all their functional leaves. At the beginning of the next growing season dormant buds open to reveal new leaves that will be used for making food by photosynthesis during a single season. By the end of the growing season the tree or shrub will have produced a crop of seeds and also laid down a new set of buds for overwintering.

Non-woody herbaceous perennials, such as the grasses and a myriad other plants including irises, daffodils, snowdrops, anemones, primroses and tulips, persist through the harsh season by having a small amount of inactive foliage above ground or by means of below-ground storage organs, such as bulbs, corms, tubers or rhizomes. Spring initiates new growth in these plant parts—which incidentally act as organs of non-sexual reproduction—with the rapid production of new leaves, stems and flowers above ground.

Although the winters or the dry season are periods of hardship for plants, they have been utilized by those same plants as essential integrators of patterns of development. In cool and cold temperate regions, for example, seeds of plants of the rose family, Rosaceae, which includes apples, pears, apricots and peaches as well as roses themselves, need to experience severe cold before they can germinate. This process is a neat and convenient protective feature in the plant's life cycle. It ensures that seeds, produced early in the growing season, are not fooled by unseasonably mild weather into germinating just before winter sets in. In a similar way, some tropical and subtropical plants have seeds that need to go through a long dry season of conditioning before they are capable of germination. ▷

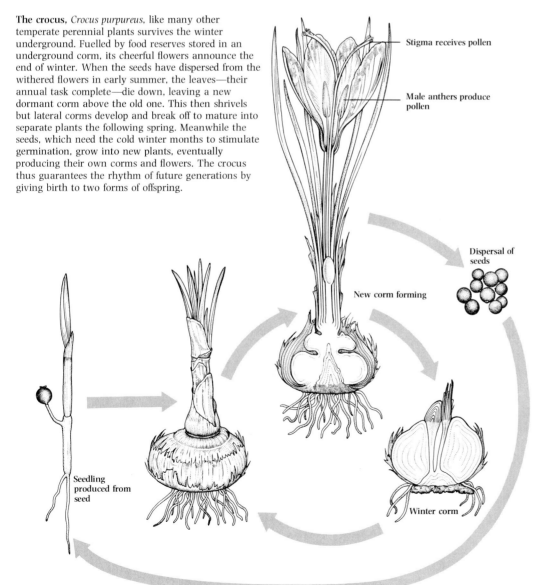

The crocus, *Crocus purpureus*, like many other temperate perennial plants survives the winter underground. Fuelled by food reserves stored in an underground corm, its cheerful flowers announce the end of winter. When the seeds have dispersed from the withered flowers in early summer, the leaves—their annual task complete—die down, leaving a new dormant corm above the old one. This then shrivels but lateral corms develop and break off to mature into separate plants the following spring. Meanwhile the seeds, which need the cold winter months to stimulate germination, grow into new plants, eventually producing their own corms and flowers. The crocus thus guarantees the rhythm of future generations by giving birth to two forms of offspring.

Stigma receives pollen

Male anthers produce pollen

Dispersal of seeds

New corm forming

Winter corm

Seedling produced from seed

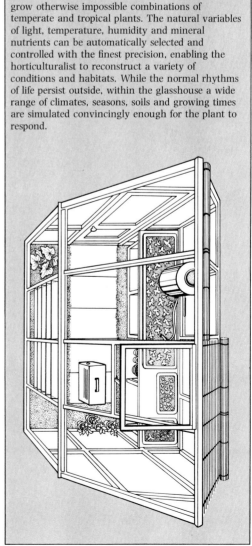

The greenhouse is the gardener's way of beating nature's order. Controllable conditions allow him to grow otherwise impossible combinations of temperate and tropical plants. The natural variables of light, temperature, humidity and mineral nutrients can be automatically selected and controlled with the finest precision, enabling the horticulturalist to reconstruct a variety of conditions and habitats. While the normal rhythms of life persist outside, within the glasshouse a wide range of climates, seasons, soils and growing times are simulated convincingly enough for the plant to respond.

▷ The seeds of flowering plants contain the embryos of new individuals and are the offspring of the plants that made them. Seeds are formed as a result of the complex process of pollination. During this process nuclei within a single male pollen grain fuse with nuclei within a female ovule of the flower. The ovule and pollen nuclei involved in this process of fertilization are made in a similar way to those of animal eggs and sperm and their fusion is analogous to the fertilization of an egg by a sperm.

There is nothing unusual about the fact that plants indulge in sexual reproduction. Plants use sex, and flowers are sexual organs. In the plant world flowers are the most sophisticated sort of sexual organs and are found in a staggering range of forms, with added layers of diversity generated by their scents and colours. Functionally, flowers can be divided into those adapted for pollination by means of physical agencies—most usually the wind but sometimes rain and very rarely water currents—and those that use animals as pollinators. Around the world, bees, wasps, ants, butterflies, moths, flies, beetles, slugs, birds, bats and even mice and marsupial possums are used by species of flowering plants as willing, if unwitting, partners in the pollination process. Urged on by some reward, which may be the gathering of pollen, guzzling nectar as food or even, in the case of some male wasps, the uncontainable urge to copulate with an orchid shaped and smelling like a mate, all these animals end up transferring pollen to new flowers, so achieving the cross-fertilization of the plants.

The male and female parts of the plant's reproductive equipment, respectively pollen-producing stamens and ovule-making ovaries, may both exist in one flower, although cross-fertilization with pollen from a different plant is usually necessary for seed to set. Male and female parts may be separate and located in different flowers on the same plant or even completely isolated from one another, in which case a plant species is divided into male and female plants. This explains why only the female plant of the holly tree, *Ilex*, can produce the showy red berries synonymous with Christmas and why only male trees of the North American shrub *Garrya* make the hanging display of eye-catching, pollen-producing catkins.

A flower is produced as a highly modified tip of a short branch, stem or twig. Indeed it is usually accepted that the petals of flowers and the sepals that surround them, which are usually green but may be brightly coloured and petal-like as in a tulip, are specialized forms of leaves, which gives the clue to the origin of the flower. In all flowering plants the growth of new stems and shoots occurs because of new cell production, concentrated into a small, but highly active, growth zone at the tip of the shoot. These points of seething plant activity are the meristems. At the end of a leafy shoot a meristem produces new cells that grow and elongate, then differentiate into particular cell types, including those that make up the tissues of new leaves.

The development of flower buds occurs when a hormone signal tells the meristems to begin making the cells of flower rather than those of a leafy shoot. It is in the control of the synthesis of such hormones—and also in more subtle ways—that photoperiodic stimuli come to exercise such a potent control over the timing of flowering. Many species of flowering plants can only begin to produce flowers, and thus begin their reproductive lives, when they experience

The flowering of most plants is triggered by a crucial number of daylight hours. The mature leaves perceive and interpret the photoperiodic changes and use the information to induce a wide range of physiological processes. Particular patterns of light and darkness stimulate internal chemical signals, which pass to the growing point of the plant and allow it to produce flowers instead of leafy shoots. While most plants cannot flower without receiving the relevant clues from the sun, some rely solely or partially upon other factors, such as temperature and humidity.

Apical meristem

Hormone signal

Long day plants such as spinach, Italian rye grass and henbane are so-called because they flower in the lengthening days of spring. Each plant has a critical photoperiod and only when the daylight lasts more than this crucial number of hours will the long day flowers appear.

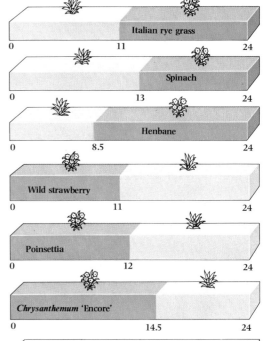

Short day plants respond to the shortening days of late summer and autumn. The poinsettia, *Euphorbia pulcherrima*, for example, only flowers when there is less than 12.5 hours of daylight. Each plant produces its seed when its pollinators are active and conditions are most favourable for its development.

Since the annual changes in daylength vary according to latitude, the actual calendar date at which a plant reaches its critical photoperiod for flowering depends on where it is growing. A wild strawberry, for example, flowers when the days shorten to 11 hours. At 50° north of the Equator (Winnipeg, Canada; Land's End, England), this occurs in October, while were it to grow at 20° north (Mexico City) it would not flower until 21 December.

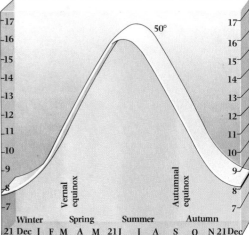

particular lengths of daylight or, more commonly, darkness. It appears that it is the leaves that react to the length of days and nights by perceiving the changing patterns of light intensity created by the Earth's rotation round the sun. Acting almost like aerials, the leaves measure the photoperiod, and once the appropriate daylength occurs hormone signals are produced which switch on flower production. Because it is daylength and not the weather that is the crucial factor in the timing of flowering, gardeners can judge almost to the day when some of their specimens will bloom, and flower shows are often planned on this basis.

Some of the crucial early experiments that demonstrated the vital role of photoperiodicity in controlling the lives of long-lived organisms were carried out on plants. In 1920 two scientists named Garner and Allard, working in the US Department of Agriculture, first realized the crucial links between daylength and plant flowering. These workers were investigating a variety of the tobacco plant, *Nicotiana tabacum*, with the impressive name of 'Maryland Mammoth'. Irrespective of the date when the seeds which gave rise to the tobacco plants had been sown, the plants that grew from those seeds would only flower in the short days of winter. In experiments that have become models for much subsequent work in this field, Garner and Allard covered up some 'Maryland Mammoths' for part of each day during the long summer days. This simple and elegant manipulation was enough to fool the plants into believing that winter had arrived and they flowered out of season.

At a stroke, the experiment proved that, at least in these plants, flowering was controlled by daylength and that daylength was the master controlling factor, dominant over all other environmental clues. The plants rooted in their summer field were experiencing summer temperatures, but this signal was ignored because the photoperiod signal was telling a different story. In other experiments it proved possible to produce the reverse effect; in the winter it became feasible to prevent the tobacco plants from flowering by extending the apparent daylength with artificial lighting. These important studies from more than half a century ago created a conceptual revolution in botanical and agricultural thought. They also had immense practical and economic significance, because they showed with startling clarity how agriculturalists could manipulate the flowering period of some crop plants with relative ease.

Further studies have revealed that flowering plants can be conveniently divided into three categories on the basis of their response, in terms of flowering, to changing daylengths. 'Maryland Mammoth' tobacco plants are an example of short day plants, because short days and long nights induce such species and varieties to produce flowers while long days and short nights have an inhibitory effect on flowering. Chrysanthemums, dahlias, asters and golden rod are other good examples of short day plants. In contrast, long day plants will only flower when days are long and nights are short. When days are short their flowering is inhibited. Examples of such plants, which only flower in the lengthening days of late spring and early summer, include commercially important crops, such as clovers and beets. The final category comprises the 'daylength neutral' plants which appear to be able to flower whatever daylengths they are exposed to. Sunflowers, dandelions and tomatoes all respond in this fashion.

Rainfall induced this Australian desert to bloom.

Once every 100 years the Chinese umbrella bamboo, *Thamnocalamus spathaceus*, flowers and dies. Other species of bamboo produce vegetation for 10, 20 or even 120 years, before all bursting into flower with an extraordinary display of species coordination. After flowering, the shoots die down, leaving roots and seeds to produce new plants. Sadly the giant panda has recently been a victim of the rare flowering of the umbrella bamboo, for its chief food source has died right down.

Flowering bamboo

During the hunter-gatherer phase of man's evolutionary development, which must be considered to stretch back from about 10,000 years ago to man's origins among the large African apes between two and three million years ago, seeds and fruits must have been a highly significant component of his omnivorous diet. Why is it that these plant structures, which from the plant's viewpoint are protected embryonic offspring, are so important as foodstuffs? The answer is that they are usually packed with nutrients for the plant embryo to use when the seed begins to germinate. In the earliest phase of seed activation and germination, the tiny growing plantlet, first within the seed coats and then bursting out of them, is in a vulnerable position when it comes to food supplies. It is liable to begin its phase of explosive growth buried in the soil or under leaf litter, in other words in the pitch dark. While it is in this state it has water, but because it has no light photosynthesis cannot take place. Unlike its much larger, established parents, it cannot produce new nutrients for itself. Equally, until it has developed a functionally effective root system, it is unable to acquire mineral salts, vital to its biochemistry, from ground water.

To overcome these two early disadvantages, and to give the minute seedling a chance of becoming established, most seeds contain large stores of concentrated nutrients, including starches, fats and oils and rich protein mixtures, such as are found in peas, beans and other pulses and in nuts. Therefore, as long as it has a complement of biochemical catalysts or enzymes available within it, so that it can digest and mobilize these foods, plus some water, a seed is able to grow and develop with no sunlight and little, if any, access to mineral salts. During the first few days of life the seedling's nutritional organization slowly and progressively switches from dependence on these internal structures to nutritional independence. This final break from the food contained in the seed—which by this time is largely exhausted—comes with photosynthesis in the first seed leaves or cotyledons, and later in the first true leaves, which can fuel the plant's nutritional needs with the help of mineral salts and water drawn up by the newly formed roots.

Just as they attract animals to themselves to bring about the process of pollination, so plants also seduce creatures into activity as seed distributors by packaging small, tough, inconspicuous seeds in showy, colourful, juicy and sweet-tasting fruit flesh. As in pollination there is an evolutionarily generated plant-animal trade off. The animal is rewarded by fruit nutrients and vitamins for its role in dispersing the resistant, indigestible hard-coated seeds which are thrown aside, spat out or, if swallowed, pass through its gut undamaged. The seeds that travel the length of an animal's intestine before being deposited not only get a free ride to a new growing environment but are also deposited in their own patch of nitrogen-rich organic fertilizer—the animal's faeces. The efficacy of this partnership between animal and plant kingdoms is revealed in the rich crops of tomato plants that spring up every year in the sludge beds of sewage works.

About 10,000 years ago, in a number of sites around the world, societies of men and women took the first steps in the transition from being nomadic hunter-gatherers to forming settled agricultural groups. Instead of relying on their knowledge of the fruiting times of the natural plant forms in their habitats and, by their wanderings, making use of a wide range of plant food sources,

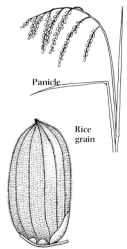

The rice harvest, at the beginning of the dry season, is perhaps the most significant event of the tropical year, as it feeds over half the world's population. In 1973, 320 million tons of rice were produced of which 15 million came from Thailand, *left*. The safe gathering of the crop is celebrated by many Asian rice growers as the beginning of the new year. Some perform a ritual symbolizing the successful union of two rice grains.

Panicle

Rice grain

Rice develops from a grain to a flowering panicle in about four months. Each panicle bears many tiny rice grains which mature 30 to 40 days after the flowers appear.

they began to plant their own crops. For each of the hundreds of plant species that has become domesticated by man and used for food, the story is roughly the same. A wild natural plant was originally gathered from its undisturbed setting and used as a source of nourishment. Then these completely unaltered plant types were grown together near human habitation. This was the crucial step, because once man started to select the seeds, suckers, tubers or bulbs that were to produce the next season's crop, an unconscious selection pressure was applied.

The human-induced selection pressure of selective breeding is immensely more powerful than the forces that operate in natural ecosystems and takes effect far more quickly. A woman living 10,000 years ago in an agricultural tribe, who, from a cluster of wild fruiting plants, took the seeds from only one plant with the biggest, juiciest fruits for sowing as the basis of the next year's crop, was changing the genetic structure of her domesticated stock. Modern geneticists and plant breeders have a much more sophisticated grasp of the processes underlying their activities and experiments, but the bulk of their work is conceptually the precise equivalent of that tribal woman's choice. So dramatic are the alterations brought about by just a few generations of domesticated breeding of plants—and animals too—that they were one of the key strands in Charles Darwin's web of interlinked evidence for evolutionary change.

The nutrition of modern man is completely dependent upon his ability to grow these radically altered, highly productive plant crops. Man's population is still increasing at an alarming rate, and the food requirements of the world's four billion people depend for their fulfilment on the evermore efficient husbandry of evermore productive crop plants. The phase of the agricultural process that conveys, above all others, the productivity of plant life and its significance to human societies is harvest. Harvesting is carried out when the cropable part of the domesticated plant has developed to a suitable and appropriate state for storage. Vital grain crops—wheat, rice, barley, oats and rye, which are all seeds of members of the grass family, Gramineae—we harvest when the seeds have dried out at the end of the growing season. Harvesting of soft fruit crops is timed to ensure the maximum survival time for the soft, edible part of the fruit, be it banana pulp, mango, strawberry or apple flesh.

In temperate climates one thinks instinctively of late summer and autumn as the time of harvest for, even though food crops are produced and mature all the year round, this is the time when the wheat is gathered in and most fruits plucked from their branches for winter storage. In tropical, subtropical and warm temperate climates harvest usually takes place at the end of the rainy growing season or at the beginning of a following dry season. The actual months depend, of course, upon exactly when the rainy and dry seasons occur. The sugar-cane harvest, upon which the economy of Cuba depends, takes place in the dry season from January to May, while groundnuts in the Kano region of Nigeria are lifted, and this lifting celebrated, after the rainy period in October and November. The completion of the harvest of any crop important to the survival of a community, even if the crop is poor, is usually a signal for a celebratory harvest festival.   ▷

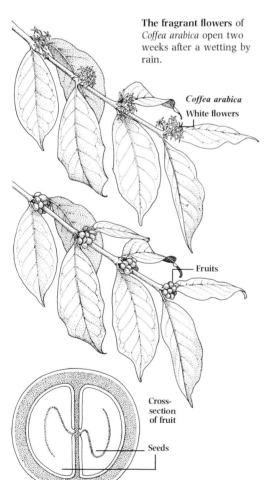

The fragrant flowers of *Coffea arabica* open two weeks after a wetting by rain.

Coffea arabica
White flowers

Fruits

Cross-section of fruit

Seeds

The evergreen coffee tree, *Coffea arabica*, which thrives in sub-tropical and tropical areas of Africa and America, produces 90 per cent of the world's coffee beans. The clumps of fruit, which develop 7 to 9 months after flowering, ripen over several weeks from a dull green to crimson, when they are ready for harvesting. Each fruit contains two seeds which are dried in the sun, and roasted to form coffee beans.

All wine-producing vines belong to a single species, *Vitis vinifera*, which prefers a temperate, Mediterranean climate. Left to itself, the vine grows rampant, sending out vigorous woody tendrils and poor fruit. In commercial vineyards the vine is regularly pruned, which channels the plant's energy into grape production. Most cultivated vines grow from grafted cuttings, and in the northern hemisphere sprout in April and May.

Tender vine shoots grow quickly into leaves and tendrils in mid-May.

Flowers appear in early June as soon as the days warm to 18°C (65°F).

Grapes form in June and begin to ripen in August for an October harvest.

▷ In Europe and North America the traditions and superstitions woven into the fabric of the harvest period provide some very powerful images. Many of these traditions, although now incorporated into notionally Christian ceremonies such as the autumnal harvest festival, undoubtedly stem from pre-Christian practices. In nutritional terms the grain harvest was often one of the most vital because, when properly dried, grass grain has excellent long-term storage properties. It is, perhaps, a mixture of unconscious racial memory of the importance of a fruitful harvest and the traditions that have been directly passed on down the generations that makes the image of the grain harvest so potent. Harvest time is linked in our minds with pictures of golden ripe corn rustling in the wind, of loaves plaited into sheaves, of corn dollies and of children bearing gifts of apples, pears and pumpkins.

It is easy to understand why spring and autumn, the transitional seasons of the temperate zone, in which the year's most dramatic changes take place, should have strong plant-related connotations. The most powerful images pass indelibly into the language; thus for North Americans, autumn is the fall and the fall is the annual, autumnal shedding of the leaves of deciduous trees. Leaf dropping is not, as might be imagined, the direct effect of the damaging environmental influences of the approaching winter. It is nothing to do with leaves being killed by declining temperatures or dropping nutrient levels in the soil. Leaf fall is a spectacular adaptation by plants, performed actively by the plants themselves, as a preparation for the damaging season which is soon— and predictably—to occur. This time-lag is crucial to an understanding of the whole process. The fact that the loss of leaves is a precautionary strategy means

that it cannot be triggered in time by the external environmental factors it is designed to avoid.

During the separation, or abscission, of leaf stalks (petioles) from tree branches, a layer of corky tissue is laid down between the end of the petiole and the branch. This abscission layer creates a seal to prevent the loss of vital sap from the branch and, once complete, allows the leaf to stay in position until only a breath of wind is needed to blow it off and carry it to the ground. The timing of abscission demands an elaborate chronological sense on the plant's part. At each moment it must know where it stands in the 12-month sequence of seasonal change and act at the appropriate time to drop its leaves before the onset of winter. As one might expect, the trick of perfect timing is accomplished by a photoperiodic response. Reacting to reducing daylength, rather than declining temperature, the temperate deciduous tree initiates the complex physiological process which, when complete, will leave it as a bare skeleton throughout the winter months. In the cool temperate subzone almost all the non-coniferous trees do shed their leaves before the onset of winter. The exceptions are so conspicuous by their unorthodoxy that they have often become endowed with supposedly magical powers. The laurel, holly and ivy are all excellent examples of broad-leafed evergreens with a firm place in folk culture and medicine.

The visual impact of the fall is not principally the result of leaf loss itself, impressive though that change may be. It is more the gloriously vivid shades of gold, red, orange and brown of the dying leaves that are unforgettable. Arguably, the eastern fringes of the world's major temperate continents have

Every year the Bactrian, or two-humped camel, grows an immensely thick winter coat, which allows it to travel for many days in freezing temperatures beyond the endurance of most large mammals. This makes it an invaluable beast of burden in its native Asian territory north of the Himalayas. The coat, which has a shaggy layer of long outer hairs and a lining of fine protective underwool, having protected the camel from the severe winter undergoes a dramatic moult in the spring. A fine film of summer hair replaces the woolly bulk, apparently reducing the animal to half its winter size, giving it a naked, shaven look and revealing the two fat-storing humps.

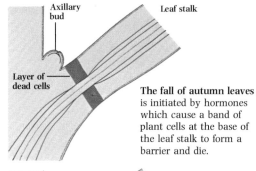

Seasonal changes in coat colour provide camouflage for many mammals living in the higher latitudes. The stoat, for example, adopts its white 'ermine' form during the winter, to hide from its prey in the snow.

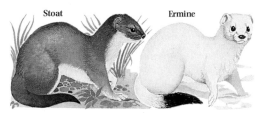

The fall of autumn leaves is initiated by hormones which cause a band of plant cells at the base of the leaf stalk to form a barrier and die.

The leaf stem soon breaks off from the plant, at the layer of dead cells, allowing the gentlest breeze to blow the old worn leaf to the ground.

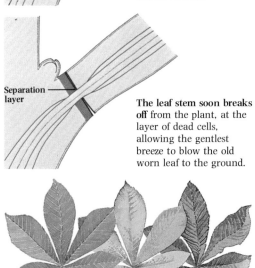

The autumnal changes of colour in these horsechestnut leaves reflect the internal chemical changes which allow the tree to survive from year to year. During the winter when the soil is cold, it becomes more difficult for the plant roots to absorb water as quickly as the leaves transpire. So deciduous trees shed their leaves, having first stored the nutrients contained in them, and their functions slow down. Sugar, protein and the green-coloured chlorophyll are withdrawn, leaving carotenoid pigments, which produce the familiar autumn tints before the leaf falls.

the most splendid autumnal leaf displays. Certainly New England and the eastern edge of mainland China are noted for their abrupt and vibrant spectacles of fall. This is partly the result of meteorological factors which make the transition from summer to winter much more rapid in these regions. In the United Kingdom the autumn period is, in contrast, spread over several months often beginning in late August and persisting until as late as November or early December.

To understand why colour change is associated with leaf fall, it is essential to make a closer examination of the physiological processes that bring it about. In relation to the initiation of flowering, leaves are the aerials that detect changes in daylength. During the shortening days of autumn, leaves, which recognize this change, set in motion a series of alterations within themselves that bring about a form of voluntary euthanasia of the leaf.

The loss of all the leaves in their active, functioning state would represent an enormous loss of past investment on the part of the tree. It would mean the sacrifice of its 'stocks and shares'—great quantities of sugars, proteins and precious minerals such as magnesium, all of them garnered and constructed from raw materials at enormous energy cost. To avoid losses of this magnitude, the leaf cashes in its investments and is stripped of its removable nutrient assets before being relinquished for ever by the plant that built it. A wide range of nutrients is extracted and passed back into the framework of the tree. In particular, the principle complex molecule concerned with photosynthesis, namely the green pigment chlorophyll, is reabsorbed and removed from the leaves. It is this salvaging process that changes the leaf colour. Golds, reds and

oranges are not pigments added to the leaf at fall time; they have been there all the time and are a variety of pigments related to carotene, the orange substance that gives carrots their colour. In summer the carotenoid pigments of most leaves are masked by a preponderance of green chlorophyll, and only when the chlorophyll is extracted, immediately before leaf fall, does the pattern of the other hues emerge.

Plants, especially trees, provide the most unmistakable and extensive colour changes that occur in living things on a seasonal basis, but animals also change their appearance with the seasons. Alterations of this sort may be brought about directly by changes in temperature or other external stimuli: lizards and amphibians become lighter in colour, for instance, as the temperature of their surroundings increases. This colour change is an adaptation which enables these cold-blooded creatures, which have no physiological machinery for controlling their body temperature from within, to reflect more heat when it is sunny and, conversely, to absorb more when the environment is cooler. Other animals change their colour or the nature of their outer covering to prepare themselves for a predictable, seasonal change and, as with leaf fall, this means that the changes must be triggered indirectly. Such changes are best seen in a range of adaptations employed by birds and mammals, which enable them to survive through the coming cold winter months.

The adaptations of birds and mammals may entail the swapping of dark plumage or fur for a white, camouflaging alternative for the snowy winter, and in the case of the ermine stoat, *Mustela erminea*, the white, dark-spotted winter fur has become highly prized by man and symbolic of aristocracy. The physical ▷

Apical bud bursting

Leaf stalk scar

The scars of the previous year's leaf fall can be seen on the budding horsechestnut twig, *above*. The old leaf's food and water-conducting vessels appear as 'nails' in the protective horseshoe-shaped cork seal.

This golden autumn in the Utah valley is typical of the spectacular fall in North America, where the transition from summer to winter is more rapid than in Europe. The endless revitalizing rhythm of the seasons is frozen in this view of millions of dying leaves.

▷ insurance premium for lasting out the winter may consist of a change in the thermal insulating properties of fur by an increase in the length or thickness of the hair. Changes of this sort, which pre-date the coming of the snow, are usually a mixture of inbuilt annual rhythms operating via hormonal changes and kept locked to real external seasonal time by photoperiodic clues. Only when daylength is short enough will the correct hormones be released to initiate the appropriate alterations in body structure. With the arrival of spring and increasing daylength, extra fur is moulted and colours change back to their summer shades.

Trees that are unable to escape the rigours of a snowy, cool temperate winter make use of leaf fall to protect themselves in the frigid months of the year. A very few vertebrate animals can make use of a protective method of coping with winter cold which is every bit as drastic as the defoliation of ashes, oaks, maples and the like. This method—hibernation—is used by only a small minority of warm-blooded vertebrate animals and all but one of these hibernators are mammals.

Some hedgehogs, woodchucks, ground squirrels, hamsters, bats and dormice hibernate, and the last of these examples provides a linguistic clue to the essential nature of the hibernation process. The dormouse is not, in strict zoological terms, a mouse at all. Its popular name comes from the French word *dormeuse*, meaning a sleeper. In the winter months, in the harsh conditions of the wild, hibernating animals change from the normal activity of warm-blooded creatures and enter a cold, sleep-like state. A single bird is known to undergo the same type of alteration, namely the North American 'poor-will'

nightjar, *Phalaeonoptilus nuttalli*. Many animals commonly supposed to hibernate, including bears, squirrels and badgers, do not, in fact, do so. Instead these animals enter a state of torpor.

What, then, is the difference between hibernation and torpor? Strictly, hibernation is the process confined to animals which can normally maintain a high and consistent body temperature close to 37.8 degrees centigrade (100 degrees F) against the ever-changing pattern of ambient temperature. Under some circumstances, however, the true hibernator can give up this controlled maintenance of body-warmth. Instead it behaves, at one level, like a cold-blooded reptile or amphibian by allowing its body temperature to fall to values close to those of its environment, which may be as low as freezing point, 0 degrees centigrade (32 degrees F).

For nearly all mammals and birds that do not hibernate body-cooling of this magnitude is rapidly lethal. In these typical forms a whole range of vital processes, such as the muscle contraction needed to maintain the heartbeat, collapses at temperatures well above freezing. In a hibernating animal, however, everything carries on working in the cold state. Breathing continues, but it is slow and shallow. The heart carries on beating and pushes chilled, thick blood around the body, but it beats much more slowly than at normal body temperatures. It is as if all the creature's life processes have dropped down into the lowest of low gears. The significance of the tremendous slowing of metabolism is an immense saving in energy expenditure.

Many scientists think that this technique of energy conservation in winter conditions is usually employed by mammals which first evolved in tropical and

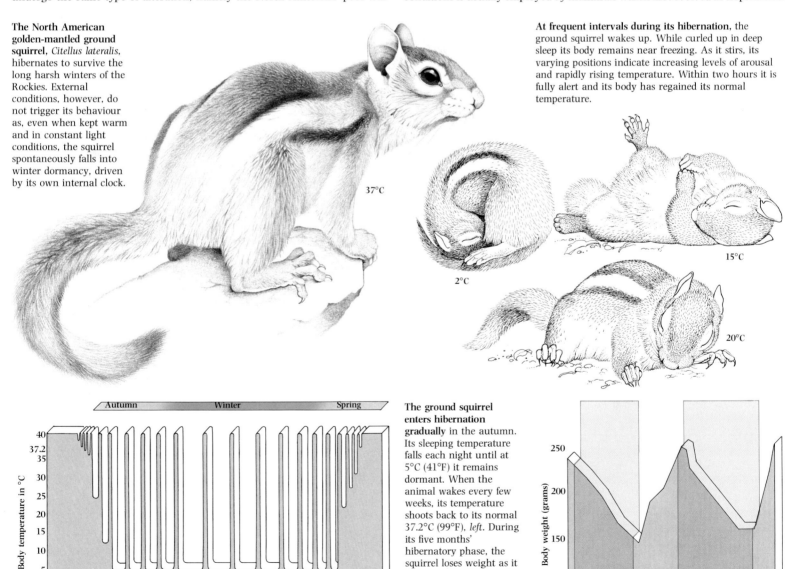

The North American golden-mantled ground squirrel, *Citellus lateralis*, hibernates to survive the long harsh winters of the Rockies. External conditions, however, do not trigger its behaviour as, even when kept warm and in constant light conditions, the squirrel spontaneously falls into winter dormancy, driven by its own internal clock.

37°C

2°C

15°C

20°C

At frequent intervals during its hibernation, the ground squirrel wakes up. While curled up in deep sleep its body remains near freezing. As it stirs, its varying positions indicate increasing levels of arousal and rapidly rising temperature. Within two hours it is fully alert and its body has regained its normal temperature.

Autumn   Winter   Spring

Body temperature in °C

40
37.2
35
30
25
20
15
10
5
0

Aug   Sep   Oct   Nov   Dec   Jan   Feb   Mar

The ground squirrel enters hibernation gradually in the autumn. Its sleeping temperature falls each night until at 5°C (41°F) it remains dormant. When the animal wakes every few weeks, its temperature shoots back to its normal 37.2°C (99°F), *left*. During its five months' hibernatory phase, the squirrel loses weight as it draws on the fat store it accumulated in late autumn, *right*.

Body weight (grams)

250
200
150
100

Jan   May   Oct   Jun

subtropical climates, but which have later colonized cool temperate zones. The patterns of physiological adaptations acquired in the warmer initial habitats seem to have made adaptation to a life of high activity in cooling conditions difficult or impossible. Instead, the hibernator opts out of any active existence during the winter period. For animals, such as hedgehogs or bats, which feed predominantly on insect prey, winter is also a time when their essential food items are hard to come by, so hibernation gives them an added advantage in the survival game.

The two conditions of torpor and hibernation are similar but hibernation is, in all respects, more extreme. Hibernation involves lower body temperatures and greater metabolic slowing and lasts for much longer periods than typical torpor. In most respects these differences are those of degree rather than of type. Thus a torpid bear does have a reduced heart rate and breathing rate. Its metabolism is slowed and it may stop urinating for several months, but the torpid bear's body temperature is only a few degrees, at most, below that of an active summer bear and never approaches freezing point as that of a hibernating ground squirrel or dormouse does.

Within the entire animal kingdom there are a number of other changes and protective metabolic states which help the animals that employ them to survive in harsh conditions, occurring in a seasonally regular fashion. Many insects go into a switched-off state of suspended animation or diapause, often in response to daylength clues which signal that harsh environmental conditions are on their way. This is part of the answer to the question, 'Where do flies go in winter?' In the larvae of the Khapra beetle, *Trogoderma granarium*, a native of

India which infests stores of wheat, maize and other grains, diapause induced by low temperatures, desiccation or starvation, as well as changing daylength, may last as long as eight years. This long diapause has made these larvae a menace in the holds of grain-carrying cargo vessels because they are almost impossible to eradicate.

Some tropical vertebrates, living in habitats that cycle between intense dry heat and a flooded rainy season throughout a 12-month period, are able to protect themselves by entering a state of aestivation which is almost exactly the tropical equivalent of hibernation. During the dry season, when the availability of water drops to dangerously low levels, aestivating animals sit out the conditions in some protected location, using levels of metabolism that will only slowly consume nutrients stored within their bodies. The African lungfish, *Protopterus*, for example, burrows into the mud at the bottom of its swampy home in advance of the approaching dry season. When the dry season arrives in earnest, the swamps become completely dried up and the fish survive in mud cocoons beneath the baked mud surface. Months later, returning rains stimulate the fish back into activity, and they wriggle from their cocoons and escape from their burrows. Villagers in some parts of the Sudan use a novel method for discovering these tasty underground fish as the rainy season approaches. Village women walk over the hard, dried mud flats rattling their fingers loudly on gourds to make a sound like rain beating on the ground. Hearing the sound, the aestivating lungfish think the rains have arrived and start to emit grunting noises—at which moment they are promptly dug up and eaten.

Lack of insect food and an inadequate coat make hibernation a useful survival strategy for the hedgehog, *Erinaceus europaeus*. Sheltered in a hedgerow or under a pile of leaves, the hedgehog rolls into a tight prickly ball for the winter, relying on its thin spiky skin to deter predators and its store of fat to sustain it.

When milder weather returns in March, the hedgehog reinjects life into its sleepy body. Like other hibernators, it activates energy-rich deposits of brown fat to raise its temperature from near 0°C to 37°C (98.4°F). Within hours, enough energy is generated by the fat to return the body's physiological functions to normal.

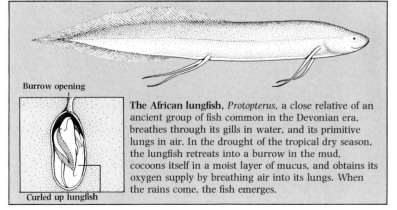

Burrow opening

Curled up lungfish

The African lungfish, *Protopterus*, a close relative of an ancient group of fish common in the Devonian era, breathes through its gills in water, and its primitive lungs in air. In the drought of the tropical dry season, the lungfish retreats into a burrow in the mud, cocoons itself in a moist layer of mucus, and obtains its oxygen supply by breathing air into its lungs. When the rains come, the fish emerges.

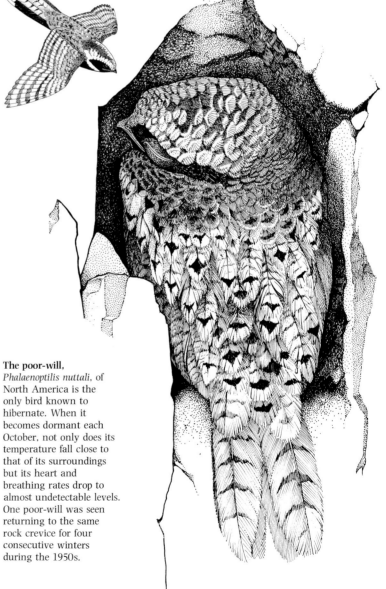

The poor-will, *Phalaenoptilis nuttali*, of North America is the only bird known to hibernate. When it becomes dormant each October, not only does its temperature fall close to that of its surroundings but its heart and breathing rates drop to almost undetectable levels. One poor-will was seen returning to the same rock crevice for four consecutive winters during the 1950s.

Rhythms within rhythms

Why is it that birds only sing in chorus at dawn? Why do crabs manage to hide under stones before low tide exposes them to the air? And why do cuckoos always arrive in the north in April? These and countless others, are examples of the way in which nature keeps abreast of the Earth's timetable of days, tides and seasons. The astronomical condition of living on a rotating planet which, together with its moon, circles around a star—the sun—places all of Earth's organisms in an environment whose time structure is composed of these three inescapable rhythms.

Night and day, high and low tide, summer and winter all involve spectacular differences in the amount of sunlight, dryness, temperature and many other environmental elements to which a creature is exposed. But few species are adapted to withstand the extremes that result from these rhythms. This problem is perhaps most obvious in tidal animals whose environment changes every six hours or so from being fully aquatic to being, in effect, terrestrial, but the rhythms of days and seasons can be equally dramatic. Animals are active either during the day or at night but usually not both, because to be both diurnal and nocturnal poses conflicting physiological demands—for example the ability to see well by night and also by day. While some creatures, such as cats and foxes, do have the ability to see and hunt well night and day, the structure of the eye of a nocturnal animal is actually different from that of the eye of a diurnal animal. Likewise for season: almost perversely it seems, snowdrops and crocuses come into flower in the depths of winter yet, by doing so, avoid direct competition from the quicker-growing but frost-sensitive annuals that smother them in summer.

These advantages of adjusting to nature's rhythms seem logical and reasonably easy to understand. But how and why did they arise, for surely primitive life had no need to be so sophisticated? The answer to this question may well lie in the composition of the Earth's early atmosphere. Four or five billion years ago, when life was struggling to assemble itself from the ingredients of the primeval soup, our planet was not protected from the sun's highly destructive ultra-violet radiation by an upper atmospheric screen of ozone as it is today. Thus the environment fluctuated violently between the searing radiation of the day and the freezing darkness of night, and it seems probable that primitive life would not have survived for long had it not developed some kind of timekeeping ability that restricted its more radiation-sensitive processes to the nighttime hours. That timekeeping ability ultimately became built into the genetic specification of early plants and animals, so that it not only had far-reaching implications for their lives but was passed on to the countless generations that followed.

All the many kinds of biological rhythms can be divided into two quite different and distinct categories. On the one hand there are all those biological functions which, especially in animals, just happen to work by oscillating—heartbeat, breathing, transmission of nerve impulses, and so on. These rhythms are not related to external time in any important sense, so that heartbeat and breathing, for example, vary mainly in relation to the demands of the body tissues for oxygen. Quite different are the rhythms by which animals and plants make the vital adjustments in their lives to the three basic periodicities of their environment—daily, lunar-tidal and annual. These three rhythms are specifically and exclusively concerned with environmental time and this chapter is devoted to them alone.

In an accurate but involuntary response to the daily, tidal, lunar and annual rhythms of the world around them, animals and plants have become adapted to their environment. These adaptations evolved so that organisms could order their lives according to the chronological arrangement of the world, and each species has its own characteristic relationship with the day, the tide or the year. From the opening of flowers to the mating behaviour of animals, all the members of a species must be active at roughly the same time if they are to breed and reproduce their kind successfully. Moreover, each individual of that species must have an inbuilt physiological timetable so that it can make suitable internal preparations in its body chemistry for the different activities which it performs around the clock.

This timetabling lies at the heart of all biological rhythms, whether daily, tidal, lunar or annual, and consists of two quite separate but closely interacting parts. One part is the organism's physiology, organized so that it cycles at approximately the right speed—once every 24 hours, or once each tide, or month, or year. The other part of the timetabling mechanism involves the environment and the responses the organism makes to the relevant cues, such as sunrise and sunset, which it receives from its surroundings about the passage of time. In this way the two parts interlock to keep the organism adjusted to local solar time, and they act together as if they comprised the mechanism of a clock—not a mechanical one, but a biological clock. We will come to the physiological reality and possible composition of such clocks later, but first it is necessary to be clear about some of the terms that are used to describe and to analyse them.

Both the rhythm of the organism and the rhythm of environmental change have some starting point—for example sunrise and waking up—and end up back at the same point 24 hours later, each having passed through one complete cycle of day and night, or activity and sleep. A rhythm is, thus, a continuous series of repeated cycles, each one very like the last. The result of this series of cycles is an oscillation, whether in the environment's light-dark cycle or in an animal's waking and sleeping behaviour. A pendulum does the same thing, repeatedly swinging through one full traverse and back again to complete each cycle of its oscillation, so providing a most useful and appropriate model rhythm with which to examine the terms that biologists have borrowed from physicists and engineers in order to study life's rhythms.

Imagine a pendulum set up with a pen fixed to it so that with each swing it traces an arc on to a piece of paper placed against it. If the paper is now moved downward at a steady speed, while the pendulum keeps swinging, the pen will trace a curving sinusoidal wave-form on to the paper. Turned on its side this wave will look like a smoothed-out version of the record of the 'sleep' movements of the leaf of the *Mimosa pudica* plant illustrated below, as it folds down its foliage at night and opens it up again during daylight. The up and down displacement of the line will show the amplitude of the pendulum's oscillation. The horizontal axis drawn along the paper represents time, and the distance between successive peaks of the trace is known as the period of the oscillation. The period of a rhythm is, thus, the time taken for the completion of one whole cycle, but here this is represented as distance along the paper—the actual distance depending on the speed at which the paper was moved.

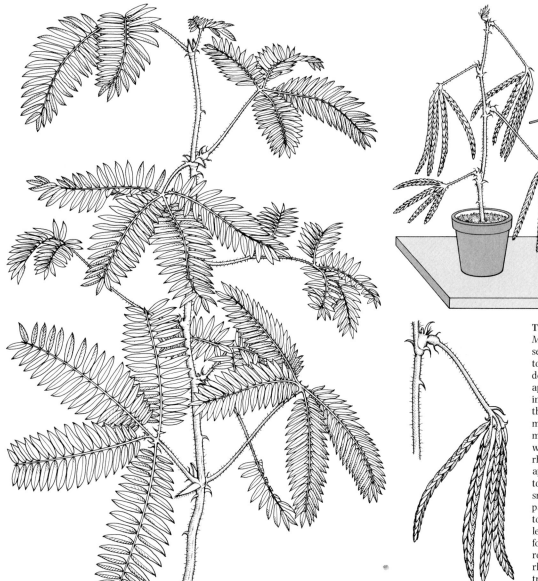

The so-called 'sensitive' plant is a small species of the *Mimosa* genus that folds its leaves within a few seconds of being touched. The tiny leaflets close together against the leaf's midrib which then folds down at a joint in its stalk, *near left*. The plant also apparently 'goes to sleep' since at night the leaves fold in the same way, whereas by day they spread out in the sun. Few plants show the touch response but many show the daily rhythm of these sleep movements, especially those like *Mimosa pudica, far left*. which belong to the bean family (Leguminosae). This rhythm may be readily demonstrated using the apparatus, *above*. The leaf's midrib is joined by a thread to a light, pivoted arm at the other end of which is a small pen. The pen presses lightly on to a sheet of paper wrapped round a cylinder driven by clockwork to make one full revolution in a week. Thus, when the leaf rises during the day the pen moves down; when it folds down at night the pen moves up. The result is a roughly snake-shaped or sinusoidal record showing a rhythm with seven nighttime peaks and seven daytime troughs.

Another measure of the speed of a rhythm is its frequency, that is, the number of cycles it completes in a given time. Thus the frequency of a rhythm such as a human heartbeat is about 70 cycles a minute. This is, of course, a high frequency rhythm compared with the very low frequency of the environmental rhythms of day and night, moon and tides and the changing seasons: the frequency of the sleep-waking cycle is once every 24 hours, of leaf fall and renewal once a year. Mathematically frequency is the reciprocal of—or, colloquially, 'one over'—the period, so that as the period of a rhythm increases its frequency decreases and vice versa. And the pendulum shows this too, because the longer its arm, the longer its swing takes, so the greater its period and the lower its frequency.

Any particular point in a rhythm is a phase. It is essential to have some definable phase of a rhythm to know where the phase is in relation to the whole cycle and that of the rhythm itself. The period of the pendulum's rhythm, for example, can be identified by the positions of the peaks in its pen record. The peak is one convenient phase point to take but, equally, so is the trough or any other point in its oscillation. In theory, every cycle of a rhythm goes through an infinite number of phases, but in practice it is sensible to select easily identifiable ones. In the movement rhythm of the mimosa leaf, for example, any position of the leaves could be used as a reference phase, but it is easier to refer to the obvious phases of maximum daytime extension or maximum nighttime folding.

The phase of a rhythm is not only useful descriptively, but also brings out an important concept that ties the two parts of the biological clock mechanism together—the physiological and the environmental. Mimosa has its leaves open by day and closed at night, and that is the phase-relationship between the plant's physiological rhythm (leaf position) and the rhythm of the environment (day/night). This phase-relationship stresses the significance of all such environmentally timed rhythms, whether they are daily, tidal, lunar or annual. There is a constant phase-setting of the organism's rhythm in relation to that of its environment. Thus nocturnal animals are active only at night, diurnal ones by day—the function of their biological timekeeping is to regulate this cycle of activity.

If the timing of the environmental cycle is changed, then the organism's cycle will be upset too, but it will, in due course, reacquire its old phase-relationship with the environment. In man the temporarily debilitating phenomenon of jet lag, after long-distance air travel across several time zones, sums up the problem admirably.

When you travel by jet aircraft from London to San Francisco your environment is changed so that it gets dark about eight hours later than you are used to. After a day or so, however, you adopt San Francisco time as your new temporal environment, although your underlying physiological rhythms may take several more days to catch up with your social schedule. In the laboratory it is easy to carry out equivalent experiments on animals or plants by changing their lighting schedule, and then seeing how they phase-shift to adapt themselves to the new conditions.

Except for organisms that live in tidal waters, and those such as bacteria that have a very short life-span, the most obvious expression of biological ▷

The main characteristics of any rhythm are illustrated on a record of the daily movement of the mimosa leaves, *top right*. The obvious **oscillation** is the daily repeated **cycle** of the up-down-up sequence of leaf movement. Each cycle is the same and lasts for 24 hours, which is the **period** of the rhythm. The rhythm's frequency is the number of cycles it completes in some chosen length of time, eg once in 24 hours. The amount by which the rhythm moves up and down is its **amplitude**, ie the maximum distance between the leaf's day and night positions. Any particular point in a rhythm is a **phase**: the phase of lowest leaf position is at midnight, for example, and of highest leaf position at noon. Each phase occurs only once per cycle and has a specific relationship with external solar time. This relationship can be changed experimentally by altering the cycle of light and dark. In this instance, *below right*, the first two days duplicate the external conditions of 12 hours light and 12 hours dark. The third dark period is extended by an extra eight hours and the plant's rhythm is temporarily suspended. When the lights are turned on again, the leaves resume their movement, but the phase of maximum leaf position has now shifted with respect to the record above it and hence to the real world.

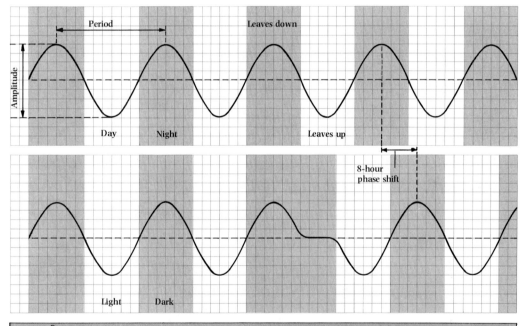

Varying rhythms allow different species to share out the environment's resources by using them at different times. The chart shows the flying times of five closely related male ant species of the genus *Dorylus* occupying the same habitat in the Ugandan jungle.

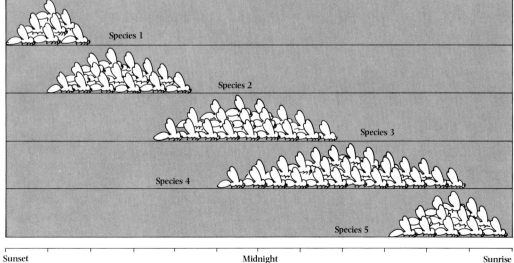

▷ timekeeping for virtually all organisms is their daily 24-hour rhythmicity. This property influences almost every aspect of their lives, making them hatch, grow, respire, feed, manufacture food products, move, mate and even die more at some times of day than at others. Even the simplest single-celled animals and plants have daily rhythms just as striking and almost as accurate as those of the mostly highly evolved mammals and the most complex flowering plants such as trees.

Because organisms tend to exploit their environment's resources at particular times they are firmly tied, in terms of both their behaviour and their physiology, to particular conditions of temperature, humidity and, most important of all, light. A simple answer to the question of why each different species has a particular phase-relationship with its environment could be that each responds to a particular phase of the daylight cycle—say dawn—and only becomes active because it is nudged by the environment to do so: just as a man can set his alarm clock to wake him up at the same time each morning. If this were so, then the logical implication would be that rhythms are nothing more than direct responses to environmental time cues and, therefore, no more inherently rhythmic than are the hands of a clock when deprived of a pendulum or hairspring to drive them.

There is, however, more to animal and plant rhythms than mere immediate responses to signals from their surroundings; organisms do seem to possess some form of internal timekeeping mechanism. While the animal or plant is in conditions of natural periods of daylight and darkness this inner mechanism may be impossible to detect. The solution is to bring the organism into the

laboratory, subject it to experimental light schedules and then measure its activity.

First an organism must be chosen for experiment—say a golden hamster or a flying squirrel since both these mammals have clear activity rhythms. Next the creature must be placed in an artificial light cycle, set at a slightly different time from the natural day-night cycle outside the laboratory. When this is done one finds invariably that the animal takes a few days to settle down, while its rhythm gradually phase-shifts to meet the new timing of its environment, as it happens in a man who has just stepped off a transatlantic flight. The creature then adapts its normal phase-relationship to the light cycle. As all the other major environmental conditions, including temperature, humidity, noise and so on are kept constant throughout the experiment, two things are clear: one, that the rhythm does continue in a highly artificial situation with the only time cue present being that from the light cycle and, two, that the rhythm phase-shifts to a new setting if the timing of the light cycle is changed in any significant way.

This gradual phase-shifting over a few days strongly suggests that the animal may have an underlying internal rhythmicity. For if the activity rhythm were merely a direct response to the light cycle it would have changed immediately to the new setting, on the very first day of the experiment. Even so, the presence of the light cycle prevents us from knowing for sure that the animal's rhythm is not due, in large part, to the fact that the light rhythm is driving it.

The second part of the experiment must be to test what happens when the

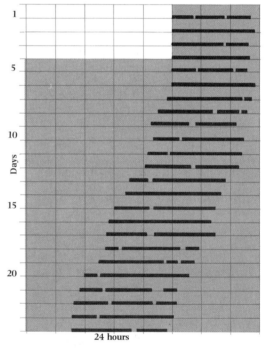

The rhythms of locomotor activity of small mammals and large insects can be measured in the running-wheel apparatus, *below*. The wheel is joined to a pen-recorder so that each revolution completes an electrical circuit and makes a lateral 'blip' on an ink trace drawn on a piece of continuously moving paper. An animal, such as a flying squirrel or a cockroach, is placed in the wheel and left to run according to its own behavioural demands. The apparatus records the animal's activity as groups of 'blips' that signify bursts of running.

The North American flying squirrel, *Glaucomys volans*, is a nocturnal tree dwelling mammal. Folds of skin joining its front and hind legs enable it to glide from tree to tree.

Running wheel experiments ensure that the animal has ample food and water and is left undisturbed for several weeks in a strictly controlled environment. To analyze the record of its running activity, the paper is cut into lengths representing 24 hours, and each day is pasted up beneath the preceding one as shown *above*. The heavily inked parts of the trace correspond to the time when the animal was active. In this instance, a flying squirrel was used and for the first four days its activity coincided with night in the natural environment. From the end of the fourth night the squirrel was kept in constant darkness for the next 19 days so that it no longer had any 24-hour time cues from the light. Its rhythm nevertheless continued, slowly drifting with respect to external solar time. Still active for the same length of time, it began its activity about half an hour earlier each day. Its rhythm was thus free-running according to a $23\frac{1}{2}$-hour day, which is the squirrel's inherent circadian periodicity.

animal is deprived of all environmental time cues by placing it in constant and complete darkness. When this is done, two significant facts are revealed. First, the rhythm continues unabated, which means that its driving force certainly does not depend on a light cycle. Second, and even more interestingly, the rhythm does not continue at exactly 24 hours per cycle, but drifts relative to external, solar time as measured by the hands of the clock on the laboratory wall.

Had the rhythm of rest and activity continued in constant darkness with the same precise 24-hour period it showed in the 24-hour light cycle, it could conceivably have been due to the animal responding to unsuspected external time cues that were not controlled during the experiment—for instance, barometric pressure or magnetic field. But that did not happen. Instead, the rhythm ran at a frequency that differed slightly, but consistently, from 24 hours per cycle, whereas all the likely—and indeed unlikely—uncontrolled time cues from the environment run at the rate of exactly 24 hours because, like light and darkness, they arise from the effects of the earth's daily rotation about its axis.

This free-running drift of the 24-hour rhythm is strong evidence in favour of the proposition that the rhythm is internally controlled, that is, driven by a physiological 'clock' system from within the organism. The same drift is, in fact, characteristic of almost all the daily rhythms in plants and animals that have so far been studied. Moreover, the period of all such free-running rhythms in constant conditions rarely differs by more than an hour or two from 24, typically falling within the range of 22 to 28 hours per cycle. It is this narrow range of period that gives such rhythms their technical identifying description

of circadian, a word derived from the Latin words *circa* meaning about and *diem*, a day. Nearly all tidal, lunar and even annual rhythms that have so far been tested in constant conditions also free-run and are thus described as circatidal, circalunar (or circumlunar) and circannual.

Although circadian rhythms free-run in the laboratory it is obvious to even the most casual observer that they do not do so in nature. What happens, of course, is that each individual's rhythm is set so that it has a constant phase-relationship with the environment according to the day-night cycle. And this brings us full circle to the dual nature of biological rhythms: the rhythmicity of the environment is said to entrain the internal rhythm of the organism to an exact 24-hour periodicity.

It is just such a process of entrainment that goes into action when a plant or animal is subjected to a changed light cycle—as by a transatlantic flight or a laboratory experiment—and takes a few days of phase-shifting to get back into its normal phase-relationship with the new light cycle time. Circadian rhythms are normally entrained most strongly by the light intensity changes of the day-night cycle, particularly by dawn or dusk when light levels are changing fastest. But in the absence of a light cycle many organisms will entrain themselves to other 24-hour cues, for example to a temperature cycle, and for man, social time cues, such as the habit of working from nine to five and eating meals at regular times, may well be paramount in entraining his circadian rhythm.

While many plants and animals organize their activities around the cycle of day and night, the organisms of the seashore arrange their actions around the ▷

The morning glory, *Ipomoea*, takes its name from the fact that each pink flower, *above right*, opens in the morning and then withers away by early afternoon. Few flowers survive so briefly, although many species open only at fairly specific times of the day.

Carl Linnaeus, the 18th-century botanist, grew a floral clock outside his window, the successive opening and closing of different flowers marking the hours. The Edwardian postcard, *above*, shows how this can be done. Such specific flowering times seem to raise the chances of pollination by bees, whose superb time sense helps them learn when nectar will be available.

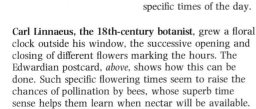

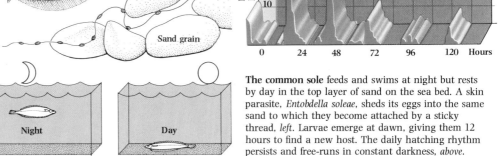

The common sole feeds and swims at night but rests by day in the top layer of sand on the sea bed. A skin parasite, *Entobdella soleae*, sheds its eggs into the same sand to which they become attached by a sticky thread, *left*. Larvae emerge at dawn, giving them 12 hours to find a new host. The daily hatching rhythm persists and free-runs in constant darkness, *above*.

▷ebb and flow of the tides. Of all the habitats on Earth the one that exists between high and low tide levels is perhaps the most difficult and demanding. At high tide, although submerged and thus more or less typically marine, the creatures that live on the shore are subject to violent battering by the waves and a perpetual shifting of the sand, pebbles and rocks that form the substrata for all the habitat's plants and non-swimming animals. At low tide—a mere six hours later—the same site has become, to all intents and purposes, terrestrial, so that it is subject to rapid changes in salinity from rain, to freezing temperatures in winter, shadeless drying radiation in summer and, at all times, to merciless predation from sea birds. Small wonder, then, that the organisms which have evolved to exploit the riches of food in the intertidal zone, and which make the investigation of the beach at low tide such a pleasurable experience, have adopted sharp timekeeping, with the result that they can anticipate and avoid the worst of its radical, twice-daily changes in condition.

Just as the denizens of land, fresh water and open sea share out their habitat by using it at different hours of the day and night, so the intertidal habitat is divided up. But the division is not nearly as equal as the day versus night share-out. Nearly all intertidal animals are essentially aquatic and so breathe with gills, structures specifically designed to extract oxygen from water. It is, therefore, hard for these creatures to use the terrestrial low tide niche, and few species do so.

A rare exception to this intertidal rule is the fiddler crab, *Uca*, which comes out of its burrow to scavenge at low tide. More common are those species that are active within the advancing or receding tide front. Plants, of course, just

have to sit it out and wait; probably their only concession to the tidal cycle is to be rhythmic in the release of their reproductive spores, so that they shed these into the water only when they are submerged. Nearly all tidal plants are seaweeds (algae), and most live at the lower shore of the beach, so that they are uncovered for only brief spells at each low tide.

What we must now discover is whether these tidal rhythms are simply direct responses to the presence or absence of seawater. There are exceptions, such as barnacles, which will open up their shells at any time when they are under water, but, apart from these, all the other animals that have been tested in the laboratory show a typical free-running circatidal—that is, roughly 12-hour—rhythm when placed in constant conditions. They, therefore, drift in their timing relative to the tide times on the beach that was their home and from which they were collected, just as circadian rhythms drift relative to solar time.

The average duration of one tide cycle is 12.4 hours. This figure is rather close to half the 24-hour circadian period, so the activity rhythms of intertidal animals are very similar to the twice-daily activity patterns many terrestrial species adopt by being active at dawn and dusk, that is, crepuscular. Moreover, 24.8 hours, or two tide cycles, is well within the range of free-running circadian rhythms, so there is every reason to suppose that, even though they may have evolved independently, circatidal and circadian rhythms must operate along similar lines and by similar mechanisms. Certainly several animals, such as the common shore crab, *Carcinus maenas*, show a roughly 12-hour, free-running circatidal rhythm in the laboratory when they have been collected from a tidal

Tidal waters contain a rich array of plants and animals whose lives have to adjust to the twice-daily hazard of being alternately submerged and stranded by the high and low tides, *right*. Few animals can withstand being exposed for long at low tide and are thus rhythmic in their activities, burying themselves in the sand or under stones at roughly 12.4-hour intervals, as if in anticipation of the receding tide. The European shore crab, *Carcinus maenas*, shows a tidal rhythm of activity as it hides under stones at low tide, coming out to scavenge only when it is covered by water, *below right*.

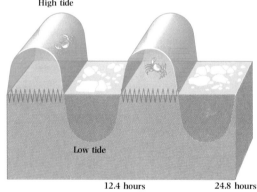

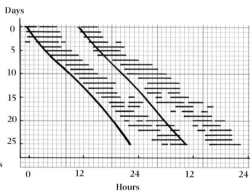

The danger of being eaten by birds, such as this crab plover, is a major hazard facing animals at low tide.

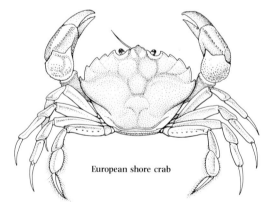

European shore crab

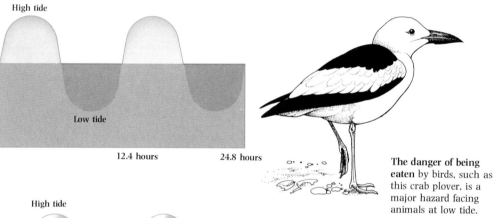

Days

Hours

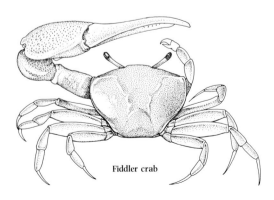

Fiddler crab

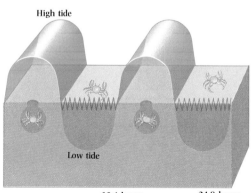

A rare example of an animal that is active at low tide is the fiddler crab, *Uca* sp, which comes out of its burrow into the open air to look for food, *left*. Much research has been done on this crab in order to discover if its activity rhythm is triggered by the water or is governed by an internal ability to measure tidal time. The record, *above*, is the result of an experiment in which the crab was kept in constant conditions for 25 days. The twice-daily horizontal lines show the crab's activity; the sloping lines, the high tide times on its home beach. These drift by 50 minutes a day since the tidal 'day' is 24 hours 50 minutes long. That the crab's rhythm persists and also drifts faster than the tides shows that it is spontaneous and internally timed.

shore, but a roughly 24-hour, circadian periodicity when they have been gathered from a tideless area such as a dock. This shows that the rhythm can become adapted to suit different conditions.

Apart from its shorter cycle time and the violence of its periodicity, the tidal habitat differs from all others by changing rhythmically with the phases of the moon. Twice in every lunar month of 28.5 days, when the moon and sun are in line—at the full and the new moon—the sun's gravitational pull on the sea augments that of the moon to cause tides considerably bigger than average. These are the spring tides, whose name often causes confusion because they are nothing to do with the seasons. The in between phase of smaller neap tides occurs when the sun's pull is at right angles to that of the moon.

The result of the moon's influence on the tides is that all coastal areas include three rather different habitats. There is a large central zone uncovered at every tide, a lower zone uncovered briefly twice a day only at maximum low-water spring tides and an upper zone that is covered briefly only at maximum high-water spring tides. The lower zone is almost always fully aquatic, the upper one almost always fully terrestrial, and there are animals that exploit both these features. A famous example is the grunion fish, *Leuresthes tenuis*, whose males and females swarm up Californian beaches to spawn, on the crest of high spring tides. There the fish bury their eggs, leaving them to develop in the warm sand until the next high spring tides wash them out to hatch in the open sea.

At the lower extreme of the tidal zone there is a highly unusual insect, a midge, *Clunio marinus*, whose larvae feed on the red algae growing at the most

seaward part of some European beaches. The adults live for only two hours, and must, therefore, emerge, mate and lay their eggs on the algae during a single low-water spring tide. Fascinatingly, when reared in the laboratory in a constant 24-hour light-dark cycle, the midges emerge from the culture in bursts at roughly 15-day intervals and thus demonstrate a clear, free-running semi-lunar rhythm.

Clunio is one of the few organisms in which a twice-monthly rhythm found in natural conditions has been shown to persist and free-run when transferred to the constant conditions of the laboratory. Another good example involves a plant, the large brown seaweed *Dictyota dichotoma*. In similar constant conditions this alga sheds its egg cells into the sea with a clear 17-day free-running rhythm. Both *Clunio* and *Dictyota*—and presumably most other inhabitants of the extreme tidal zones—thus possess the internal ability to measure out approximately 15-day intervals, presumably by means of a semi-lunar physiological clock mechanism.

There is one more aspect of lunar periodicity that affects at least some organisms, namely that of the full 29-day lunar month. Around coral reefs in the Pacific there is a particularly spectacular expression of this every year for just a few nights during the last quarter of each moon in October and November. During these nights, and only then, the surface of the sea swarms with an astronomical number of mating palolo worms of the species *Eunice viridis*. A more familiar example occurs in man—or rather in woman—as the menstrual cycle. Although in effect it free-runs, since it is not directly related to the timing of the real lunar cycle, this cycle can still be described as circalunar. ▷

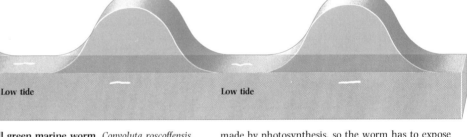

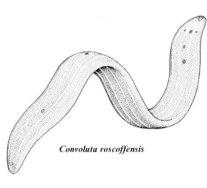

Convoluta roscoffensis

The small green marine worm, *Convoluta roscoffensis*, which lives just below the neap tide high-water mark in Brittany, is part animal and part plant. Its greenness comes from an alga that grows symbiotically in its tissues and from which it derives carbohydrate. This is made by photosynthesis, so the worm has to expose itself to sunlight. At high tide and at night it lies buried in the sand but at low tides during the day it crawls out into the light, *above*. Thus it possesses a combination of both a tidal and a circadian rhythm.

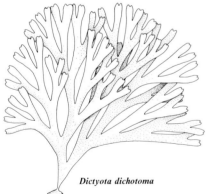

Dictyota dichotoma

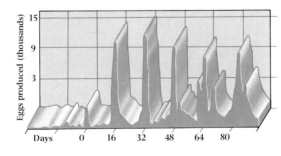

Every full and new moon the high tide is higher and the low tide is lower than usual. The 15-day rhythm of these spring tides causes the upper extreme of the tidal habitat to be covered for a few days and the lowest extreme to be uncovered. The brown alga, *Dictyota dichotoma*, lives in this lowest spring tide zone, where it discharges its eggs and sperm to fuse in the sea every 15 days during the smaller neap tides, presumably to avoid their being stranded by the low spring tide. In tideless conditions the rhythm of egg discharge persists, and free-runs every 17 days, *left*.

The aquatic larvae of the midge, *Clunio marinus*, feed on red algae growing in the low spring tide zones of many European shores. When the moon is full and new, the zone is exposed to the air for a few hours at each low tide. As waves break over them the larvae pupate, and when the tide is right out, quickly emerge as midges. The adults emerge, mate, lay eggs and die all within the few hours afforded by a single low spring tide. A larva that fails to pupate must wait two weeks before it has another chance. The timing of adult emergence with low spring tide is shown, *right*. In constant laboratory conditions the larvae persist in their rhythm, emerging as adults on one of four evenings at roughly 15-day intervals.

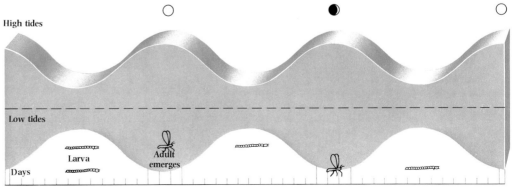

▷ There are several well-documented 29-day rhythms, clearly coupled to the moon's phases, to be found among invertebrate organisms, not only in marine species, where such behaviour might be expected from the cycle of the spring tides, but also in terrestrial species. For example, the emergence of some mayflies occurs in a peak just after the full moon. Most of these lunar rhythms are known only in their natural environment, but a few have been studied in the laboratory. A particularly good case is that of *Platynereis*, a European relative of the palolo, which swarms almost exclusively around the new moon in natural conditions, but shows a free-running circalunar rhythm when it is removed from its natural habitat and placed in an unchanging 24-hour light-dark cycle in the laboratory.

For the mayflies, palolo worm and *Platynereis* the function of this monthly swarming is presumably to bring the sexes together. But some tidal species show non-reproductive lunar rhythms, and it is difficult to see why they should have evolved this periodicity when the 15-day semi-lunar spring tide cycle must be much more important to their day-to-day survival. Nevertheless, all the cases so far investigated show free-running circalunar rhythms when they are placed in constant conditions, so they must all have evolved the internal ability to measure out a circalunar period, an ability indelibly inscribed in their genetic material.

Clearly none of these moon-caused rhythms, whether 12.4-hour, 15-day or 29-day, is free-running in nature. Just as with circadian rhythms they are kept to the correct, relevant environmental time—that is, they are entrained—by external time cues. In the few organisms that have been fully studied, semi-

lunar and circalunar rhythms are susceptible to entrainment by moonlight. Thus the rhythms of *Clunio*, *Dictyota* and *Platynereis*, for example, can be made to toe the environmental line by giving the organisms a few consecutive nights of artificial moonlight once a month in their otherwise unrelieved 24-hour cycles of light and dark. Experimentally this is done simply by replacing the total darkness of the relevant nights with dim light, controlled so that it is not confused with daylight.

For the twice-daily tidal rhythms, matters become a bit more complicated. The 24-hour day-night cycle that entrains circadian rhythms, although clearly discernible to most tidal species, is misleading because of its 0.8-hour (48-minute) difference from the 24.8-hour tidal 'day'. If tidal organisms used light cues to entrain their rhythms they would certainly get wholly out of phase with the tides in a few days. What these organisms use as tide-time cues are, therefore, the mechanical disturbance of the water as it rushes over them, or changes in temperature, salinity or pressure, or any combination of these reliable markers of the tidal cycle.

In the short term, the tidal environment creates the Earth's most demanding rhythm, but in the long term the changing year can be at least as exacting. Higher latitudes become totally uninhabitable in winter unless creatures stay hidden under sea ice, or go into a deep hibernation. And even in temperate latitudes, plants lie dormant and the food supply of many animals disappears. Anticipating this predictable annual crisis is, therefore, vital to survival for both plants and animals.

There are three ways in which one could imagine this adjustment being

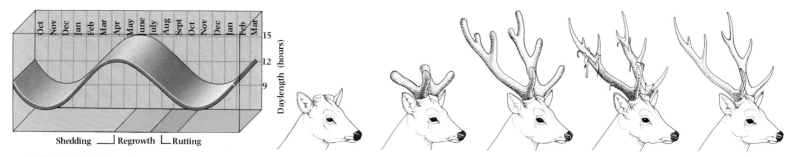

Shedding ⌐ Regrowth ⌐ Rutting

Most deer shed and regrow their antlers in an annual cycle. Like this American wapiti, *left*, they often add a branch or two over the years, so the complexity is a reflection of the stag's age. Generally antlers are shed in late winter or in early spring and then regrown in summer. Before the rutting season occurs in the autumn, the soft velvet under which the antlers develop must be lost. The annual antler cycle of the oriental sika deer is shown, *top left*, and the sequence of regrowth, *above*. If sika deer are kept in various lighting conditions, the antler cycle can be shown to be due to two factors. Mainly it is a response to the seasonal changes in daylength, which varies between 9 and 15 hours. It responds also to a weak, internal and roughly annual rhythm in the deer's physiology.

accomplished. First, organisms could merely switch into their winter dormancy or migratory phase when they first feel the pinch—say, from the first frost. Since frost, rain, drought and other such natural disasters arrive at the whim of the weather, however, they are unreliable markers for the change of season. The second possibility is that animals and plants use the totally reliable annual change of daylength as a predictor of seasons. This ability to measure out the number of hours of daylight is photoperiodism, and it is certainly used by many organisms, but to do so they must possess a sophisticated ability to measure daylength.

The third possibility would be for living things to have an internal annual clock—a circannual biological clock analogous with the circadian kind. Although this may sound improbable, such clocks do apparently exist in a wide variety of organisms. Circannual clocks were probably first noticed when European varieties of tree, normally exposed to changing daylength with the march of the seasons, were taken to the tropics by nineteenth-century colonists. There in an environment with virtually unchanging daylength, the trees continued to pursue their usual cycle of leaf opening, flowering and so on, although they tended to lose synchrony with the local year; they in fact free-ran in this natural laboratory.

More rigorously, although not very often because of the necessary experimental time involved, several animals and plants have now been shown to have free-running circannual rhythms in unchanging 24-hour light-dark cycles. The carpet beetle, *Anthrenus verbasci*, even shows a persistent circannual rhythm of adult emergence when kept for years in constant total darkness. In all these instances, the annual event that is actually timed in the free-running situation, such as mating, migration or moulting, can occur in the laboratory at any time of the year, and, therefore, quite out of phase with the real seasons outside. The experimental animals and plants are, therefore, not picking up uncontrolled cosmic cues, and must possess truly in-built circa-365-day clock systems within their bodies.

In natural surroundings, what is it that entrains these circannual rhythms to a precise year? The answer is not entirely clear, but is probably mainly a change in daylength. Certainly species such as deer and willow warblers show both circannual rhythmicity and responses to change in daylength. This sounds like a confusing mixture between photoperiodism, which is a direct response to daylength, and an organism's internal annual timing mechanism. Although quite different, the two phenomena do serve the same function in nature and go hand in hand. The circannual rhythm presumably makes the physiology of the animal, such as its production of sex hormones, change at roughly the right time of year so that it can correct its seasonal response when the appropriate daylength arrives.

All the observations and laboratory experiments performed on animals and plants leave us in absolutely no doubt that daily, tidal, lunar and annual rhythms are all controlled by underlying physiological timekeeping—that is, by physiological clocks. Such clocks clearly cannot be anything like man-made clocks, but any clock, whether biological or the product of human engineering, must, in principle, consist of the same five basic components. Number one constitutes the 'hands', the visible parts that indicate the clock's underlying ▷

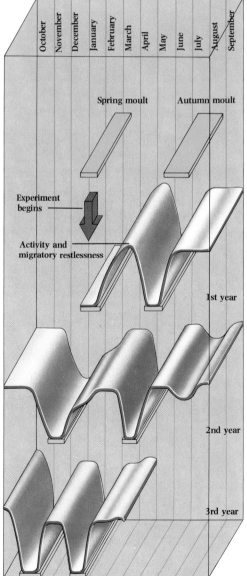

The willow warbler is a common European migrant, usually arriving in Britain in April. Like most other birds it moults twice a year: once in early spring before it migrates north, and once in late summer before it migrates south. When it moults its activity and migratory restlessness is reduced to a minimum. These moults are in part due to a response to seasonal changes in daylength. But if the warblers are kept for more than two years in a stable cycle of 12 hours light and 12 hours dark, they continue to moult roughly twice a year. The results of these experiments show that the interval between the spring moults and the interval between the autumn moults are both about 10 months. The birds must, therefore, have an internal and roughly annual clock to time the moults.

The European garden warbler has a circular migration route. In spring it migrates north across the Sahara and the eastern Mediterranean from its winter home in central Africa. In autumn it returns by flying over the Straits of Gibraltar, then southeast over the Sahara.

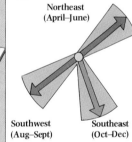

Northeast
(April–June)

Southwest
(Aug–Sept)

Southeast
(Oct–Dec)

Garden warblers know when to migrate and also the direction they must take. Built into the bird's behaviour is an ingenious annual schedule. Caged warblers in a migratory mood perform many repeated hops in the direction of their normal migratory flight path. These hops are easy to record, so for long periods the bird's attempts to migrate can be monitored. When the days are kept the same length so the birds have no clue as to the time of year, their hops point in three directions, *above*, corresponding to their annual behaviour in the wild.

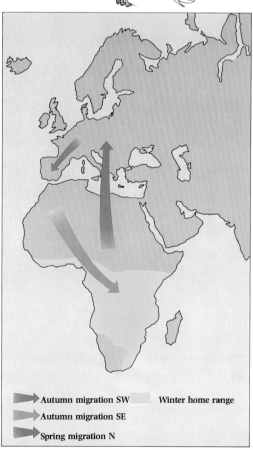

Autumn migration SW ▬▬ Winter home range

Autumn migration SE

Spring migration N

▷rhythm and show what time it is. Number two is a 'mainspring', the source of energy that drives the clock's mechanism. Three is an 'escapement' device, the basic regulating oscillator which, in mechanical clocks, governs the oscillation of the pendulum or combination of balance wheel and hairspring. Number four is the 'cogs', the couplings that link the escapement to the hands. Number five is an 'adjuster', the mechanism by which the clock can be corrected to tell local time.

In biological clocks, the 'hands' are all the visible, measurable rhythms, from running activity and leaf movement to migration and leaf fall. These observable rhythms are not an intrinsic part of the clock mechanism, but merely indicators of the state of the clock at any given moment. In theory the 'mainspring' should be the energy demands of the physiological mechanism, but these are so minute that they cannot be detected against the organism's general metabolism which uses comparatively huge amounts of energy. The 'escapement' of a biological clock is the key, unseen, driving oscillation, the essential timekeeping pacemaker.

The 'cogs' of the biological clock are those physiological coupling mechanisms such as hormones and nerves through which the 'escapement' drives the 'hands'. Although the environmental entraining time cue of the 'adjuster' mechanism is well known for many biological clocks—involving the perception of night and day by the eyes in circadian rhythms, for example—little or nothing is known about how this information is used within the body to reset the 'escapement' device.

For circadian rhythms, the two clock components about which most is

known are the 'escapement', or driving oscillator, and the coupling 'cogs'. Three plausible theories have been proposed for how the basic oscillator might work. The first theory places the clock mechanism in the cell's nucleus, the part of the cell in which hereditary information is contained in the chromosomes. These threadlike structures are composed of functional units, the genes, made up of the chemical deoxyribosenucleic acid (DNA) which acts as a coded series of instructions to control all life processes. The theory proposes that the circadian period is measured out by continuous reading of the genetic code along a very long loop of DNA in a chromosome. This idea has several attractive features and is supported by some experimental evidence, but it also has some flaws.

The second theory assumes that the clock has its roots in the organism's oscillating biochemistry. The problem is that it is theoretically difficult to construct a 24-hour clock from the known biochemical oscillations because these all have a high frequency with the time-span of each cycle measured in seconds or, at best, in minutes. The theory thus proposes that circadian rhythmicity is generated by an interaction of many of these high frequency chemical oscillations in such a manner that one inhibits another to create a net slow output. As yet, however, there is no concrete physiological evidence that this actually occurs.

The third theory of the biological escapement device envisages that the circadian oscillation occurs as a result of slow changes in the membranes within a cell. As a result of these changes, which affect the permeability of the membranes, ions, such as potassium, leak passively through a membrane from

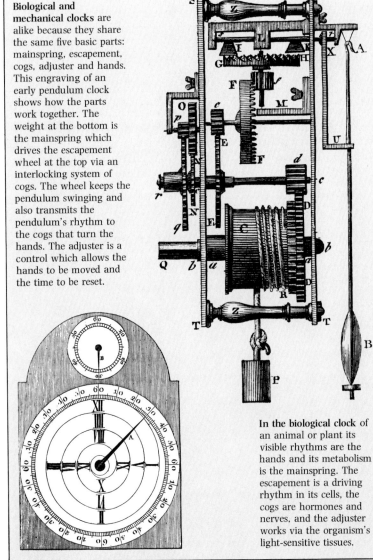

Biological and mechanical clocks are alike because they share the same five basic parts: mainspring, escapement, cogs, adjuster and hands. This engraving of an early pendulum clock shows how the parts work together. The weight at the bottom is the mainspring which drives the escapement wheel at the top via an interlocking system of cogs. The wheel keeps the pendulum swinging and also transmits the pendulum's rhythm to the cogs that turn the hands. The adjuster is a control which allows the hands to be moved and the time to be reset.

In the biological clock of an animal or plant its visible rhythms are the hands and its metabolism is the mainspring. The escapement is a driving rhythm in its cells, the cogs are hormones and nerves, and the adjuster works via the organism's light-sensitive tissues.

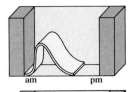

Hyalophora cecropia

Tissue transplants are relatively easy to perform on these silkmoths because they are large and their body cavity is an open blood system without arteries or veins.

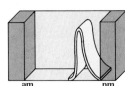

Antherea pernyi

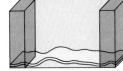

Moths of *Hyalophora* normally emerge from their pupal cases in the morning while those of *Antherea* emerge in the late afternoon.

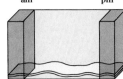

am   pm

This circadian rhythm is lost if the brains are removed from the pupae; ultimately they still emerge normally, but at random times.

am   pm

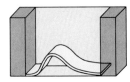

The normal daily rhythm of emergence can be restored if the removed brain is replaced in the abdomen of the brainless pupa.

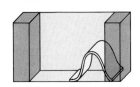

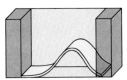

Rhythms are exchanged if the brain of one species is given to the other: *Hyalophora* then emerge in the late afternoon and *Antherea* in the morning.

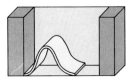

An adult silkmoth emerges from its pupa at a particular time of day and, as the above series of experiments has shown, this is under physiological control. This emergence—the splitting and shrugging off of the cocoon—will occur at any time during the day if the brain is removed. But since the rhythm is restored when the brain is replaced in the pupa's abdomen, the brain is needed for a specifically timed emergence. A hormone must control this timing since the transplanted brain has no nervous connection with the abdomen but is only bathed in blood. This brain hormone switches on the circadian emergence rhythm.

The fact that the rhythm can be transferred from one moth species to the other when their brains are exchanged means the brain itself has a timing mechanism. Thus there is a clock in the pupa's brain that produces a hormone at a particular time of day and triggers the adult to emerge.

one part of a cell to another, and are then actively pumped back again in the reverse direction when a particular imbalance in their concentration is reached. There are good non-circadian models for this kind of thing in the membranes of nerves and other sorts of cell. Moreover, while most chemical treatments either have no effect on circadian clocks or put them permanently out of commission, chemicals such as alcohol and heavy water, which are known to alter the permeability of cell membranes, actually slow down or phase-shift circadian rhythms in both plants and animals. This implies that these substances are directly affecting the basic oscillators contained in the organism, so it looks as though the membrane clock does have a sound experimental basis.

Despite the attraction of a cell membrane 'escapement', it is still too early to dismiss the DNA theory completely. The reason is that one or two antibiotic drugs that selectively inhibit the manufacture of proteins in cells—a process that can only occur through the active participation of DNA—have also been shown to phase-shift circadian rhythms. The current view of the biological clock is, thus, that it may involve both slow changes in the permeability of cell membranes and also the nucleus, via its role in synthesizing membrane proteins.

So what of the 'cogs' that link the escapement device of the biological clock with its 'hands'? It is a basic assumption that circadian rhythmicity evolved in step with primitive life, and hence that it is an essential property of all cells. Certainly single-celled algae and protozoa, that is, the very simplest plants and animals, have circadian rhythms, and so do isolated cultured cells taken from higher plants and animals. But if each cell of the millions from which higher organisms are built contains a circadian clock mechanism, why do some specific operations on animals, such as removing the brain of a silkmoth, the pineal body from a bird or the suprachiasmatic nuclei from the hypothalamus of a rat, make these creatures arrhythmic? The answer is that these operations only stop overt *behavioural* rhythms. And that gives the lie to the whole circadian organization of these animals. There are clocks everywhere in the body but some cells, such as those in the silkmoth's brain, bird's pineal body and rat's suprachiasmatic nuclei, have acquired specialized timekeeping abilities and, as a result, act as driving clocks for other tissues throughout the creature's body.

The body of an animal or plant, thus, consists of a panoply of clocks at many levels of organization—in cell, tissue and organ—and in the whole organism they must all tick together. It seems that all these clocks are arranged in a hierarchical network, mutually entraining each other to some degree, but with the brain clocks that control behaviour being, in effect, the primary driving oscillators. In this hierarchy, all the other clocks are subservient to the brain clocks, but the clock in the adrenal gland seems to be higher on the organizational ladder than those in other hormone-releasing glands. Similarly, these hormonal clocks appear superior to those in individual cells. The brain clocks take on their dominant role because they control behaviour, but also because they are the only clocks directly connected to the eyes and, therefore, the only ones with a direct entraining influence from the powerful signal of daylight.

The pineal gland at the top of a bird's brain is probably the location of the clock controlling the bird's circadian rhythms. This 'third eye' is not used for seeing but is, nevertheless, sensitive to light. Its removal makes the bird's behaviour arrhythmic and its replacement restores the rhythms. If pineal glands are exchanged between two birds on different time schedules the recipient assumes the schedule of the donor. The pineal seems to be a circadian clock and also a rhythmic source of a behaviour-regulating hormone.

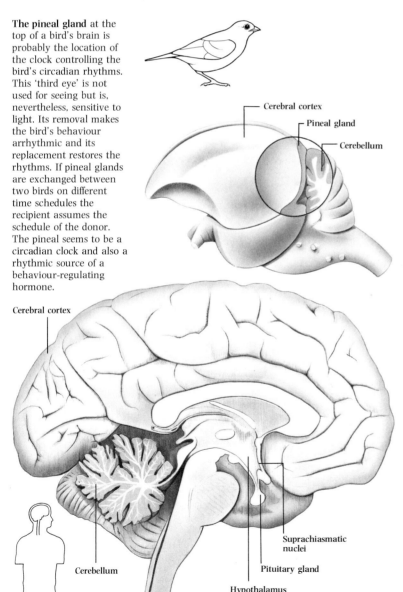

Cerebral cortex

Pineal gland

Cerebellum

Cerebral cortex

Suprachiasmatic nuclei

Pituitary gland

Cerebellum

Hypothalamus

Optic lobes

The brain of an insect, such as a housefly, is the largest nerve centre in its body. Groups of brain cells seem to act as circadian clocks, which in some insects have been found in the optic lobes either side of the brain.

The daily activity rhythms of mammals and presumably, therefore, of man are driven by a clock in the hypothalamus—a key area of the mid-brain that controls body functions such as temperature, rate of metabolism, appetite and thirst. The actual clock cells are in two tiny groups called the suprachiasmatic nuclei. They are entrained and adjusted to external daylight by a specific nervous input from the eyes that has nothing to do with normal vision. Unlike the bird's pineal and the silkmoth brain, these cells exert their influence by a direct nervous route rather than via hormones.

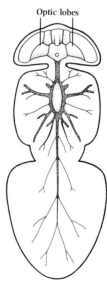

Many biological functions of animals such as their behavioural activity show clear circadian rhythms. There are also rhythms in excretion, hormone production and in the metabolism of single cells. Normally these functions keep in time with each other and work in harmony. There is probably a hierarchy of clocks, perhaps as shown *below*. In a mammal, the hypothalamus is entrained by daylight via the eyes, and itself probably entrains the clocks of other organs, which in turn adjust the clocks of individual cells.

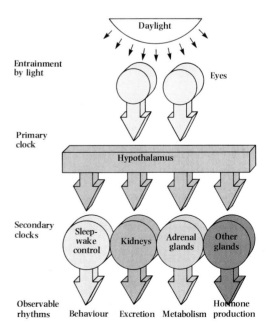

Entrainment by light

Primary clock

Secondary clocks

Observable rhythms

Daylight

Eyes

Hypothalamus

Sleep-wake control | Kidneys | Adrenal glands | Other glands

Behaviour | Excretion | Metabolism | Hormone production

Rhythms of sex

To ensure the continuance of their kind, animals must have a means of reproducing themselves. This process of reproduction does not, however, go on all the time and there are no species of animals capable of continuous reproduction. Some creatures are able to reproduce on many occasions during the year but others—the majority—are able to breed on only a few days in each 365. The reason why the annual rhythm of reproduction is so restricted is that it is essentially a process of development. A single egg cell is shed by the female which, after fertilization by one sperm from the male, eventually develops into a new, whole animal. Most invertebrate animals, the creatures without backbones, produce their eggs and sperm in batches and the processes of fertilization and egg development take place outside the body of the female, but in mammals, including humans, the egg is held inside the female's body. Following fertilization, a great deal of development occurs before birth so that reproduction requires a much longer involvement of the whole animal than, say, an act of movement, breathing or digestion.

Because of the inherent restrictions on continuous reproduction, there is a paramount need for the reproductive rhythms of males and females to be synchronized so that the male has ripe sperm ready just when the female has produced her eggs. If the rhythms of one sex are out of phase with those of the other then fertilization will be impossible and the species will not reproduce. The forces of evolution have seen to the attainment of synchrony, but simultaneous production of mature sex cells is not always enough. This is because many species—particularly among birds and mammals—must undergo a time-consuming change in behaviour prior to mating. Without courtship, the mutual suspicion and aggression between the sexes may prevent mating from taking place. So courtship must precede mating and must itself be triggered by the body's changing chemistry as the process of sexual maturation progresses.

The actual cues that animals obey to time their breeding seasons are many and varied. Some creatures respond to the lengthening days of spring, others to the shortening days of autumn, yet others take their cue from the amount of rainfall or from the gradual change in temperature related to the seasons. But the timing of mating is always geared to the timing of birth or hatching, for if the young are to have a high chance of survival they must be born when the food and other resources they require are at their most abundant. This may mean that the adults have to make immense breeding migrations: some birds fly half way round the globe to breed and some whales cruise the length of an entire ocean.

Giving birth at the very best time for juvenile survival is the key to many of the fascinating aspects of reproductive rhythms of large, terrestrial animals. But in the sea, where there are few seasonal influences other than daylength, the chief problem is for the sexes to find one another, in order that fertilization can take place. Most marine animals mate in a rudimentary fashion: the females shed their eggs into the water all around them and the males do the same with their sperm. Therefore, considering the vastness of the oceans, it is absolutely imperative that eggs and sperm should be shed simultaneously if fertilization is to take place at all. It is the moon that is the all-important coordinator of these events because it exerts an influence on the sea through the tides, and those tides cause rhythmic changes in water pressure which synchronize and impart a rhythm to reproduction in marine animals.

The time of year during which an animal is fertile, and so ready and able to reproduce itself, is its breeding season. But although reproduction is an animal's most important function—the reason for its existence—many species have breeding seasons lasting only a few days. The exact timing of the annual rhythm of the breeding season depends upon a variety of environmental cues which change the animal from being sexually quiescent to being sexually active. Once the process of change has started, the sexual cycle begins its course.

In mammals, including humans, the cycle of sexual events is the estrous cycle of the female and is composed of two distinct phases, the follicular phase and the luteal phase. The follicular phase is the first of the two. During this part of the cycle developmental changes take place in the ovary of the female and some eggs ripen. The whole cycle is controlled by hormones, and the most vital are those that come from the front or anterior part of the pituitary gland, a small but influential organ situated at the base of the brain. The cycle starts when the anterior pituitary produces a hormone which acts specifically on the capsules, or follicles, in which the developing eggs are enclosed. This hormone, which is called the follicle stimulating hormone—or FSH for short—brings about an expansion of the follicles and an enlargement of the eggs within them. But FSH does more than just this, and its other actions sow the seeds of its own destruction. It causes the whole ovary to start secreting the female sex hormone, estrogen, which is soon sufficiently abundant to feed back to the anterior pituitary gland via the bloodstream and reduce the production of FSH.

The rise in the estrogen level and the fall in FSH marks the end of the follicular phase of the estrous cycle. Next, under the influence of estrogen, and as a result of the absence of FSH, the anterior pituitary starts secreting a second hormone, luteinizing hormone (LH), which now acts on the ovarian follicle in a rather different way. The first effect of this hormone is to bring about the process of ovulation or egg release. Acting on the follicle walls it makes the follicle burst. The egg is shed into one of the Fallopian tubes that join the ovaries to the womb or uterus. The follicle, which nurtured the egg for so long, now has a new role to play. Under the influence of LH the follicle grows and fills up the space that was once occupied by the egg. The empty follicle, now yellowish in colour, is given the name of corpus luteum or yellow body.

This second phase of the estrous cycle, dominated by the emergence of the corpus luteum, is the luteal phase. The corpus luteum starts to secrete progesterone, the hormone necessary for maintaining pregnancy. If the egg or eggs from the female are fertilized just after their release from the ovary, then the corpus luteum left behind by each egg will continue to grow and pour out its life-supporting chemicals. But if the eggs are not fertilized, the corpus luteum gradually starts to shrink and its production of progesterone wanes. During all this time the output of estrogen has been declining, and when the corpus luteum is finally extinguished the production of progesterone stops. The cessation of progesterone production marks the end of one complete estrous cycle. Free from the restraints imposed upon it by estrogen and progesterone the anterior pituitary is now able to secrete another batch of FSH and so a new cycle begins.

Some species of mammal, such as the dog, are described as monestrous

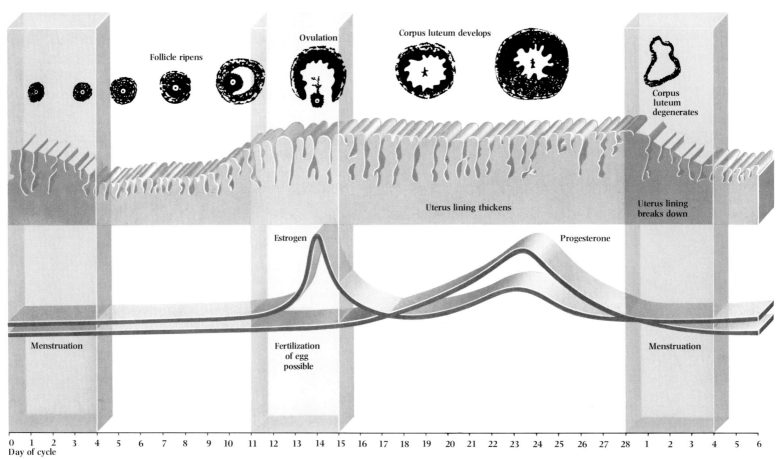

Follicle ripens

Ovulation

Corpus luteum develops

Corpus luteum degenerates

Uterus lining thickens

Uterus lining breaks down

Estrogen

Progesterone

Menstruation

Fertilization of egg possible

Menstruation

```
0   1   2   3   4   5   6   7   8   9   10  11  12  13  14  15  16  17  18  19  20  21  22  23  24  25  26  27  28   1   2   3   4   5   6
Day of cycle
```

A woman's menstrual cycle is one of the most familiar human rhythms, with the same events taking place each month unless altered by pregnancy. An egg ripens in the ovary under the influence of follicle stimulating hormone (FSH), produced by the pituitary gland. Midway through the cycle ovulation occurs—an egg bursts from the ovary and travels down the Fallopian tube to the womb (uterus). The level of FSH drops and the level of luteinizing hormone (LH), also produced by the pituitary, rises. This hormone helps the development of the yellow body (corpus luteum) in the ovary, which in turn produces progesterone. Under the influence of LH the egg, if fertilized, will implant itself in the uterus lining. If the egg is not fertilized the yellow body breaks down, the uterus sheds its lining and the progesterone level drops. Associated with menstruation is a temperature rhythm, *right*: the woman's body temperature rises with ovulation. This signal of fertility is used in the 'rhythm' method of contraception and is important in fertility therapy.

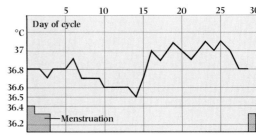

```
        5      10      15      20      25      30
°C   | Day of cycle
37
36.8
36.6
36.5
36.4
36.2   |—Menstruation
```

because they have just one estrous cycle during each breeding season. Others, such as the cow, mouse and rat, have several cycles in each breeding season and so are polyestrous. The smaller of the polyestrous species may be able to produce several litters of young during a single breeding season, a capacity for which mice are notorious.

Man, and a few species of ape, have no rigid breeding season. Instead, estrous cycles follow one another continuously throughout the year. The human cycle is known as the menstrual cycle, a name that comes from the Latin *mens*, meaning month, because it has a periodicity of approximately 28 days. Although the hormonal basis of the human menstrual cycle is identical with that of the estrous cycle, the two are different in one important respect. During the estrous cycle of mammals, apart from man and other menstruating species, there is a period of 'heat' during which mating normally occurs and this coincides with ovulation. In addition there are a number of behavioural changes, many of which are easily observed.

The hallmark of the human menstrual cycle is the periodic bleeding that normally occurs for between three and five days midway between ovulations. For a long time it was thought that menstruation was the equivalent of estral heat, and accompanied ovulation, but this is now known not to be so. In menstruating species, one of the effects of the hormone progesterone is to cause a thickening of the wall of the uterus in preparation for receiving and nourishing a fertilized egg should one become implanted in it. If fertilization does not occur, and the corpus luteum starts to degenerate, there is no support system of hormones left for the thick, blood-rich tissues. Sometimes slowly, and

sometimes with a rush, the uterine wall sloughs off its newly acquired thickness and prepares itself for the next cycle. Bleeding during the sexual cycle is not an exclusive attribute of man and some of his primate relations. A few species with estrous cycles bleed shortly before ovulation, but this bleeding is never as heavy, nor as long-lasting, as in menstruating species.

The rhythmic monthly changes in a woman's hormone levels are so great that they affect other parts of the human body, apart from the ovaries and uterus, in a regular fashion. Under the influence of progesterone a woman's breasts swell slightly, reaching their maximum size just before menstruation, and they may also become more sensitive to the touch and tingle slightly. The nose, too, suffers from the hormonal surges. During the time of ovulation—roughly 14 days after the start of menstruation—the sensitivity to musks and musklike odours increases tenfold or more, only to decrease sharply during the bleeding phase of the cycle. Progesterone also affects the tiny blood vessels in the nose, and many women suffer from nosebleeds and blocked noses during the latter part of their menstrual cycle. And because progesterone (produced now by the placenta) is secreted throughout pregnancy, some women suffer from mild nasal irritation at this time. The effect of the cyclic production of progesterone on the nervous system is not well understood, but it is well documented that many women suffer from premenstrual tension before their periods, a time during which they are abnormally anxious and tense. There are other strange effects: it has been found that toward the end of the cycle, women singers with the ability to pitch notes perfectly sometimes find their ability impaired.

▷

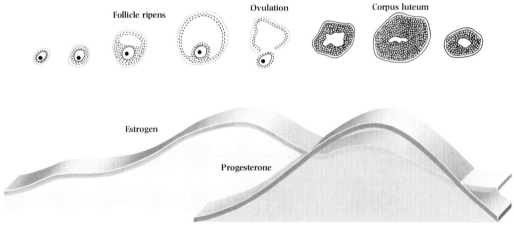

Follicle ripens   Ovulation   Corpus luteum

Estrogen

Progesterone

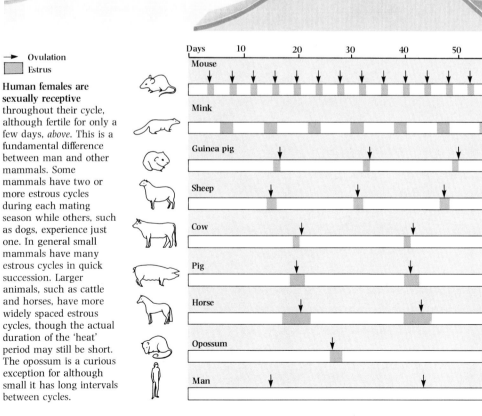

→ Ovulation
▩ Estrus

Human females are sexually receptive throughout their cycle, although fertile for only a few days, *above*. This is a fundamental difference between man and other mammals. Some mammals have two or more estrous cycles during each mating season while others, such as dogs, experience just one. In general small mammals have many estrous cycles in quick succession. Larger animals, such as cattle and horses, have more widely spaced estrous cycles, though the actual duration of the 'heat' period may still be short. The opossum is a curious exception for although small it has long intervals between cycles.

Days	10	20	30	40	50	60
Mouse						
Mink						
Guinea pig						
Sheep						
Cow						
Pig						
Horse						
Opossum						
Man						

Most mammals are sexually receptive and fertile for a restricted number of short periods in each year and do not have a monthly cycle. During these periods of sexual activity an animal is 'on heat' and many behavioural changes accompany the physiological developments in the reproductive system. True estrus is marked by ovulation—the shedding of one or more eggs from the ovary—and is controlled by the same hormones as control the human menstrual cycle. In the few days immediately before estrus, external signs of fertility appear, such as changes in body coloration or the swelling of the vaginal area characteristic of some primates, and the production of a powerful body odour. After estrus the body is becoming prepared for pregnancy and this broody state of pseudopregnancy is typical of many species. If the animal is not pregnant the womb gradually shrinks back to its resting state to await the next breeding season.

Ground squirrel

Rabbit

Mink

The physical stimulus of copulation is needed by some animals to bring about ovulation. These 'induced ovulators' experience a regular estrous cycle but only release eggs when mating occurs, thereby running no risk of wasting eggs. The complete mechanism is not yet understood, but it seems that the stimulus of mating releases the hormone oxytocin from the pituitary gland, and triggers ovulation. Badgers, mink, rabbits, ground squirrels and many small rodents are all induced ovulators.

▷ In many species of mammals estrous cycles can be synchronized by certain odours or pheromones that emanate from both males and females. It has been suggested that human cycles are regulated by various odorous substances found in male urine and sweat, or even in the sweat of certain 'driver' women, but much more experimentation is required before the precise mechanism of this phenomenon is known—or is proven to be non existent.

Merely coming into estrus at a particular time is of little use if a potential mate is unaware that the time for mating has arrived. To overcome the problem of an overlooked estrus the females of most species have developed some means of signalling their readiness to mate. Since the signal is tied to the underlying sexual cycle it, too, makes itself apparent on a regular, cyclical basis. Animals use a wide variety of sensory cues to advertise their state. Deer and antelope, for example, make strutting movements which act as visual signals, while whales and dolphins sing complex songs. Very many mammals, including cats and dogs, produce specific sexual odours when they are ready to mate.

Among the most bizarre sexual advertisements in the whole animal kingdom are those of man's primate relatives. In many species, such as the chimpanzee, baboon, mangeby and macaque monkeys, the onset of estrus is heralded by a huge swelling of the vaginal lips. In some chimpanzees they not only swell up to huge spheres more than a foot (30 cm) across but also change colour, becoming deep pink or even crimson. The odour of the vaginal secretions is, however, able to influence the quality of the visual signal. At the end of her cycle, after conception has occured, the swelling of the female baboon is as large as before, but even so the male shows little sexual interest in

her. Clearly, then, the secretions bear a powerful olfactory (smell) message which cancels out the visual stimulus that was earlier of overriding importance.

In many of the more lowly primates, such as lemurs and tarsiers, the vaginal aperture is completely sealed off during the non-breeding season, so that mating is physically as well as physiologically impossible. As estrogens start to course through the bloodstream, the thin membrane which closes the orifice breaks down and the rim of the vagina starts to swell and flush with colour. At the peak of estrus the vagina is slightly everted, but quickly returns to normal at the end of the cycle. As if rear-end display were not enough, the gelada baboon, *Theropithecus gelada*, and a few other primates enhance its effectiveness with another form of display. During estrus the upper part of the female gelada's chest (her under or ventral surface), becomes slightly puffy and assumes a pink flush. Furthermore, a peripheral necklace of white promontories (tubercles) focuses attention on the pink skin within. Presumably the mimicking, ventral signal increases the chances that the message, carried primarily by the anal 'face', will get across to a willing, fertile male.

Clearly a message can only be understood—and acted upon—if the perceiver can read it. There is evidence that males of a number of animal species can only read the message telling them that a female is sexually prepared when they themselves are in breeding condition and ready to mate. Outside the breeding season both male and female sticklebacks, *Gasterosteus aculeatus*, look so much alike that it is hard to tell them apart. But as the male's testes develop, the hormone cycle induces a rich red flush to spread over his whole belly area.

When she is ready to mate, the female gelada baboon, *Theropithecus gelada*, signals her receptiveness by means of a thick, purplish-red swelling of her vagina and strong-smelling vaginal odours attractive to the male. Her chest also becomes pink and puffed up and surrounded by a 'necklace' of white promontories. These changes are brought about by a surge of the hormone estrogen which is essential to the release of an egg for fertilization

In the mating season, the belly of the male stickleback, *Gasterosteus*, turns bright red due to the influence of the hormone testosterone, made by his testes. This coloration deters other males from entering his breeding territory. The hormone also makes him extremely aggressive and he will attack any red object.

The female golden hamster, *Mesocricetus auratus*, sends out the message that she is ready to mate by marking her territory with strong-smelling vaginal secretions. Once an egg is released the number of markings drops rapidly.

▲ Days of estrus

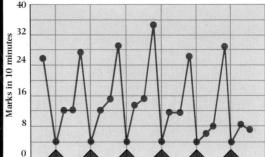

When the male sets eyes on a silvery female he starts a courtship dance and she responds correctly, triggering off a chain reaction of events which ends in mating. A male in breeding condition will respond to anything coloured red, fearing a rival, but outside the breeding season, he is unaffected by red objects even when they are dangled close to him.

Other animals broadcast the news of their sexual readiness by applying special scents to various parts of their territories. Glands on the necks and chests of many species of primate are highly sensitive to sex hormones. As the sexual cycle progresses these glands pour out increasing quantities of odorous chemicals. This reaction is coupled with a cyclical variation in the frequency of scent-marking behaviour. Sometimes the scent is directed toward other individuals of the same sex—males of many big cats, for example, heavily mark an area that contains their females. With the golden hamster, *Mesocricetus auratus*, however, scent serves to attract the sexes to one another, for these animals are solitary creatures, coming together only for a brief period for mating. The female has a four-day estrous cycle but only during a receptive period of 20 hours will she allow a male to come close to her. To attract the male, the female indulges in frenetic scent-marking shortly before estrus. The odour of the vaginal secretions conveys a strongly attractive message, and the female spreads this widely about her home burrow, using a bobbing movement during which her vaginal orifice is dragged along the ground. The frequency of marking may be 30 times higher on the day before estrus than on the estral day itself.

Although humans do not have a breeding season, nor any obvious signal indicating ovulation, some research workers claim to have shown a rhythm in sexual activity. In Europe and North America most conceptions take place in spring and early summer, making January to March the favourite birthday months. In the southern hemisphere the same seasons are popular, even though the months are half a year apart. But there is no simple explanation to embrace the switch that occured in Puerto Rico between 1940 and 1960 from the typically northern pattern to the southern one. Other workers have tried to show that the frequency of intercourse is highest at a woman's mid-period point, and again shortly before menstruation, but there are great problems surrounding the collection of all such data.

In man, and in all other species in which the male inserts his penis into the female during copulation, the sex act itself is a rhythmic activity. The repeated physical thrusting of the human male induces waves of involuntary contractions in the smooth muscle of the vagina, which may assist in transporting the sperm high up into the uterus and on to the Fallopian tubes for fertilization to take place. These orgasmic waves which, contrary to popular belief, are not essential to conception, are approximately 0.8 seconds apart and closely follow the ejaculatory rhythm in the male during which the sperms and accompanying fluids are pumped out.

Since most species of animals are seasonal breeders, it is pertinent to ask what initiates the breeding season. The factors that trigger the beginning of breeding behaviour have long puzzled zoologists. In some northern species of birds and mammals the date of the start of the breeding season can be predicted with great accuracy, for example, the hare always mates in late March. In other ▷

A welcome summer resident of the northeastern oak forests of the USA is the scarlet tanager, *Piranga olivacea*. When males and females arrive after their winter in South America, both are the same drab, brownish-yellow colour, *right*, but as the days lengthen and the testes of the male develop, his drab plumage is moulted and replaced with resplendent scarlet cape and breast feathers, *below*. The pair build a nest of sticks and twigs high up in the trees and the brightly attired male defends his domain from pirates. He plays no part in incubating the eggs laid by the female but does help to fetch insect food for the hungry youngsters. After the breeding season the male sheds his bright plumage and returns to the nondescript coloring of his mate. As winter approaches the pair migrate south and return the following spring.

Non-breeding coloration

In an attempt to analyze human sexual behaviour in relation to the events of the menstrual cycle, 40 women recorded the timing of sexual intercourse and their periods over several years. The results, *right*, show a hint of maximum frequency of intercourse and orgasm around mid cycle, (at ovulation). Both fall to a minimum during menstruation.

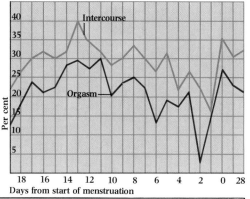

Copulation is prolonged in large predatory mammals such as the tiger, *Panthera tigris*, that do not risk being attacked during intercourse. Until she is in estrus, the female repeatedly pushes the male away. Then her powerful sex drive overcomes her fear and she allows him to approach and to mount. The first mating may last only a minute or two, but the pair will mate up to 20 times each day for the 3-week duration of the female's estrous period. In domestic cats it has been found that the friction of copulation stimulates the release of the egg from the ovary and the same is probably true of the tiger. Prolonged copulation also occurs in dogs and in some large herbivores.

▷ species the timing is more variable. Probably no single factor functions as the switch that turns on the breeding season. Instead it is more likely that the interaction of two or more environmental factors, such as daylength and temperature, start off a chain of physiological and behavioural events that leads to sexual union.

Many birds, including members of the duck family such as the shelduck, *Tadorna tadorna*, have a distribution range covering more than 15 degrees of latitude. Breeding starts earlier in the more southerly regions because here the daylength increases and the temperature rises considerably earlier than in the north. But to confuse the issue, there is some experimental evidence to suggest that there are no environmental switches at all, and that the initiation of the reproductive season is governed by some biological processes within the animal.

Studies have been performed on some male ducks in which the gradual, and cyclic, increase in the size of the testes, typical of the beginning of the breeding season, occurs in the total absence of light and of cues concerning temperature and other factors. The same cycles are seen whether the ducks are kept in total darkness or constant light, but it is apparent that the amplitude of these cycles is smaller than during normal seasonal changes, so that the testes do not become as big as they would under natural conditions. Also the periodicity of the changes is shorter, so that the breeding season does not last as long as usual. Other studies, in which sheep were kept in continuous light for three years, revealed that the time of the onset of breeding was no different from that of sheep kept in natural light conditions. Thus it has been speculated that some

animals have a built-in or endogenous sexual rhythm.

Such a rhythm has been postulated for birds that regularly migrate great distances and experience rapid changes of daylength and temperature. The short-tailed shearwater, *Puffinus tenuirostris*, which breeds in the Australian summer and spends the southern winter in the northern hemisphere in the Aleutian Islands before returning to the shores of Australia in September, maintains its normal breeding pattern in captivity, when exposed to a completely different range and pattern of light conditions from those it experiences in the wild.

Despite these complications, there is an abundance of evidence to show that, for many species of vertebrate animals, light is the trigger which initiates breeding. Some respond to a regular spring extension of daylength and are known as long day breeders because they breed when daylength is increasing, while others—the short day breeders—respond to the autumnal decrease in daylength. Small species with a short gestation period mate and give birth within a few weeks, and it is among these species, such as squirrels and other small rodents, small insectivores like shrews and hedgehogs, and small carnivores, such as cats, ferrets, raccoons, weasels and mongooses, that the breeding season is cued by an increase in daylength. By breeding in spring these species ensure that their young are produced in summer when food is most abundant and of the highest quality, and when environmental temperatures are higher, which gives offspring the best start in life and the best chance of surviving the winter when food is scarce.

Large species, with a gestation period of six months or more, cannot use

The key to successful breeding in the animal kingdom is to produce young at the time of year that will give them the best chance of survival. In the tropics breeding is often timed to coincide with the start of the wet season. In temperate latitudes, where winters are harsher, it is more advantageous for animals to give birth to young at the start of spring when food is plentiful. Large species with gestation periods of more than six months, such as goats, sheep and deer, must thus mate in autumn. These short day breeders mate when days are shortening.

Goat

Cat

Hare

Deer

Sheep

Ferret

Small mammals, such as the cat, hare and ferret, have short gestation periods. They can thus mate towards the end of winter and give birth to their young in spring. These species are known as long day breeders because the cue that switches on the breeding cycle is an increase in daylight. It is not yet clear how the daylength cue actually triggers the hormonal mechanisms that allow breeding to begin, but the pineal body, in the brain, seems to be involved. Most rodents are long day breeders but in a mild autumn will produce young well into the winter.

The precise timing of the start of the breeding season depends on local climatic conditions. Species tend more often to breed in synchrony as they move to higher latitudes because the weather dictates that matings and births must be squeezed into a short period. The graph shows the number of species of birds producing eggs at the same time in various latitudes. The birds included the song thrush, *Turdus ericetorum*, the prothonotary warbler, *Protonaria citrea*, and the Californian white-crowned sparrow, *Zonotrichia leucophrys*.

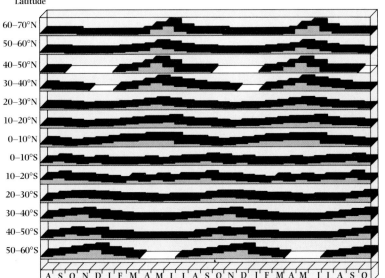

Latitude

60–70°N
50–60°N
40–50°N
30–40°N
20–30°N
10–20°N
0–10°N
0–10°S
10–20°S
20–30°S
30–40°S
40–50°S
50–60°S

A S O N D J F M A M J J A S O N D J F M A M J J A S O

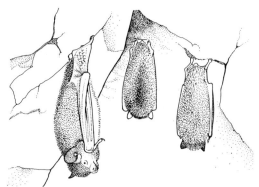

Bats spend the daylight hours inside dark caves and so cannot use daylength cues to trigger their breeding cycles. Despite this, bats such as *Miniopterus australis* always mate in September. It may be that the changing time of twilight, which marks the start of the bat's active period, is enough to begin the cycle, but the mechanism is still a mystery.

increasing daylength to start their surge of sexual activity. If they did, their young would be born in the autumn or early winter and would stand little chance of survival. So such animals are tuned to respond to decreasing daylength. Some short day breeders are sheep, goats and deer which all give birth in spring when food stocks are plentiful. Larger species, such as the horse, which have gestation periods of 10 months or more, respond to springtime cues of lengthening days, so timing the arrival of their young for the start of the following spring.

How do daylight cues actually influence the hormonal changes that are associated with the reproductive cycle? This is still a largely unanswered question, although significant progress toward understanding the mechanisms involved has been made. What is known is that no response in the gonads—the ovaries and testes—to daylength (photoperiod) occurs in animals which have no anterior pituitary gland. This is hardly surprising since it is the anterior pituitary that is responsible for the production of FSH early in the breeding cycle. What is surprising is that the eyes are not always essential to a gonadal response. In rats and ferrets, for instance, the eyes—and the optic nerves that link them to the brain—are necessary, whereas in ducks they are not. If a tiny glass fibre rod less than one twentieth of an inch (1 mm) in diameter is inserted into the brain of a male duck, and light is passed directly to the part of the brain adjacent to the pituitary gland, an increase in testis size takes place.

A largely mysterious and little-researched brain structure, the pineal body, has been implicated in the light-hormone pathway. In some lizards the pineal body actually pokes up through the skull, looking like a third eye on top of the head; in fact the pineal body is actually pigmented like a true eye and also has the ability to perceive light. Removal of the pineal body in male chameleons kept under natural light conditions results in enlargement of the testes. Thus, normally, the pineal seems to act in an inhibitory way, probably by secreting various substances into the blood which serve to regulate sexual development. In man and other mammals the pineal body does not seem to have a photoreceptive function because it is buried deep within the brain and light rays cannot reach it, nevertheless it appears to have some influence on sexual function. In one series of experiments the pineal bodies were removed from some female rats and the animals were then kept in complete darkness. It was found that these rats had larger ovaries than those not operated on and allowed normal access to light. Such experiments suggest that light, perceived either by the eyes or by the pineal body, acts on the pituitary gland and associated regions of the base of the brain, stimulating them to produce the various hormones that rule the sexual cycle.

The interrelationship between daylength and temperature in controlling the sexual cycle is clearly seen in a number of species. As far as the sex glands are concerned the breeding season consists of an enlargement phase and a shrinking phase. Like most small mammals, the American thirteen-lined ground squirrel, *Citellus tridecemlineatus*, starts to breed in spring. In midsummer the testes and ovaries start to shrink and by autumn are fully regressed. But if the squirrels are kept at a constant temperature of 4 degrees centigrade (39 degrees F), and under normal light, sex gland regression does not start until the end of autumn. This sort of link between sexuality and temperature does not ▷

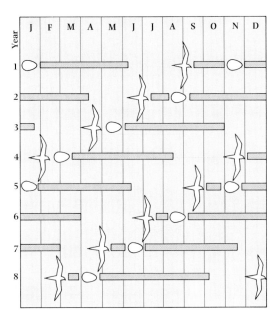

Most of the world's species of terns are migrators, covering incredible distances between their winter and summer quarters. Some tropical species do not migrate and one truly stay-at-home tern is the sooty tern, *Sterna fuscata,* which lives on the tropical Atlantic island of Ascension. The sooty tern breeds in a 9½-month cycle. Since the seasonal differences in daylength and average temperature are so slight it is not clear what triggers the breeding cycle, although it may be due to a cyclical variation in the abundance of fish food. Another explanation is that the sooty tern has an internal clock programmed to start the breeding cycle every 9½ months, allowing enough time for pairing, nest building, mating and moulting before the cycle restarts.

Flocks of sooty terns living on Ascension Island feed exclusively on fish. Each pair of terns lays its eggs in a crude nest on a crowded breeding ground where there is much squabbling, complaining and even open piracy. Despite this, enough chicks hatch each year to maintain the population.

71

▷ occur in all mammals, as studies on the European hamster, *Cricetus cricetus*, show. This creature normally hibernates in autumn, and before it does so its testes and ovaries are fully regressed. But if the hamsters are kept at 20 degrees centigrade (68 degrees F), so that they do not go into hibernation, their testes still regress in the usual way. So as for light, no simple rules about temperature apply to the regulation of mammalian breeding seasons.

Among the fish, cyclical temperature change is definitely an important cue for the sexual cycle. The environment enjoyed by fish has a relatively constant temperature, but this means that the smallest changes have significant repercussions on food supplies and, like mammals, fish also need to ensure that their young hatch out when food is available. In the male killifish, *Fundulus heteroclitus*, sperm formation, the process of spermatogenesis, starts when the water temperature reaches 10 degrees centigrade (50 degrees F), but is quite unaffected by daylength. Minnows, *Phoxinus phoxinus*, come into breeding condition as the water temperature rises, but full maturity depends upon the additional stimulus of long days. Interestingly, it has been shown that if minnows are kept under conditions of artificially short daylength, warm water inhibits sperm manufacture rather than stimulates it. Temperature also cues reproduction in certain lizards and amphibians, but again there is no universal rule governing this particular environmental influence.

A great many species of animals live either in the tropics, where light and temperature regimens undergo little annual variation, or in harsh environments where the likely food supplies for newborn young are unpredictable. Such species are seasonal breeders but even so they rely on certain environ-

mental triggers to set their sexual cycles in motion. Tropical animals rely on food availability, and the presence of certain compounds in fresh food seems to activate the cycle of sexual activity. In areas such as East Africa, where the year-round rainfall is relatively high, large mammals calve in every month of the year. This has been dramatically demonstrated in the case of the giraffe which has a gestation period of 15 months. In other parts of Africa, however, where rainfall is more seasonal, the timing of births corresponds more to the rainy season than to the dry season. Thus in the Kruger National Park in South Africa giraffes have two calving peaks corresponding to the periods of early summer and late summer rain.

Kangaroos are cued into breeding condition by daylength, and mate within a few days of becoming fully fertile. At ovulation, up to 20 eggs may be shed from the female and fertilized, but only one actually becomes implanted in the wall of the uterus and starts to develop. The others are held in a state of suspended animation, and may never be required. If there is adequate rainfall during the short gestation and throughout the first part of the youngster's pouch life, development will continue until the joey is weaned. The reserve eggs will then break down and disappear. If conditions of drought prevail, however, and the mother cannot find enough green vegetation sufficiently high in nutritive value to support both herself and her young, the embryo is expelled from the pouch before it has grown too large.

The cue that initiates this expulsion of the female kangaroo's burdensome embryo is the protein level of the forage; when it falls too low, the milk supply system is disturbed, resulting in the death of the young. Immediately this

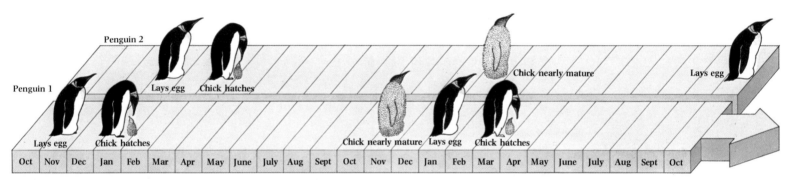

Oct	Nov	Dec	Jan	Feb	Mar	Apr	May	June	July	Aug	Sept	Oct	Nov	Dec	Jan	Feb	Mar	Apr	May	June	July	Aug	Sept	Oct

King penguins, *Aptenodytes patagonica*, must raise large chicks within one short Antarctic summer. They lay one egg at a time and can breed twice in every three years. In each season there are two peak periods of egg laying: one in November the other in January. Penguin (1) lays in November the first year and this chick is sufficiently independent for the parent to lay again in the January period of the next season. The third year, however, this second chick still needs its parent's attention and breeding is missed that year. Penguin (2) lays in January the first year. This chick is not fully grown the following season so the parent misses breeding that year but is free to lay early in the third year.

happens the corporea lutea in the ovary, which until this time have been suppressed, start secreting the hormone progesterone and one of the fertilized eggs from the reserve pool becomes implanted in the uterine wall. If the food supply does not improve this egg will go the way of its predecessor. This system of an environmental cue influencing the development of a new embryo is an excellent means of ensuring survival of a species in unpredictable surroundings, although it may seem wasteful, for in a year of intense drought many eggs will come to nothing.

Frogs and toads are other creatures whose reproduction is linked to water supplies. These animals need water in the form of a pool or puddle in which they can lay their eggs. Responding quickly to the sound and the feel of rain they actively seek out a mate and spawn before the water dries up. In some arid areas of the world tadpoles are dependent upon standing water for only a few days. A small African plains dwelling bird, the quelea, also needs rain before it will breed and in bad years no breeding will take place. Just the sound of rain falling on the parched earth is sufficient to induce gonad development, courtship and egg laying, although the precise mechanism by which the acoustic signals actually influence the hormonal balance is still a mystery.

The reproductive cycle of animals consists of more than just the physiological sexual cycle. Successful breeding embraces many behaviour patterns, including the courtship that precedes mating, and the eventual production and care of young. Perhaps one of the best-known behaviour patterns associated with breeding is the annual song cycle in birds. There are probably no birds whose song is absolutely constant throughout the year. In the northern hemisphere, most small garden birds do not sing at all during the period from midsummer to late winter, although the red breasted European robin, *Erithacus rubecola*, is a tuneful exception to this rule. As the days begin to lengthen in February and March, the enlarging testes of the males of all the species produce increasing quantities of the male hormone testosterone, the same hormone that is made by the testes of mammals, including man. The birds' tendency to sing has been found to be closely correlated with the level of testosterone circulating in the blood.

As the mating season progresses—in April, May and June—and the testes approach their maximum size, the amount of song reaches its zenith. Experimentation has revealed that song production can be carefully controlled by artificially regulating daylength, and the songs of the birds can be kept at maximum level if the amount of light remains constant at its late spring or early summer level. This trick has long been known by bird fanciers who seek to lure migrating song birds to the ground. A caged member of the same species is kept at maximum singing level by means of controlled lighting, and is used as bait for its kin passing overhead.

Superimposed on the annual rhythm of bird song is a 24-hour daily rhythm. Most species sing less during the middle of the day than they do in the early morning or late afternoon. In hot, dry places the midday lull seems to be inextricably associated with heat and wind, but this cannot be the full story because the same phenomenon occurs in northern temperate areas. Certainly in arid regions any increase in humidity is usually associated with a renewed burst of song. Shortly after first light many male birds produce a burst of song, ▷

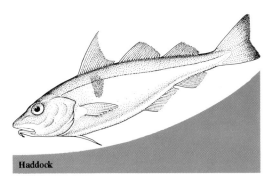

Haddock

The onset of breeding in fishes seems to be related to variations in water temperature, although the level of illumination may also play a part. Most fish species have a restricted range of temperatures within which they can breed. Giraffes, like man, are aseasonal breeders. Calves are born every month of the year with a slight increase coinciding with the onset of rains. Spadefoot toads, *Scaphiopus bombifrons*, inhabit the unpredictable environment of shallow pools which often dry up. During periods of drought the toads dig themselves into the mud and wait for rain. The arrival of the rain not only frees the toads but also triggers the development of the gonads and the production of spawn from which tadpoles emerge.

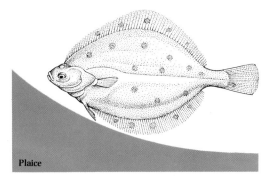

Plaice

▷ which lasts for about an hour. On wet, windy mornings it may be shortened or even absent, but on fine sunny mornings this dawn chorus can be dramatic. It seems to need just one particularly enthusiastic songster—often a blackbird—to start the entire choir into full voice. A few species of song birds sing only in the dawn chorus, but most sing later in the day as well, although the evening melodies are never as loud or as long as those of the early morning. Most ornithologists agree that the waves or bursts of song that punctuate the daily song period of any one bird are controlled by some innate factor, which is itself influenced by the level of production of testosterone.

There is certainly something spectacular about a great many animals all doing the same thing at the same time, and nowhere is mass activity more breathtakingly orchestrated than in the displays or leks of many game birds. Timing their activities for immediately after dawn, groups of male grouse and related species, such as prairie chickens, heath hens and blackcock, adopt the most ludicrous postures and swagger about in front of one another with a view to psychological intimidation. After many months of parading, the dominant cock emerging from this charade mates with the greatest number of females so that his traits of display are passed on to the next generation. The regularity of the lekking behaviour is remarkable, becoming earlier each day as daylength increases. Leks of a slightly different nature can be observed in certain species of bat, for example *Hypsignathus monstrosus*, the hammerhead bats of the Zaire Basin, which foregather in huge assemblies each evening just before twilight and sing loudly to one another.

Urban birds such as the starling, *Sturnus vulgaris*, perform the most intricate

aerial manoeuvres just as dusk is descending. Gathering in immense swarms of up to several thousand individuals, the entire mass rises into the air as one, wheeling and turning as if performing a carefully choreographed ballet. Just as suddenly as the aerial display starts it ends, with the dispersal of the flock. The function of these aggregations is not wholly understood, although it appears to be closely related to the prenuptial period and the assessment of the population size. It is not known how each individual actually responds to the assessment, although in certain years only a fraction of the adult population takes part in breeding. It may be that the stress of the display, or the activity it involves, tells more heavily on the weaker than on the stronger individuals in the flock and makes them less able to take, or to defend, a mate. What is particularly impressive, however, is the timing of these mass phenomena. Since they serve the important role of helping to prevent the population from becoming too large, they must occur at a clearly defined time of day. And the times of day at which light conditions are changing fastest—dawn and dusk—are clearly defined circadian landmarks.

The rhythmic movements of courtship often look like those of a ballet that has been professionally choreographed. In the crested grebe, *Podiceps cristatus*, for example, the male and female birds perform a sequence of courtship movements without which mating will not take place. As the male approaches the female he advertises his presence by raising the feathered, tuftlike crests on the top of his head and expanding his feathery ear flaps. As he gets nearer he adopts the 'cat' posture, hunching his neck and, at the same time, spreading out both wings to expose their white markings. The two birds then wag their

In the highlands of New Guinea young, unmarried tribesfolk take part in a type of mass 'courtship'. Couples traditionally rub their heads together and sing to one another before having sexual intercourse. Relationships forged in this ritual mating do not, in fact, lead to marriage. Instead, 'suitable' marriages are arranged by the families of young men and women. The rituals and rules of human courtship are different the world over and in the West are

undergoing rapid alteration related to more permissive attitudes and the changing role of women in society. In the animal world, matters are more rigid because courtship behaviour is largely governed by the levels of sex hormones in the blood. Closely related to courtship is aggressive behaviour between males with several rival males competing for the favours of one female. Paradoxically, behaviour such as fighting intimidates other males but may also attract a female.

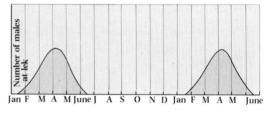

Prairie chickens

Before mating, males of many species of game birds, such as prairie chickens, *Tympanuchus cupidio*, strut about in front of one another for weeks or even months. Because it occurs in a traditional display place, or lek, this behaviour is described as lekking. When the few days for mating arrive, the highest ranking males will mate with the most females.

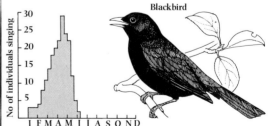

Blackbird

Like most garden songbirds, European blackbirds, *Turdus merula*, do not sing with the same intensity all year. From a slow start in January or February the number of individuals singing in early morning rises dramatically and reaches a peak in May. This peak corresponds with the nesting season. By July the number of birds singing has dropped and no song is heard from August to January.

heads at one another, pointing their bills rhythmically up and down.

After the head-wagging ceremony, the grebes may rise up out of the water and perform a 'penguin dance'. Both partners dive to the bottom of the pond, grab a piece of weed and rear out of the water face to face. Following this dance, or after head-wagging, one bird may retreat. It then faces its partner again and re-adopts the 'cat' display posture. The whole cycle of events is repeated several times before the pair go off to build their nest. Mating does not take place until the nest is finished and a platform of reeds has been built for the male to stand on. The male then drops his head to signal that he is ready to mate and the female stretches out her neck and flattens her crests to convey her readiness. A noisy copulation, with much wing-flapping by the male, is the culmination of the courtship ritual.

When it comes to courtship there are few animals that display a behaviour as rhythmic as that seen in fireflies. In many species of these nocturnally active beetles, which are members of the family Lampyridae, only the male is winged. The wingless female, who looks like a caterpillar and remains on the ground, is also commonly known as a glow-worm. Specialized light-producing or photogenic organs found on the undersurface of the abdomen in both sexes are used to help bring them together. The light from each male, which may be as strong as a fortieth of a candlepower, is flashed on and off rhythmically, with a frequency of about 5 seconds, each flash lasting about 0.2 seconds. The flashing, which is not like that of a lighthouse with a shutter temporarily obscuring the beam, but is the product of a controlled chemical reaction, continues as the flies rise and fall over the vegetation. Once a female on the ground flashes her light in reply, the male will fly down and copulate with her, but a male may flash his light rhythmically for many hours before he finds a mate and, because many males prefer to search together in a swarm, the twinkling effect is a delightful spectacle.

Although at first glance human courtship might appear to be devoid of any rhythmic or cyclic components, tribal peoples do still adhere to an ancient system of regular gatherings. The famous sun dances of the North American Indians, and the corroborees of the Australian Aborigines, although they have largely died out, were once gatherings of immense social significance, not least of which was to encourage courtship between members of different groups. They may also have allowed a population census to be made and this certainly happened in southwestern Asia during the era of the great nomadic caravanserais. Rather as the leks and massed swarms occur at the time of day at which they are most easily detected, early human social gatherings normally took place around the summer solstice, when the sun was at its highest in the sky. Regularly every year the tribes would gather to provide a formalized framework for intergroup trials of strength and stamina, as well as to encourage courtship dalliance. The reason these congregations took place in midsummer rather than in midwinter, which would have been equally distinctive, was because food was abundant in summer, so time could be spend on non-productive activities with equanimity.

Such social gatherings, tied to the calendar of the seasons, are a sort of social migration. Throughout the animal kingdom migration is a common phenomenon, although the reasons for its often vast scale of operation are sometimes ▷

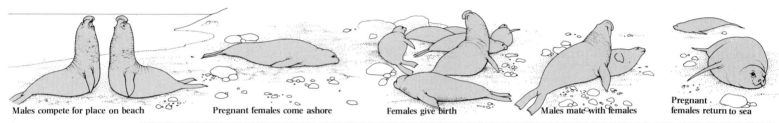

Males compete for place on beach   Pregnant females come ashore   Females give birth   Males mate with females   Pregnant females return to sea

The breeding season of the southern elephant seal, *Mirounga leonina*, occurs on dry land. The first individuals ashore are the old bulls which squabble with one another over the division of the shore into territories. As the smaller females come on to land they are herded into harems by the bulls and many females are stolen by rival bulls. Within a few days any pregnant females give birth and spend much time suckling the young on rich, fatty milk. Meanwhile the bulls are busy inseminating the non-pregnant cows, all the time fighting off interested neighbours. About 10 days after giving birth the cows come into estrus again and they, too, are mated by the harem master. Soon afterwards they return to sea. The fertilized egg remains dormant for 3 to 4 months before development begins, and the new pups will be born after the adults have spent 11 months at sea.

▷ obscure. Defined correctly, there are two components to a migration—an outward journey and a return journey. The irregular explosions of populations of Arctic and subarctic birds, such as waxwings, crossbills and sandgrouse, are one-way mass movements and certainly not predictable. In the truly migrant species the return and departure dates can be guaranteed as regular, rhythmic occurrences.

Breeding and feeding are, broadly speaking, the two reasons why animals adopt an annual migration pattern, but the two purposes may not be distinct. Subsidiary reasons for migration may include travel to a less hostile climate in which to give birth, so a species may undertake a migration to ensure that its young are born in an area in which the quantity or quality of food is optimal or the environment least damaging. Indeed there are few species that migrate for breeding or feeding only.

By definition, long distance migration must be a phenomenon restricted to those species that are highly mobile, the most dramatic examples being found among the birds, fishes and marine mammals and reptiles, and the large swift-footed land mammals. Small terrestrial mammals, reptiles and amphibians do not have the ability to move far but many, nevertheless, show a pattern of migratory movement. Thus the common toad, *Bufo bufo*, leaves the sanctuary of its home among vegetation every spring and searches for a pond—probably the one in which it was hatched—in which to spawn. After a few days, spent seeking for a mate and spawning, it returns to its former abode. In its own way this pattern of movement, in which the animal may travel only a few hundred yards, repeated year after year, is every bit as spectacular as the 17,000 mile

(27,000 km) breeding journey of the Arctic tern, or the 3,800 mile (6,000 km) trek of the caribou. Even rabbits, among the most sedentary of creatures, show a burst of activity annually at the start of the breeding season and build new sets of underground delivery rooms in which the mothers-to-be will incarcerate themselves to await the birth of their litters. Compared with the incredible breeding journeys of some species such activities differ only in scale—their purpose is the same.

A small number of species shows an annual pattern of migration that is associated only with feeding. This may be because of the seasonal pattern of rainfall, as it is with the annual east-west migration of herds of wildebeest, impala and zebra in Tanzania. Associated with these movements are migratory groups of hunting dogs, jackals and hyenas which prey upon the newborn and the weak. In these tropical areas breeding continues throughout the year with animals showing no preference for a special time or place. In North America huge herds of bison, *Bison bison*, once roamed the plains from northern Alberta to New Mexico, and from Oregon to Pennsylvania. They moved northward in spring and gave birth soon after the young, nutritious grass had started to sprout. In autumn the herds migrated southward again to avoid the harshness of the northern winter in which food was hard to find. Although the migration was primarily for feeding purposes, breeding regularly occurred at the start of the trek.

Depending for their survival upon the bison herds were the plains Indians, who regularly migrated northward and southward, following their quarry. Thus the Arapaho from central Colorado annually trekked a couple of

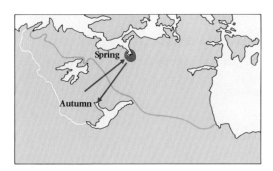

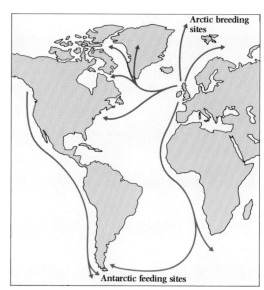

Migration means that an animal can have the best of both worlds—constant temperature conditions throughout the year and no shortage of food. Caribou, *Rangifer tarandus, below,* spend the winter in the forests of Alberta, Saskatchewan and Manitoba in Canada, *left,* and start their trek northward to the Barren Lands in spring. The deer head northward for almost 4,000 miles (6,400 km), lured by the prospect of lush lichens and spring grass during the Arctic summer. On the journey the calves are born and from their first day are initiated into a life of regular migration. Mating occurs in the summer pasture before the caribou begin their long march south.

thousand miles through Wyoming, Montana and into Alberta in pursuit of bison. Humans, as we have seen, have no set breeding season, and births are recorded in every month of the year, so as far as the Arapaho people were concerned the annual migration was for feeding purposes alone.

The most astonishing breeding migrations are displayed by the birds that breed in the northern parts of the world in the summer months and spend the winter in southern latitudes. One such is the Arctic tern, *Sterna paradisea*, a marine bird which breeds in the summer as far north as Baffin Island and northwest Greenland, then flies off to Antarctica. By undertaking such vast migratory journeys from one polar region to the other, this bird is able to spend its whole life in more or less constant climatic conditions. During August, when its chicks are fully fledged and able to fend for themselves, the terns start to fly southward, following the northern coast of Labrador, and then off southeastward across the Atlantic.

Once they pick up the west coast of the Iberian peninsular, the migrating terns may be joined by others of their kind who have bred in northern Scotland and Scandinavia. As far south as the Azores the flocks tend to split up, one part following the west coast of Africa, the other the east coast of South America, until the Antarctic continent comes into sight. The return route is the same. The birds feed little on either their way south, or on their way back, although short resting periods do occur. A similar pattern of migration is seen in the European swallow, *Hirundo rustica*, which breeds in northern Europe and overwinters as far south as South Africa.

Why do these birds undertake such incredible journeys attended by dangers of every conceivable kind? There is no simple answer to this question, but the food resources at the breeding site certainly play a vital role in maintaining the migratory behaviour. Because the bulk of the earth's land mass lies in the northern hemisphere the amount of shallow seawater, in which the Arctic tern finds the whitebait on which it feeds, and the number of flying insects, upon which the swallow depends for sustenance, are far higher in the northern than in the southern half of the globe. The cost, in terms of both energy and life, of undertaking a huge migration, is more than offset by the riches of food to be found in the northern summer at journey's end.

For some terrestrial species the lure of such banquets is sometimes like a two-edged sword. The caribou of North America, *Rangifer tarandus*, which is identical with the European reindeer, spends its winters in the Canadian forests of Alberta, Saskatchewan and Manitoba. Caribou mate in late summer, just when they are preparing to leave the lichen-rich heaths of the northern Barren Lands, but the females do not feel the first birth pangs until some eight months later when they are migrating northward, again seeking nutritious lichens. But if the early spring weather is particularly inclement, the caribous' northward progress may be much slower than normal and the young may be born in a most inhospitable environment. Not only may the young perish in intense cold and blizzard conditions, but their mothers may also be unable to make sufficient milk to feed them during their first few shaky days of life. And like the hyenas following the wildebeest in Africa, there will be wolves eagerly awaiting their chance of an easy meal. Enough caribou survive, however, for there still to be large stocks of this courageous species. ▷

Man is frequently a migrant species and his migrations are sometimes related to the movements of herds of large ungulates. In northern Europe the Lapps, *right*, have for countless generations followed the migration of vast herds of caribou (known in Europe as reindeer) as they seek out the rich plant growth of the high Arctic summer. Human reproduction is not linked to any phase of the migration—the association seems to have developed so that man can exploit his food resources more efficiently.

Among the most astounding of all the regular migrations undertaken by animals is the annual 17,000 mile (27,000 km) pilgrimage made by Arctic terns, *Sterna paradisea, left,* from their Arctic breeding sites to their Antarctic winter feeding sites. Flying over the open ocean as well as hugging continental coastlines, the birds use Atlantic winds to help them on their way. They stop infrequently, but spells of bad weather may force flocks to seek refuge for several days at a time. The tremendous effort of the journey is finally rewarded with a rich supply of whitebait.

▷ A number of truly outstanding breeding migrations involving fish, reptiles and mammals occur in the sea. The largest of all mammals, the whales, live mostly in the southern oceans where the continental shelf of Antarctica causes a vast upwelling of nutrients from the cool, deeper layers, which mix with the relatively warmer surface waters. Here microscopic plants and animals, which collectively compose the plankton, thrive as nowhere else on earth, and the whales cruise slowly through the dense plankton swarms, their mouths agape like the cutters of a combine as it passes through a field of wheat.

Although warm enough for well-insulated adult whales, the surface waters of the southern oceans are much too cold for newborn calves, which explains why many whales undertake breeding migrations towards the Equator, away from their feeding grounds. Thus the humpback whales, *Megaptera novae-angliae*, travel northward to the coastal regions of South America and southern Africa and there, in the warm and shallow waters, produce their calves. But while the diversity of planktonic organisms may be high in these regions, the sheer bulk of them is lacking. So once the calves are big enough to travel, the whales set off southward once again in search of the rich plankton pastures. During their migration the whales eat little, relying on their stored reserves of blubber to supply them with energy for movement and the nutrients to make milk for their calves. Humpbacks live in the north Atlantic as well as in the south, and the northern stock migrates to the warmer waters of the Caribbean shortly before giving birth. Since each stock travels toward the Equator in its own particular winter, the two communities do not meet at the breeding grounds.

Salmon, and a few related game fish, lay their eggs in fresh water, choosing fast-flowing rivulets with sandy or gravelly bottoms, high up among the headstreams of major rivers in Europe and North America. After hatching, the young fish spend the first two years or so of their lives in this relatively secure environment before embarking on a downstream adventure. With a re-markable physiological accommodation to the problems posed by passing from fresh to salt water, the young smolts head out into the Atlantic or Pacific oceans. There, at the edge of the continental shelf, they feed voraciously on the rich plankton whose pink colour is transmitted to the fishes' flesh. For one or two years they remain at sea then, as grilse, start the long haul back to their natal streams.

It is still not known how the grilse find their way back from 1,500 miles (2,400 km) out at sea to the coastline they left two years before, but once in the vicinity of the river mouth they can recall the odour of their home stream and, jumping huge waterfalls and negotiating tumultuous rapids, the adult salmon press on their way. They enter the rivers in August and September when in some early-running fish the gonads are not fully developed, but by October/November both males and females are fully mature.

Reaching the spawning grounds the fish court briefly and, as the female extrudes her eggs, the male sheds his sperm all over them. Huge stretches of the river may turn white because of the millions of sperm suspended in it. For the Pacific salmon—sockeye, chinook, coho, chum and humpback, all species of *Oncorhynchus*—spawning marks the end of their lives. The spent fish lie listlessly in the shallows, providing abundant food for foxes, wolves, brown

Because they encounter no physical barriers, water dwelling animals can migrate with comparative ease. Some migratory aquatic animals have to contend with physiological barriers, such as the change from fresh to salt water or vice versa, but migrant species have evolved means of overcoming these problems. Migration in the sea has the same function as migration on dry land, that is, it allows a species to breed in one place and to feed in another. How migration evolved is still a mystery, but it may be that the drifting apart of the continents in past eras was significant. Whatever the reason, the migrations of whales, eels, turtles and salmon are spectacular.

The world's two migratory populations of humpback whales, one in the northern hemisphere, the other in the southern, never compete for food at the breeding grounds because there is a six-month interval between the winters at the North and South Poles. The whales return to colder waters after calving.

Every eel or elver, *below,* found in European or North American waters, was born in the Sargasso Sea off the southeast coast of Florida, *left.* The 8,000 mile (12,800 km) journey to European waters takes up to three years. After several years the eels set out to retrace their journey and breed and die in the Sargasso.

Sargasso Sea

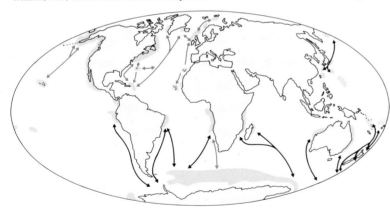

→ Migration routes of southern species

→ Migration routes of northern species

☐ Feeding and calving grounds

The nutrients in the currents of Antarctic waters support huge crops of krill, small shrimplike creatures on which the humpback whales feed. Despite this rich supply of food the whales migrate north in winter to the warm coastal waters of Africa, South America and Australia.

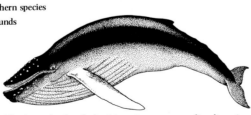

The humpback whale, *Megaptera novae-angliae,* lives in the freezing waters of the Arctic and Antarctic. Although nutrient-rich, these cold waters are no place for giving birth, so shortly before the winter the humpbacks migrate toward the Equator in search of warmer, shallower water. There the calves are born and the young suckled.

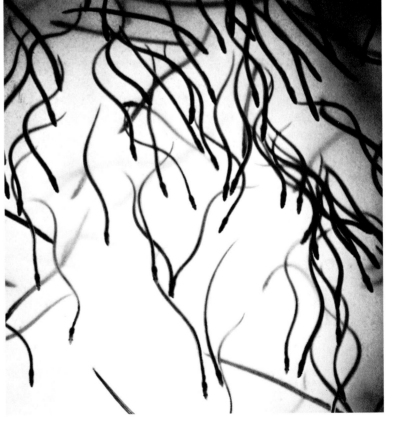

bears and bald headed eagles. But the spent Atlantic salmon, *Salmo salar*, has another chance. As thin, ragged-looking kelts they drift and float downstream, and while many perish from exhaustion on the way—for adult salmon do not feed from the moment they enter fresh water—a few manage to return to the oceanic feeding grounds. There, fortified with a rich plankton diet, they quickly rebuild their strength in readiness for the next spawning run. The timing of the run varies only slightly from year to year, so anglers can book salmon fishing holidays for the late summer with great confidence.

The annual breeding migrations of the European and American eels, *Anguilla anguilla* and *A. rostrata* respectively, are just as spectacular as those of the salmon. These fish are described as catadromous because they breed in the sea but feed in fresh water, in contrast to the anadromous salmon which behave in the opposite way. (Both types of creature have similar physiological problems to overcome when they pass from salt to fresh water and back again.) For almost three centuries zoologists have been actively trying to understand the biology of the eel, because sexually mature adults are never found in fresh water. Even Aristotle noticed this in the fourth century BC, declaring that eels arose from the 'entrails of the earth'. What appears to happen is that eels lay their eggs at great depth in the Sargasso Sea, some several hundred miles east of Florida. After hatching into tiny, flattened larvae, the baby eels begin their long swim toward fresh water.

For the American eels living in the eastern United States the journey takes only about a year, but the young European eels take three years to reach fresh water. By this time they are unmistakably small eels, or elvers, about 4 inches (10 cm) long. After some years in the rivers, the fully grown—but not sexually mature—eels set off again out into the ocean, heading for the Sargasso Sea. They swim at the great depth of 1,700 feet (500 m) or more, and on this journey their ovaries and testes develop. When they finally reach the Sargasso they spawn and die. No eel has ever been recorded as completing more than one spawning migration although American eels, because they have a far less strenuous journey than their transatlantic cousins, may spawn twice in a lifetime.

No account of breeding migrations could be complete without mention of the extraordinary travels of the green turtle, *Chelonia mydas*. Huge numbers of these large reptiles feed on rich growths of 'turtle grass' that fringe the eastern coast of Brazil. Green turtles breed on Ascension Island—a tiny outcrop of rock some 870 miles (1,400 km) distant in the south Atlantic ocean. Tagging studies have shown that young turtles hatched on Ascension return two years later to breed on the shore where they were born, but how they manage to find such a tiny target in such a vast ocean is a mystery. It is indeed a puzzle why turtles should bother to migrate at all for breeding, because perfectly adequate sites exist close to their feeding grounds. It has been argued that, before the process of continental drift separated the southern land masses of South America and Africa, Ascension Island was merely an offshore island, easily reached by the turtles in a day. Remaining faithful to their ancient breeding site as it moved eastward, the turtles were forced to make an ever more hazardous journey for breeding. That they succeed at all is a testament to their remarkable navigational ability. ▷

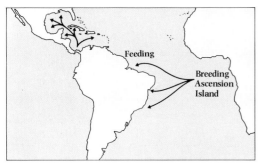

Salmon hatch in the headstreams of rivers and migrate out to sea, returning to fresh water to breed, *right*. After two or three years in fresh water, young Pacific salmon, *Oncorhynchus* spp, travel down the rivers and out to sea to enjoy the rich feeding grounds of the continental shelf. After two or three years young salmon feel the urge to breed and head back towards the land. On finding their natal stream they follow its branches until they reach the breeding shallows in which they were born. After spawning once, Pacific salmon die and provide an abundance of food for animals such as bears and bald eagles. The young hatch, develop and repeat the migration cycle.

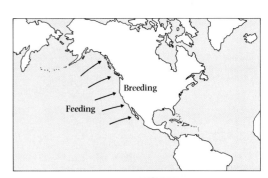

At breeding time, green turtles, *Chelonia mydas*, converge on Ascension Island 870 miles (1,400 km) from their feeding grounds off Brazil. The turtles may find their way by following the island's odour, which moves west in surface currents, and the turtles retain a memory of this odour in their brains. This memory is formed in the first days after hatching.

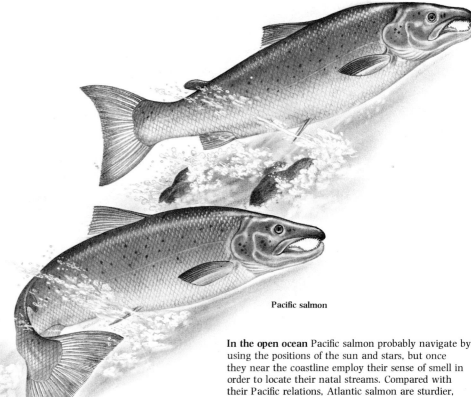

Pacific salmon

In the open ocean Pacific salmon probably navigate by using the positions of the sun and stars, but once they near the coastline employ their sense of smell in order to locate their natal streams. Compared with their Pacific relations, Atlantic salmon are sturdier, have a shorter journey, and can return to their breeding grounds to complete a second breeding cycle.

▷ The actual process of spawning, or egg laying, is also a rhythmical and carefully timed activity. An abundance of evidence indicates the rhythmic waxing and waning of the moon exerts a strong influence on spawning, either directly, through nighttime illumination levels, or indirectly, via its effect on the tides. There is a very good reason why the moon should be so influential: life in the sea is remarkably free from the pronounced seasonal effects to which terrestrial animals are exposed, and by which they synchronize their reproductive activities. And in a vast, homogeneous environment such as the sea it is absolutely crucial that the mechanism of reproduction be synchronized, so that the eggs of one individual stand a reasonable chance of being fertilized by the sperm of another.

Most marine invertebrates, such as molluscs, annelid worms, sea urchins, jellyfish and crustacea, simply cast their eggs and sperm in the waters in a haphazard fashion, so if all members of the population do this at the same time, the chances of fertilization are greatly increased. By far the most dramatic example of this is the South Pacific palolo worm, *Eunice viridis*. This worm, which is related to the common lugworm of temperate sandy shores, lives among coral reefs surrounding the Pacific islands of Samoa and Fiji. It breeds just once a year in a most curious manner. As the November spawning season approaches, the rear part of each worm becomes packed either with eggs, which impart a bluish hue, or sperm, which give a ruddy appearance. By day the worms remain curled up inside crevices in the reef, but at night they emerge to forage for microscopic invertebrates among the coral. During these forays they detect the level of lunar illumination. Then, around full moon, the rear

ends of the worms break free and wriggle to the water surface. There they burst open, and the sea turns white with the spilt sex cells or milt. The Samoans find baked palolo milt a great delicacy, so the November full moon signals an annual feast for them.

Careful experiments under controlled conditions show that it is the number of hours of lunar illumination that brings the worms to the peak of readiness—rather as the increasing daylength brings on sexual receptivity in mammals and birds, and triggers the dawn chorus—but the actual release of the sperm and egg bags is triggered by the pressure exerted by sea water in a particular phase of the tide.

A similar, though less dramatic effect is seen in several other types of worm. The Atlantic palolo, *Leodice fucata*, which lives among the coral rocks of the Tortugas islands off the coast of Florida, swarms every year within three days of the last quarter of the moon between 29 June and 28 July. A number of other species appear to relate their reproductive activity to the lunar rhythmicity, including many female 'fireworms' which flash luminescent lights to attract males during swarming.

Only one example is known of a sea urchin that responds to a lunar rhythm. The species *Diadema setosum* from the Red Sea spawns each full moon, requiring the intervening four weeks for the maturation of a new batch of eggs and sperm. Lunar periodicity of breeding is not known for any other species of sea urchins, even those that live in similar latitudes and habitats, but fishermen in the Mediterranean region still maintain that all sea urchins are best eaten at the time of the full moon.

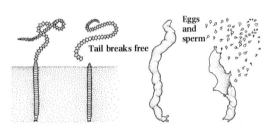

At spawning time the rear half of the palolo worm, *Eunice viridis*, thickens and becomes attracted to light. The worm's front half, *left*, stays in a crevice but the rear half breaks off, shedding eggs or sperm.

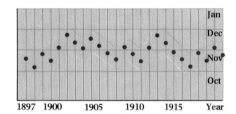

Palolo worms spawn around the November full moon, *left*. The end of each worm detects the waxing of the moon and when it reaches a maximum breaks off.

● Swarm

For many peoples of the South Pacific, including the Samoans, November full moon heralds a feast, because the palolo worms swarm at this time and can be caught in huge numbers. The South Sea islanders then bake the freshly caught morsels which are considered a delicacy. This fishery is one of the safest in the world, for no matter how many worms are caught the breeding stock is in no danger. The reason is that the parent worms, which released their bags of eggs and sperm, are safely ensconced in the dark crevices of rocks and coral reefs.

Through its action on the tides, the moon also influences reproduction in a few species of fish. In one, *Enchelyopus cimbrus*, from the east coast of Canada, the breeding season lasts several months, reaching its peak in July and August. The eggs are pelagic, that is, they float in open water toward the surface of the sea, and routine sampling of the surface waters shows a great abundance of eggs at the time of spring tides, when the moon is full. Between the spring tides of July and August the rate of egg production is almost nil. Even the neap tides—some 10 feet (3 m) lower than the east coast springs—are insufficient to trigger spawning. Since these fish live at a depth of 98 feet (30 m) or more, where temperature and illumination differences are likely to be insignificant, the actual cue that triggers off mating must be the rhythmic rise and fall of water pressure as the spring tide approaches.

Over on the west coast of North America, particularly in the San Pedro region of California, lives another fish with a remarkable sex life. This little fish, the grunion, *Leuresthes tenuis*, buries its clusters of eggs in the sand at the high water mark of the spring and neap tides. The moment the high water mark has been reached the grunion start to 'run'. They swim ashore in great numbers and can easily be observed mating and laying eggs in shallow scrapes which the female makes as the eggs pour out of her. The gentle lapping of the water covers the eggs with sand and within 60 seconds the adults have swum back out to sea. The eggs remain cool and damp, but are not exposed until the next spring tides. Then the erosion of their protective cover induces them to hatch, and the larval grunion head out to sea, starting their free life among the plankton. A much smaller run occurs at the neap tides, so there is a fortnightly cycle of

ripening eggs and sperm. It appears that the variation in water pressure associated with the tidal rhythm acts to control egg development, which continues unabated from March until June but peaks strongly in late April and May.

Synchronization of spawning with the phases of the moon may be far more widespread than has been thought up to now. It has been discovered, for example, that one species of mayfly, namely, *Povila adusta*, which inhabits the airspace over Lake Victoria and Lake Albert in East Africa, shows a lunar rhythm. The adult female mayflies lay their eggs in the water and these eggs hatch into water dwelling larvae. The emergence of swarms of adult flies from the eggs occurs about two days after the August full moon each year. The behaviour of the mayfly larvae also shows a circadian rhythm. During the day these larvae stay hidden and inactive in burrows made in the litter at the bottom of the lake, but at dusk they emerge and swim vigorously toward the water surface. Experiments in which the larvae are kept in conditions of continual darkness reveal that the rhythm is maintained, which suggests that it is inbuilt.

Future research into the rhythmic reproductive behaviour of animals may well reveal that most spawning species are affected by a combination of an annual reproductive season, the monthly periodicity of the moon and the tides, and the daily cycle of changing light intensity. The total purpose of all these rhythms is to bring the reproductive activities of a whole population into synchrony, perhaps during just a single hour, once each year. As with all reproduction, this behaviour ensures the continuance of the species.

The Californian grunion, *Leuresthes tenuis*, is a 6 in (15 cm) fish which spawns at night, either at full or crescent moon. Carried on to shore on a wave, the male and female intertwine and in a brief moment the female extrudes her eggs which are fertilized by the male and deposited about 2 in (5 cm) below the surface of the sand. The fish are carried out to sea and two weeks later, at the next spring high tide, the eggs hatch. More eggs may be laid every 15 days.

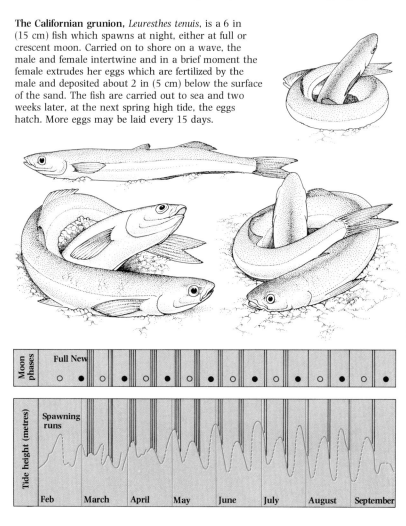

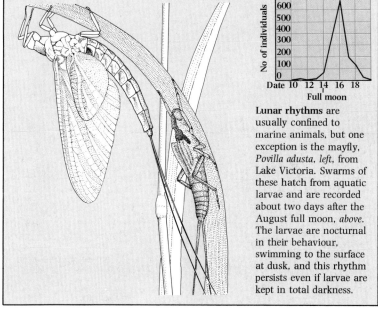

Lunar rhythms are usually confined to marine animals, but one exception is the mayfly, *Povila adusta*, *left*, from Lake Victoria. Swarms of these hatch from aquatic larvae and are recorded about two days after the August full moon, *above*. The larvae are nocturnal in their behaviour, swimming to the surface at dusk, and this rhythm persists even if larvae are kept in total darkness.

Grunion spawn from March until September but the breeding season may end in late June. The runs may be controlled by changes in water pressure due to the rhythmic cycle of the tides, and by perception of the lunar phases. They can be predicted with great certainty since they always occur at the highest spring and neap tides. Because of this predictable availability, grunion fishing must be carefully regulated.

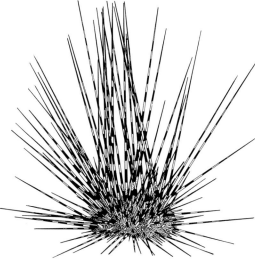

Only one sea urchin, namely, *Diadema setosum*, *left*, shows a lunar rhythm and in only one location. In the Suez region of the Mediterranean the urchins spawn at full moon, becoming ripe for further spawning at the next full moon. At Naples, Marseilles, Alexandria and on the Atlantic coast at Roscoff, there is no hint of moon-related spawning. Why this spawning should occur in such a limited way is as yet unexplained.

Population cycles

Every time a baby is born, utters its first yell and begins to breathe another human generation is complete. Both the mother and the father of this baby have brought their lives full circle and fulfilled, in reproductive terms, their *raison d'être* as part of the human race. Like man, all species of plants and animals that populate the earth have some sort of life cycle in which stages of development follow one another in an ordered sequence. The cyclical patterns have no beginning or end as long as the species persists, because an end to the cycle means extinction. For individuals the ring is broken if they do not or cannot reproduce themselves, or if they die of natural causes or are killed before they have a chance to procreate.

Often the word generation is used to mean the same thing as life cycle, but strictly a generation is the time that elapses between the birth of the parents and the birth of the offspring. The two are often synonymous but in some animals such as aphids, which infest plants, the life cycle contains several generations. Taking the egg produced in autumn and designed to last out the winter as the starting point in the aphid's life cycle, the next cycle will begin with the production of the next batch of overwintering eggs in the autumn of the next year. This annual life cycle is divided into many generations because the female aphids make many clones of female-only offspring with developing embryos of the next generation already forming inside them.

Annual breeding is commonplace in the living world. Man has extricated his reproduction from the web of the weather but for almost all other creatures the seasons have a profound effect upon their life cycles. Annual plants are a prime example, passing the winter as resistant seeds, then growing, flowering and setting fresh seed during the summer. Another big group of plants, the biennials, which includes many important root crops, such as parsnips, carrots, beetroot, swedes and sugar beet, takes two years to complete its life cycle. In their first year biennials grow from seed and lay down food stores in a swollen root or some other sort of storage organ. This food store carries the species through the winter so that it can grow rapidly in the following spring and produce flowers and seeds. Perennials—plants that live more than two years—have a persistent root system or storage structure, such as a bulb, corm or tuber, and make flowers and seeds every year.

Among the animals many soft-bodied invertebrates have annual life cycles but there may be one, two or three cycles squeezed into every 12 months. At the other end of the scale, the 17-year cicada, *Magicicada* spp, a native of North America, spends 17 years as a juvenile form or nymph, feeding underground on the sap of plant roots. Adults emerging from these nymphs in the same year spend a deafening summer of song, then mate, and the females lay their eggs. The young nymphs that hatch out burrow underground and 'disappear' for the next 17 years. This 17-year cycle is a local one so odd individuals which hatch out of phase with the majority have little chance of finding a mate.

In most populations patterns of change occur which encompass every developmental stage in a creature's life cycle. Animal and plant species usually have irregular populations which respond to changes in food or the climate, but some show definite cycles of change in which the number of individuals rises and falls. Four and ten-year cycles are common in some mammal and bird populations, while six to eight-year cycles occur in some moths, and two to four-year cycles are typical of several diseases, such as measles and rabies, which infect man and other animals.

The development of almost all plants and animals takes place as a cycle of events starting from a fertilized egg, progressing through various juvenile growth stages and ending with mature adults, which complete the cycle by producing eggs which are fertilized and give rise to the next generation. This is an organism's life cycle and, like all cyclical events, it goes through a sequence of stages before returning to its starting point. Exceptions to this basic cyclical pattern, which could begin anywhere but, purely for convenience, is taken as starting with the fertilized egg, are found throughout the animal and plant kingdoms. Many microscopic organisms, including bacteria, yeasts and a large proportion of single-celled animals (protozoa), reproduce simply by splitting into two in the process of fission, or by making buds which later break off the parent to become independent. The non-sexual life cycle is thus growth, fission and back to growth again. Occasionally these organisms have a 'sexual' stage inserted into the cycle. In this two different individuals of opposite 'sexes' or strains meet, fuse and give rise to a sort of egg, which may be useful by being resistant to specific environmental hazards.

Whether sexual or asexual, life cycles are by no means uniform and may differ in a number of distinct ways. One important variation is the length of time it takes for a life cycle to come full circle. For both animals and plants, the general rule is the bigger an organism, the longer its life cycle. So while it takes less than three weeks to increase the total cell-weight or biomass of a blowfly egg up to that of an adult, the microscopic egg of an elephant takes about 14 years to reach the 7,700 lb (3,500 kg) norm of an adult female capable of releasing fertile eggs. Most of the organisms that are a menace to man, be they

weeds, pests or diseases, are troublesome primarily because they are small and thus have a short generation time. In favourable environments brief lives allow for the rapid build up of the weed or pest, so large animals and plants can never be pests or weeds but mosquitoes and many grasses often can be.

The length of a life cycle has other important implications for the life-style of an organism. In a very short life cycle survival depends on quantity rather than quality of egg production, while for organisms with long life cycles the reverse is the case. Thus small creatures, such as insects with short life cycles, reach maturity quickly and release vast numbers of eggs, which are easily dispersed but just as easily destroyed or eaten by other animals. As a life cycle lengthens, so the time and energy invested in it by the advent of maturity is much greater, and there is, therefore, much more emphasis on covering the capital sum committed to the project by means of defence mechanisms, increased efficiency in obtaining food, and coping with competition. The eggs that are made tend to be few in number and well provided with food reserves to nourish the new individual in its first stages of growth and development. And the parents often expend considerable energy—as human parents will testify—in protecting and providing for the young of the next generation.

For some organisms habitat has a profound influence on their life cycle. Many plants and animals complete their life cycles in the same habitat. The young elephant, for example, has much the same structure and requirements as its parents, and if food or space became limited there would be direct competition between juveniles and mature adults for the available resources. In a forest this sort of competition is easy to observe. The mature trees are well

The life of the lung fluke, *Haematoloechus medioplexus,* cycles between three hosts, in water as well as on land. The adults of this common parasite live in the lungs of North American frogs, eg *Rana pipiens,* and toads, eg *Bufo americanus,* which frequent ponds and lakes. Eggs produced by the adult flukes are swept out of the lung by the action of its tiny hairs or cilia, and are subsequently swallowed, passing through the digestive system and out in the amphibian's faeces. The faeces, containing the first infective stage, are then eaten by an aquatic snail, which consumes organic

debris at the bottom of the pond. Enzymes in the snail's intestine stimulate the eggs to hatch, and the young parasites burrow through the gut wall and move into the snail's liver. Here they develop, multiply asexually and, in heavy infections, destroy the liver totally. After about three months the second infective stage, the cercariae, escape from the snail. These are able to swim, and live for only 30 hours during which time they must infect the next host, a young nymph of the dragonfly, *Sympterum.* Modified for respiration, the rectum of the nymph can pump water in and out via

the anus. Any nearby cercariae, drawn into the nymph's rectum by these respiratory currents, quickly burrow into the rectal tissue and form cysts around themselves. Frogs and toads eat either the nymphs or the freshly emerged adults which sit on leaves while their wings expand and their cuticles harden. The third infective stage emerges from the cyst in the amphibian's stomach, makes its way up the oesophagus and down into the lungs, so completing its cycle. The flukes mature in 37 days during the summer and remain in the frogs for up to 15 months.

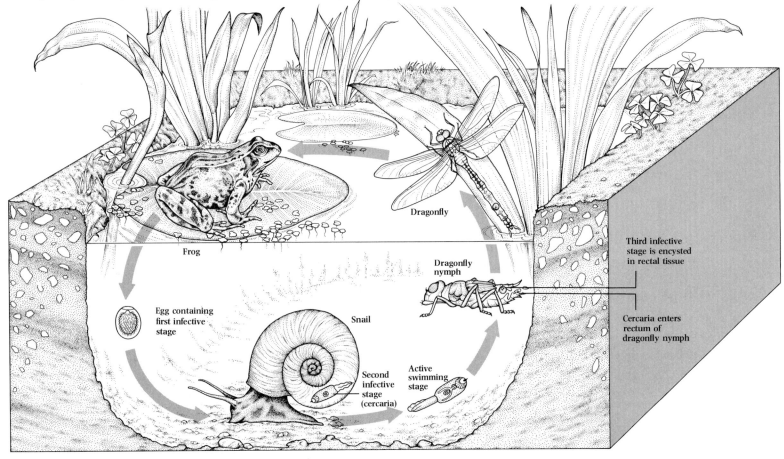

Dragonfly

Frog

Third infective
stage is encysted
in rectal tissue

Dragonfly
nymph

Cercaria enters
rectum of
dragonfly nymph

Egg containing
first infective
stage

Snail

Second
infective
stage
(cercaria)

Active
swimming
stage

spaced out with their canopies forming a continuous cover. This dense umbrella prevents sufficient light from reaching the forest floor and thus stops any tree seedlings from developing successfully. Clearings created by fallen or felled trees are, in contrast, a riot of young growth, and in time the vacant space in the canopy is filled by the most successful offspring.

To side-step the intense competition that may develop when the young and mature stages of a species both occupy the same habitat, many organisms make use of two different habitats during the course of their life cycle. Insects, for instance, which in terms of numbers of species are the most abundant of all organisms on earth, frequently double up on habitats, using one for the developing young and another for the mature adults. Most adult butterflies and moths exist primarily in the aerial habitat and feed on pollen, nectar and rotting fruits, but the herbivorous, land-locked larvae live for the most part on vegetation, chewing on leaves, stems and sometimes roots.

In fresh water, adult water beetles and water boatmen may share the available space and food with their offspring, but dragonflies, caddisflies and a host of other aquatic insects leave the water as adults to feed and live on land or in the air. Among the vertebrates those that make use of the land-water habitat duo are mostly amphibians, a group that represents the ancestors of the first backboned animals to leave the water for a life on land. Frogs, toads, salamanders and newts are all amphibians that use aquatic habitats for their juvenile tadpole stages, but exploit the land in a variety of ways as adults. These changes in habitat and feeding methods lead to dramatic alterations in shape and form—the contrast between caterpillar and butterfly, for example.

Competition between adults and their young is not the only reason for having a second home. In seasonal temperate environments, where the winter is distinct and cold, making use of two habitats may be a way of counteracting the effects of an adverse climate. In water the temperature fluctuates far less than it does on land, so an aquatic habitat has inbuilt buffers against rapid temperature change. Many insect larvae continue to grow and develop in fresh water during the winter, when the adults that gave rise to them have been killed off by the cold. Migration is another way of dealing with locally adverse conditions. Birds often use temperate food-rich regions of the world for raising their young, then move to a warmer zone to avoid the winter.

Parasites are animals or plants that live at the expense of—and often inside—other creatures which are known as their hosts. The life cycles of parasites differ widely in complexity and these differences are based on the theme of changing habitats. The body of the host is, by definition, the habitat of a parasite, but the host's own habitat may also influence the parasite. A few parasites complete their entire life cycle in a single host. For example, the pinworm, *Enterobius* sp, is a roundworm that commonly infests children, making its home in the lower section of the intestine and living on digested food. Repeated infection takes place when eggs are accidentally eaten as a result of lax hygiene. A similar situation is found in pigs who are prey to infestation by another roundworm by the name of *Ascaris*.

More commonly, parasites use two hosts so that the juvenile stages develop in one host, the mature egg-producing adults in another. The serious tropical disease bilharzia is caused by a parasitic flatworm *Schistosoma*. The adult ▷

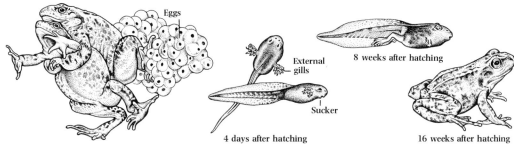

Eggs

External gills

8 weeks after hatching

Sucker

4 days after hatching

16 weeks after hatching

The larger an animal species, the longer is its life cycle. A full-grown elephant weighs 3 to 4 tons and is at least 10 ft (3 m) high at the shoulder. The foetus spends nearly two years in the womb and only one offspring is produced at a time. Each baby has the same structure and food requirements as the adult, reaches sexual maturity in 13 to 17 years and lives for up to 70 years. The elephant's immense body requires a constant intake of food so it forages for most of the day. In some parts of Africa the elephant population is so large that vegetation has been seriously damaged.

In spring, frogs journey to the nearest pond to breed. A male grips a swollen female with his enlarged thumbs and is carried on her back for a few days. In the water, the female lays a cluster of between 3,000 and 6,000 eggs encased in a protective jelly. At the same time the male produces sperm which swim to the eggs and fertilize them. The eggs hatch into tadpoles, which are omnivorous and aquatic and look very different from the adult. The tadpole uses a sucker to hold on to vegetation and breathes through external gills. It slowly loses its gills, develops limbs and absorbs its tail. Four months after fertilization the tadpole has become a fully formed frog, is carnivorous and essentially terrestrial. Four years later it has become old enough to breed.

▷ worms live in the blood vessels of the human abdomen, including those of the gut and urinary bladder, while juvenile stages use snails as their hosts. In some parasites as many as three hosts are involved, with two separate hosts for the early and late juvenile stages and the third for the mature adult. But whatever a parasite's life-style, moving house from one host to the next is inevitably a hazardous business. Where these transfers take place in water the parasites often have active swimming stages, as can be seen, for example, in the fluke *Haematoloechus* which lives a sedentary life as an adult in the lungs of frogs and toads.

In terrestrial situations the risks are potentially much greater than they are in water because of the added problem of desiccation. The complex life cycle of the sheep liver fluke, *Dicrocoelium dendriticum*, illustrates one way of making the transfer from host to host in a wholly terrestrial situation. Eggs of *Dicrocoelium*, liberated in sheeps' faeces, are eaten by snails. Early juvenile stages develop which, in leaving the snails, irritate them so that they encase the parasites in balls of slime. The slime protects the parasites from fatal drying out and is also a highly attractive food for the second hosts, ants. After the ants have eaten the slime balls, the juvenile parasites escape and invade the ants' bodies. Following further development, some of the parasites move to the heads of the ants and begin to alter their behaviour in an odd way. The ants climb up grass stems and anchor themselves with their jaws. Stuck with lockjaw at the top of a blade of grass, an ant is likely to be eaten by grazing sheep and so the parasites in the ant are able to complete their life cycles by reaching their final host.

The number of offspring a creature produces and the way in which this is achieved vary widely. The elephant only has one offspring at a time. So too do tsetse flies of the genus *Glossina* and, 98 times out of 100, humans. In all these cases the survival rate of the offspring is high because the parents shepherd the young through its development until adulthood is achieved. In contrast, a frog produces several thousand eggs in each breeding season and the oyster is reputed to liberate several million eggs a year. In these cases survival depends on the way the numbers game is played—produce enough and a few are bound to surmount the rigours of the environment and the hazards of predation.

Safety in numbers is also the motto of social insects but these creatures have added insurance policies to protect their young. In such insects the fertile females, the queens, produce many offspring by laying huge numbers of eggs—up to several thousand a day in some termites—and these are well cared for by the infertile workers of the species so they have a high chance of survival. It is difficult to trace the life cycle of a social insect, such as a bee or termite, because, although the female may lay many millions of eggs in her lifetime, few will develop into new, virgin queens. The vast majority become workers and soldiers, which do not reproduce and so are reproductive 'dead ends'. In such instances it is more meaningful to think in terms of the life cycle of the colony as a whole, which is founded, grows large and eventually gives rise to potential queens, who leave to found new colonies of their own. Colonies tend to be large and long lived and create relatively few 'offspring' in the form of new self-contained colonies.

For most plants and animals, seeds or eggs are the primary means of producing offspring and increasing the number of individuals, but some species

A sex-change commonly occurs during the life cycle of the slipper limpet, *Crepidula fornicata*. Individuals live stacked upon each other, the youngest at the top and the oldest at the bottom. The young start as males, but later their male organs degenerate and they become temporarily sexless. If the stack contains no females the sexless limpet will become a female; if there are several females then it will revert to being male.

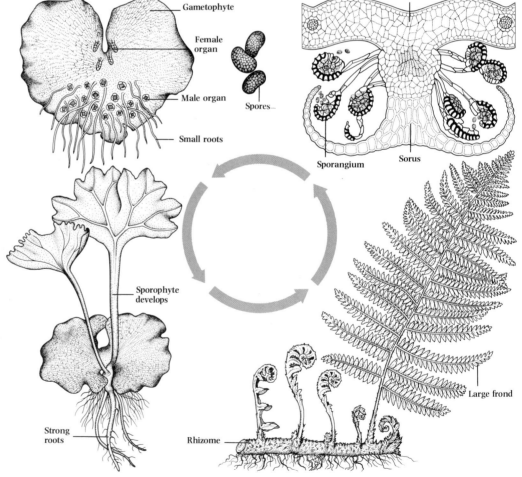

The life of a fern alternates between a sexual and an asexual phase. The large, leafy fern, familiar in woods and other moist habitats, is the culmination of its asexual generation. Starting as tight spirals growing from an underground rhizome, the young fronds unfold and expand into mature leaves, which possess small spore-containing structures or sori on their undersides. Within each sorus there are many stalked sporangia bearing clusters of asexual spores, which are liberated and dispersed when mature. On reaching a moist location they develop into small, flat, green, heart-shaped pads of tissue. These inconspicuous gametophytes are the short-lived sexual stages of the fern's life cycle. On the underside of this gametophyte are its male and female organs, and in moist conditions the male sperm swim to the female organs to fuse with the eggs. A fertilized egg develops into a young sporophyte, with simple leaves and a strong root system. The gametophyte then dies and the sporophyte grows into the large and characteristic fern, such as bracken, that often dominates the landscape.

use alternative methods of multiplication. Some species of parasitic wasp use a method which is a sort of cloning. In the body of its host—usually the caterpillar of another insect—the female wasp lays a single egg which then divides again and again to produce up to 3,000 identical wasp embryos. Although this system is rare, it can confer a distinct advantage if a female runs a serious risk of being damaged when she attempts to lay her eggs in the host.

The life cycles of many parasites contain phases in which the juveniles undergo a process of multiplication. Three of the parasites already mentioned, *Schistosoma*, *Haematoloechus* and *Dicrocoelium*, for example, all have multiplication stages, or sporocysts, in their intermediate snail hosts. This method of increasing the number of individuals in a population without sex is, in many ways, similar to the life cycle pattern with the rather confusing name of alternation of generations, which is seen in some plants and animals, notably insects. In such organisms the life cycle is divided into two parts with two offspring-producing stages or generations, one sexual, the other asexual. The confusion arises because the term generation is normally used synonymously with life cycle to encompass the time it takes for an offspring to grow up and produce new offspring. In alternation of generations a complete life cycle includes both sexual and non-sexual phases.

Reproduction by alternation of generations is well illustrated by the ferns. The familiar leafy fern frond is known as the sporophyte generation because it bears spore-producing organs on the underside of the frond. These spores are produced without sex. After they are liberated each of these spores is capable of developing into a small pad of tissue, the gametophyte generation, which bears

the male and female sex organs. Male cells or 'sperm' liberated from the male sex organs swim through drops of dew or rainwater deposited on the gametophyte to the female organs and fertilize the eggs they contain. The fertilized eggs then develop into new sporophytes, so completing the life cycle and also completing the alternation of generations between the sexually produced sporophyte and the asexually produced gametophyte.

Some of the most complex and highly variable life cycles of the living world are found among the aphids. The aphid life cycle is essentially annual and contains several 'generations', some sexual, some asexual. Different generations may be structurally and behaviourally distinct and induced by either genetic or environmental factors. Some generations lay eggs but most 'give birth' to live young. For even greater flexibility some aphid species switch hosts, either to prevent predators from congregating in regions of high aphid density, or to exploit more succulent new growth, or both. The sycamore aphid, *Drepanosiphum platanoides*, completes its life cycle on a single species, overwintering as an egg which hatches in spring as the sycamore leaves unfurl to produce the first of a series of female-only asexual generations. In autumn both males and females are produced and then mate and give rise to overwintering eggs. In contrast, the bird cherry/oat aphid, *Rhopalosiphum padi*, overwinters on bird cherry but switches to grasses and cereal crops in summer and has a wide variety of asexual generations.

Aphids, ants, humans and many other species live in groups or populations and occupy a particular situation or habitat. A population pulsates with change, often increasing or decreasing in size in naturally induced cycles. ▷

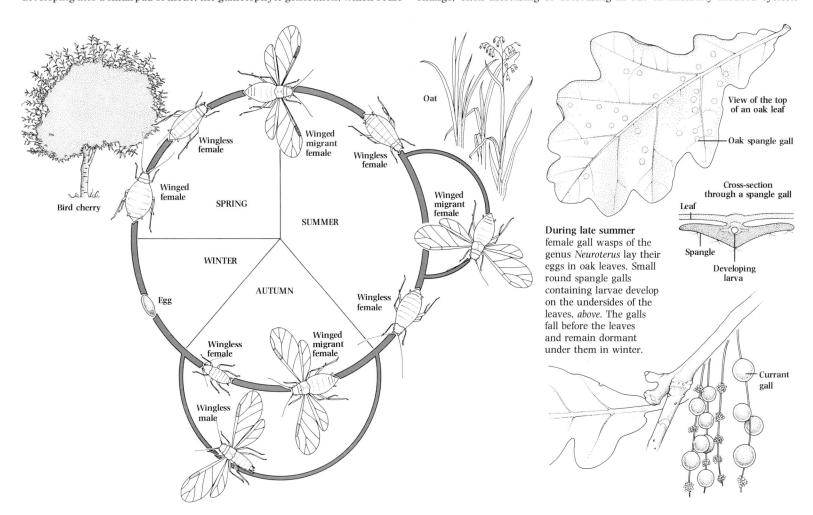

In its search for good quality plant sap the aphid, *Rhopalosiphum padi*, has developed a life cycle that not only alternates between sexual and asexual generations but also between the bird cherry tree and various grasses, such as the oat. The eggs overwinter in the bird cherry and in spring hatch into wingless females. These give birth to live wingless females which produce live winged females that migrate to the oats in a nearby field. Overcrowding can cause one of the next generations to be winged, enabling them to move to a less populated area. In autumn, winged males and females are born and fly back to the bird cherry. The females give birth to live wingless females that mate with the males and lay eggs on the tree.

In the spring only female wasps emerge, flying up into the oak tree and laying unfertilized eggs in the buds. From these eggs larvae hatch and stimulate the oak to form round currant galls, *above*. Male and female wasps emerge from these galls in midsummer and mate. The females then lay their eggs on the undersides of the leaves and the life cycle begins again.

During late summer female gall wasps of the genus *Neuroterus* lay their eggs in oak leaves. Small round spangle galls containing larvae develop on the undersides of the leaves, *above*. The galls fall before the leaves and remain dormant under them in winter.

▷ These changes in population size, which involve the processes of birth and death, emigration and immigration, occur regularly over long periods of time in any one locality.

In a perfect environment with unlimited resources, a population of almost any species of plant or animal will grow at its maximum or exponential rate. Imagine a population of bacteria, organisms consisting of a single cell, growing in a vast vat of nutrient-rich sludge. Every half-hour or so a bacterial cell will divide to give two cells. If we start with just one cell there will be two after 30 minutes, and four at the end of the first hour. After two hours the cell number rises to 16, and after three hours it is 64. At the end of the first 24 hours there will be approximately 2.8×10^{14} (280,000,000,000,000) cells. If this growth is plotted on a graph of numbers against time, the result is an exponential curve of increasing steepness and this shape is retained even if the time-scale of reproduction is altered.

In practice, however, environments are not perfect since they contain limited amounts of space, food, shelter and the other necessary requirements for the existence of a species. There is also a restriction on the volume of waste products they can accept and process. So while the population has a tendency to increase exponentially, as the population size increases the limitations of the environment's resources become increasingly apparent and exert an ever-more rigid restraint on unchecked exponential growth. Eventually a balance point is reached at which the growth and death rates of the population are equal, and this is the carrying capacity of the environment. The introduction of sheep to Tasmania in the early part of the nineteenth century is a good example

of such growth. To begin with the sheep multiplied exponentially, but then the rate of increase gradually slowed down, levelling out at a carrying capacity of 1.5 million sheep. Plotted out, this gives an S-shaped logistic growth-curve. Such a curve describes the population growth of many species in limited environments and occurs over and over again in both laboratory and natural situations.

If this pattern of growth is so widespread, we would expect the world's natural populations to remain constant. Some species do, in fact, maintain a steady rate of growth, but man's population is still enlarging exponentially while those of other species go through distinct cycles of expansion and contraction. Cyclical behaviour in population size has been extensively studied in certain Canadian populations of the lynx and the snowshoe hare, both of which have approximately ten-year cycles between peaks of population size. The cycling of lynx populations is easily explained. The lynx feeds primarily on the herbivorous snowshoe hare. When hares are abundant the lynx have plenty of food and can reproduce exponentially, or nearly so. When hares are scarce the lynx have difficulty in finding prey and so many die. Through its dependence on the hare for food, the population size of the lynx changes as the hare's population changes.

For other predators with pronounced population cycles similar explanations hold good. In the tundra regions, populations of the arctic fox show a four-year cycle linked to the population fluctuations of their principal prey, the lemming. Farther south on the forest margins, the marten and the red fox also have four-year population cycles synchronized to the changing size of the population of

A population of animals or plants will grow at its maximum rate if it lives in conditions where there is plenty of food and space and an absence of predators and disease. Humans are no exception to this, and in many countries the rate of population increase is near the maximum. In some developed countries there is a minimal increase, due to effective control measures. The global pattern of population, *below*, shows a continuing rise in numbers and also in the rate of

increase. The world's population has doubled since 1940. In natural populations the environment effects its own controls and, unless man imposes his own restrictions on human populations, one or more of these natural controls will do it for him. There will be either a massive shortage of food or space, or some disease will spread. Technological advances, better agricultural practice, medical breakthroughs and a lack of regard for natural laws have buffered the human

race from many natural population controls. When sheep were introduced into Tasmania they multiplied rapidly until they exceeded the population which Tasmania could carry. Their numbers then dropped from 2 to 1.5 million. Such an overshoot is common in natural populations and results from the time delay between the rate of population increase and the rate at which controls, such as lack of food, can come into effect. Improved farming has let numbers rise again.

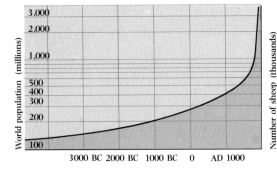

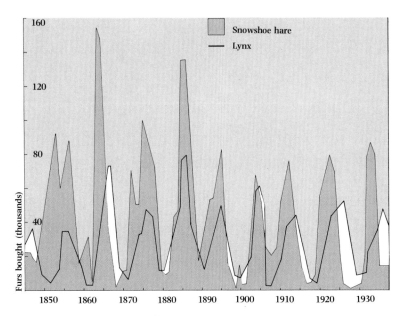

Populations of the lynx, *Felis canadensis,* and the snowshoe hare, *Lepus americanus,* reach a peak every 10 years. This regular cycle in their numbers is found in the records of furs bought from trappers by the Hudson's Bay Company in Canada. These show peak numbers of lynx are a year or two behind those of its prey, the hare. Lynx numbers rise if hares are abundant, but when most hares have been eaten, lynx numbers decline. The hare population cycle seems to be driven by climatic changes and the availability of plant food.

voles on which they feed. However, in parts of Canada the marten and the red fox feed on the snowshoe hare and so have ten rather than four-year population cycles. Similar explanations can be applied to the cyclical population changes in some predatory birds, such as snowy owls, which show cycles generated by changes in the populations of their prey.

Ample evidence exists to support the speculation that predator populations are controlled by the abundance of their prey, but what causes the cycling in the prey populations? A variety of answers has been put forward to explain the ten and four-year cycles in the herbivores. Some early analyses claimed to have found a link between the population cycles and sunspot cycles, but the fact that population cycles are not simultaneous across large tracts of land and show a huge site-to-site variability rules out any such global climatic influence. Another suggestion is that the predators have some force in driving the change. The idea is that the predator population grows in size as its food becomes more abundant, but that its demand for food eventually outstrips the rate at which the prey population is increasing, so that the prey population crashes because of over-exploitation by predators. The strongest argument against this idea is that the prey population's decline should be precipitous not gradual as, in fact, it is. Despite this objection, some insect populations seem to obey these rules of population growth and decline.

The influence of disease epidemics and changes in the physiology of individuals at high population densities have also been proposed to explain population cycles. Both do occur, but they, too, would produce more rapid reductions in population size than actually take place, so they may well be symptoms rather than the causes of change. A suggestion that fits the facts better is related to the abundance of the plant foods on which the hares, lemmings and voles feed. When population cycles are at their peak both the hares and small rodents behave as if they were searching for food. This is most noticeable among the lemmings which, at times of peak population, often embark on extensive, legendary migrations in search of new food resources. If food is scarce the juveniles are likely to be affected first so that the population decline begins with juvenile mortality, followed by the death of adults and a reduction in the numbers of offspring produced, so explaining a decline lasting several years.

The rise and fall of populations are part of a complex web of interactions that take place in an environment. Thus the population cycles of the hares and small rodents may regulate the cycles of their predators and also affect other species in the area. Several species of grouse, for example, show approximately ten-year cycles in Canada. The willow grouse has a ten-year population cycle in Canada but a four-year one in Norway, and these cycles are synchronized to the population cycles of the local hares or small rodents. Since the hares and grouse have different diets, depletion of one food source should have little effect on the other, but grouse are the favourite alternative prey for the Canadian predators of the snowshoe hare, such as the red fox and the goshawk, while in Norway the willow grouse stands in for the lemming as an alternative prey for arctic foxes and snowy owls. A decline in hare and small rodent populations, therefore, induces their predators to switch prey and so influences the population cycles of the grouse.   ▷

Fighting breaks out when the density of a vole population increases, but aggressive individuals, while able to stay alive, cannot reproduce fast and are prone to disease. A reduced birth rate and an increased death rate cause the population to decline. This cycle of rise and fall in population size occurs approximately every four years. When the population density has fallen, less aggressive but faster breeding individuals make the population rise again.

Adult blowfly

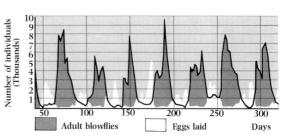

Adult blowflies   Eggs laid   Days

The ecologist A.J. Nicholson grew colonies of the sheep blowfly, *Lucilia cuprina*, in a cage with a limited but renewed supply of food, and every day counted the numbers of flies and their eggs. Flies showed a regular cycle in their numbers, proving that they were being limited by the amount of food available for their larvae. At low densities there was ample food and many eggs were laid. At high densities few eggs were laid on the food because competing females obstructed one another. The two-week time delay between the laying of these differing egg numbers and the emergence of the adults governs the blowfly population cycle.

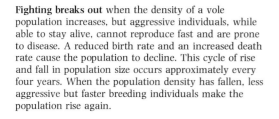

Vole

At the start of their breeding season male black grouse assemble at a certain place on the moors and make aggressive territorial displays to each other. A limited number of successful males is thus selected to mate with the females; the weaker ones are excluded and eventually killed by predators. As the population increases, these displays become more vigorous and more individuals are excluded from mating.

Black grouse

▷ If population size is controlled by the effects of another species operating in the same environment then, according to the biotic school of thought, cycles develop because there is always a time-lag between the development of a regulator and its actual effect. This is like trying to control the temperature of an oven with a slow-acting thermostat. The oven temperature will rise and fall in a cyclical way. The faster the thermostat reacts, the smaller the fluctuations in temperature will be. Many of the early ideas of the biotic school were developed by entomologists studying the possibilities of controlling insect pests by using other insect species, and this concept of biological control is used today to deal with a wide range of pests around the world.

Opponents of this sort of argument have put forward the suggestion that animals regulate their own population sizes so that they do not overstep the carrying capacity of their habitat. Several mechanisms have been proposed to explain how this self-regulation is achieved. Some birds, mammals and even insects hold territories, patches of a habitat defended against intruders. Territory holders are at an advantage because having a territory may improve the chances of finding a mate or ensure space in the habitat in which a family can be successfully reared. Only a certain number of territories can be fitted into any one habitat and this automatically restricts the number of productive matings. Unsuccessful individuals who do not manage to raise a family will die out. In any habitat the amount of food will vary from year to year so that a territorial bird will have to evaluate the quality of a territory before deciding how big it should be. The poorer the conditions, the larger the territories will need to be and the fewer breeding pairs will be supported by the whole habitat.

In this situation population size is not, therefore, self-regulated but governed by the quality of the habitat, but the existence of territories does ensure that population levels do not outstrip the habitat's resources.

For many species the regular procession of the seasons creates a regular cycle of population size. Aphids, for example, build up their numbers throughout each summer before regular autumn and winter crashes. In less distinctly seasonal situations climate can also bring about marked changes in population size, but these are irregular. Small mammals and birds of the southern Sahara have populations that fluctuate in a non-rhythmical way in response to the erratic rainfall in the area, and in Australia grasshopper populations go up and down in a similar rain-induced way.

One striking climatic effect is seen on the western coast of South America. The cold, nutrient-rich Humboldt current sweeps up from the south and provides eminently suitable conditions for the development of large plankton and anchovy populations off the coast of Peru. Huge flocks of birds, tuna, whales and porpoises feed in these rich waters, and the Peruvian people harvest the anchovies for fish oil and meal and remove the bird droppings for fertilizer. At irregular intervals the wind patterns in the area change and warm, low-nutrient tropical waters displace the Humboldt stream, with catastrophic effects on the marine life, including the anchovies. As the anchovies die, all the other species that feed on them die too or leave the area. Population cycles of anchovies and pelicans are thus linked but show irregular patterns of change because of the vagaries of the climate. The notorious unpredictability of the weather means that regular patterns of population change are not linked to the

Catastrophe strikes the marine life and the fishing industry when 'El Niño'—the Christ child—comes to the waters off the coast of Peru. Occasionally, around Christmas time, unusually strong northeast trade winds bring warm nutrient-poor waters from the Equator to replace the cold nutrient-rich Humboldt current, flowing up from the Antarctic. This warm current causes a rapid rise in water temperature which kills the plankton and anchovies and, consequently, the larger animals that depend on them. Anchovies are the main catch for the Peruvian fishermen and also provide food for many of the birds, whose faeces form guano deposits which are used as fertilizer worldwide.

Brown pelican

Anchovies are the key to the pyramidal food chain that thrives in the cold waters of the Humboldt current. They feed on millions of tiny plankton and are themselves eaten by cormorants, pelicans, whales and porpoises. 'El Niño' kills the plankton and anchovies, and many of the animals dependent on them also die.

climate. Such patterns can only be the result of density-dependent processes—factors that exert an increasingly detrimental effect on the population as its numbers increase—and time-delays between cause and effect. The climate is density *independent*: the inflow of warm waters off Peru would kill the anchovies irrespective of how many there were.

By using computer simulations almost any pattern of population change can be mimicked and many modern ecologists believe that both climatic and biological influences are important in population dynamics, but that only biological, density-dependent factors can operate to generate cycles. The caterpillars of the larch budmoth, *Zeiraphera griseana*, feed on the needles of larch trees, and in parts of the European Alps have a regular seven-year population cycle. A detailed, long-term analysis of this species has revealed a complex interrelationship of mechanisms which together produce the population cycles.

Climate plays an important role in governing where these caterpillar cycles will occur, but does not influence the pattern of change itself. Unless they experience at least 120 days of cold weather, the overwintering eggs of the larch budmoth will not hatch. This delaying tactic or diapause, a quiescent period in an animal's life cycle common in many invertebrates, is usually broken by some specific external factor. The correct cold period is only found at specific Alpine altitudes. At high altitudes the climate is too cold for too long, so the eggs fail to develop, while at lower altitudes the eggs hatch too early, before the larch needles have burst from their buds. Conditions are best at between 5,000 and 6,000 feet (1,500 and 1,800 metres) and one of the most studied areas in this

range is the Upper Engadin Valley in Switzerland.

Within the larch budmoth populations there are two physiological strains—the 'strong' race and the 'weak' race—and much of the population cycle can be explained by the dominance of one race over the other. One of the key features in this story is that in the Upper Engadin the larch budmoth suffers from a unique virus disease. The virus lives latently in the 'weak' race, which is resistant, but the 'strong' race is highly susceptible to it. Parasitic wasps also play a role—they are more successful at laying their eggs in the 'weak' race than in the more active 'strong' race.

In several other cyclical species, including the western tent caterpillar, *Malacosoma* sp, and voles, separate races have been discovered which may exert an influence on population cycles. The 'strong' form of the moth is superior in almost all respects. Its eggs survive the winter better; its larvae grow more quickly, are bigger and more active; and the females of the 'strong' race lay more eggs. The success of a particular race, whether weak or strong, depends on the population density so that the regular oscillations in population size of the larch budmoth arise as a result of the changing fortunes of the 'weak' and the 'strong' races.

In the diseases that afflict man and other animals cyclical patterns can exist at two separate levels. The organisms that cause the disease may undergo cyclical population changes within their host, in much the same way as free-living species may rise and fall in numbers in a cyclical way in their own habitats. The prevalence of a disease among the individuals in the population that it attacks may also show cyclical patterns. These patterns are particularly ▷

Defoliation and extensive browning of larch trees occurs regularly in some parts of Switzerland, such as the Engadin Valley, *below*. Larvae of the larch budmoth, *Zeiraphera griseana*, feed on young larch needles; they cause most damage when their population reaches a maximum every seven to ten years.

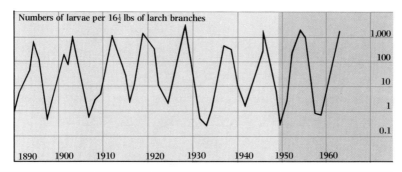

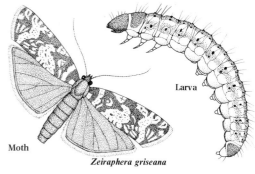

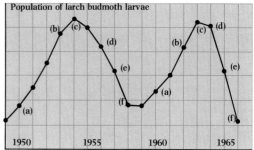

Zeiraphera griseana

Moth

Larva

A virus, parasitic wasps, competition between individuals, and a lack of tender larch needles all contribute to one complete cycle of the larch budmoth population, *above*. Two races make up its total numbers; a strong race, which is prone to the virus, dominates a weak race which carries the virus but is more often parasitized by the wasps. The population (*a*) grows as the strong race proliferates, and damage to the trees is noticeable. At high population densities (*b*) larvae compete for larch needles, and damage to the trees is most severe. Population growth slows (*c*) as female moths find it harder to discover fresh tender needles on which to lay their eggs, and the virus begins to spread. Population falls (*d*) as the virus kills the strong race, leaving the weak race to predominate. Population declines further (*e*) as many of the weak race are killed by the parasitic wasps. As population reaches its lowest density (*f*), the virus becomes latent in the weak race and so allows the strong race to proliferate once more.

▷significant to the study of epidemics and may be useful to us all if they help scientists to plan preventive strategies for the future.

The life cycles of organisms that cause disease are often complex and they contain specific phases in which the organism is transmitted from one host to another. This may involve another species—a vector—or a new host may be infected directly. Transmission via a vector demands highly accurate synchronization between the behaviour of the virulent organisms and that of the vector species. In the tropics blood-sucking flies, such as mosquitoes, blackflies and tabanids, are frequently disease vectors, and throughout every 24-hour period they tend to have distinct feeding times. The disease-causing organisms these vectors transmit have behaviour patterns which ensure that forms of the organism suitable for transmission will be present in the blood vessels near the skin surface of an infected animal and will, therefore, be most likely to be taken up as the vector feeds. Loaiasis, for example, is an African disease, caused by the parasitic roundworm *Loa loa*, which is spread from human to human by several tabanid vectors which suck human blood during daylight hours. Transmission of *Loa loa* is, therefore, maximized because the greatest concentration of the infective stages in the superficial blood vessels occurs during the day in humans.

In other cases of diseases caused by roundworms, such as river blindness, or onchocerciasis, and elephantiasis, the infective stages also migrate to the superficial blood vessels to rendezvous with the biting vector. With malaria the situation is rather different. The single-celled protozoan *Plasmodium*, the cause of malaria, has a characteristic pattern of development in humans. Depending on which species of *Plasmodium* is responsible for the disease, a periodic fever typically comes on every third or fourth day.

As part of its life cycle *Plasmodium* invades the red blood cells that are responsible for carrying oxygen round the body. When a parasite invades a red blood cell it divides four times to give 16 new infective stages or merozoites. The red blood cell then disintegrates, liberating the merozoites. Depending on the species it takes 48 or 72 hours for the 16 merozoites to be formed, and the fever is caused by the release of cell debris and waste products into the blood as the red cells disintegrate. What is particularly interesting is that the parasites are synchronized so that the merozoites are all released together. And the timing is so precise that the merozoites are liberated at a particular time of day, which explains why the tertian fever, typical of *Plasmodium vivax* infection, occurs every 48 hours in the early afternoon.

The fever period is not, however, the time at which the mosquitoes that transmit malaria bite and suck blood, nor are the merozoites infective to them. The timing is more subtle than this. Some of the merozoites are destined to become sexual cells, or gametocytes, designed to continue the infection in the mosquito. These gametocytes take between 30 and 36 hours to develop from the merozoites. Thus the very regular asexual cycle producing merozoites in the early afternoon ensures that the gametocytes will be at the correct stage of development some 30 to 36 hours later, that is, during the night of the following day. This fits perfectly with the nocturnal feeding habits of the mosquito vectors.

This system of disease transmission takes its time cues from the host. In most

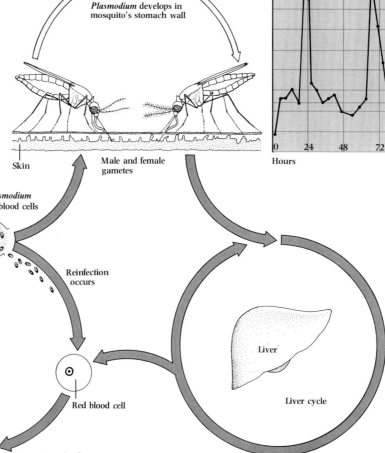

Plasmodium develops in mosquito's stomach wall

The mosquito's need for human blood has caused millions of people to suffer from malaria. When it bites, it introduces into the wound asexual stages of the protozoan *Plasmodium*, present in the mosquito's salivary glands. These quickly make their way to the liver, where they multiply and live for many years, causing bouts of the disease. From the liver they enter the bloodstream periodically, invade the red blood cells and here develop and multiply. When mature they rupture the cell, and thousands of cells rupturing at the same time cause a fever. Most of the liberated spores invade other cells, so beginning the red blood cell cycle again. Others become male and female gametes, which will only develop and begin the sexual stage of the life cycle if they are sucked up in a mosquito's meal of blood.

Skin

Male and female gametes

Many *Plasmodium* cause red blood cells to rupture

Reinfection occurs

Brain

Spleen
Kidney

Colon

Red blood cell cycle

Red blood cell

Liver

Liver cycle

Plasmodium divides

Hours

Temperature (°F)

Periodic fevers are the classic symptom of malaria but different *Plasmodium* species produce different patterns of fever. In all cases the fever occurs when the infected blood cells rupture and parasites, cell debris and waste products are released into the blood. *Plasmodium vivax* and *P. ovale* both take 48 hours to complete one blood cell cycle, and cause tertian fevers every third day. *P. malariae* takes 72 hours, causing quartan fevers every fourth day. The fevers begin in the middle of the day with a feeling of cold, and chattering teeth. The body temperature rises to 40°C (104°F) and headaches, vomiting and delirium may occur. After about three hours profuse sweating occurs, the temperature drops to normal and a feeling of well-being returns.

Symptoms of malaria include kidney failure, enlargement of the spleen, disorders of the colon and, in fatal cases, brain damage.

The mosquito swallows the male and female gametes of *Plasmodium* and, when they enter its stomach, parasitic sexual reproduction occurs. The resulting parasites penetrate the stomach lining where they grow and eventually produce hundreds of cells which migrate to the salivary glands of the mosquito.

Once these cells are in human blood they can be transmitted by a mother to her foetus across the placenta causing abortions, stillbirths and infant mortality. Even in blood bank conditions *Plasmodium* cells remain viable for several weeks so that malaria can also be transmitted by transfusions.

animals, including man, there are 24-hour cyclical changes in several physiological activities—temperature, blood acidity and hormone levels all fluctuate regularly during waking and sleeping. Night workers with malaria would, thus, experience the start of their fever just after midnight, that is, about half-way through their waking cycle. From experiments with malaria in animals it has been established that the *Plasmodium* organism makes use of slight temperature changes to synchronize its life cycle. If monkeys with malaria are cooled down under anaesthesia then their fever is delayed. *Loa loa* also employs temperature changes as cues for its timing, while *Wucheria bancrofti*, which causes elephantiasis, orders its activities according to changes in the oxygen levels of the lungs between day and night.

The cycles of human disease are frequently linked to the seasons but their periodicity may be greater than a year. Measles, for instance, has a two-year cycle, rabies a four-year one. It is unlikely that climate alone could produce such regular cyclical changes. More probably density-dependent factors, such as the proportion of susceptible people available, influence the populations of disease organisms, and so regulate their cycles. Season is important to several diseases. In temperate zones the crowded, intimate association of school-children during winter and spring allows measles to spread easily among susceptible children. The virus is helped by the lowered resistance of children during the winter because of respiratory tract infections, and the typically warm, moist, enclosed classroom atmosphere is ideal for the transmission of airborne virus particles. Summer peaks occur in many diseases spread by accidental ingestion and via faeces. Salmonella food poisoning, cholera,

infective hepatitis and poliomyelitis are all such diseases. Summer peaks are also common in diseases spread by biting insects or ticks, whose populations are largest in the warmest months of the year.

Plague is a disease with both summer and winter peaks. The disease is spread by fleas which become increasingly active as the weather warms up. In late autumn small rodents, the natural reservoirs of the disease, tend to move into human houses to avoid the winter cold and this migration enhances the risk of cross-infection to man. Plague can be spread from man to man in air breathed out and there is a greatly improved chance of this causing infection in the enclosed atmosphere of winter accommodation.

In foxes, rabies shows a seasonal cycle linked to the fox life cycle. Male foxes are territorial creatures who aggressively defend their areas against intruders. Fox territories are established during the mating season which results in a considerable increase in the degree of contact between individuals. The prevalence of rabies in the fox population peaks in January or February, just after mating. Both rabies and foxes also show striking longer-term cycles of three, four or five years. Mathematical simulations have shown that these may result from a time-delayed density-dependent influence of rabies on foxes. Rabies reduces the fox population to such a low density that there are not enough individuals to spread the disease. This point of lowest density seems to be reached when the fox population drops below a single animal per square kilometre. The time-lag is determined by how long it takes the fox population to pick up and increase its numbers enough for the rabies virus to establish itself once more.

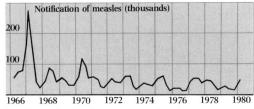

Measles epidemics show an annual cycle in temperate countries as the number of reported cases reaches a peak in spring. The incidence of the disease is markedly annual in North America but shows a two-year cycle in England and Wales, *above*. This biennial cycle persists even though vaccination, started in the late 1960s, has reduced the total number of cases.

Rabies attacks the nervous system causing excessive salivation, a dread of water, paralysis, madness and eventual death. The red fox is the chief reservoir of rabies in Europe and a three to five-year cycle in the prevalence of rabid foxes has been found. In Africa the dog is the chief reservoir and in South America, the bat, although many animals can become infected. Virus particles are found in the saliva and one bite from a rabid animal is enough to spread the disease. The pitiful sight of a docile red fox whose hind legs lack coordination may inspire sympathy, but if approached it is liable to bite viciously. Change of temperament is an early symptom of the disease, normally docile animals becoming aggressive and vice versa. This is always followed by a phase when the animal bites readily and foams at the mouth.

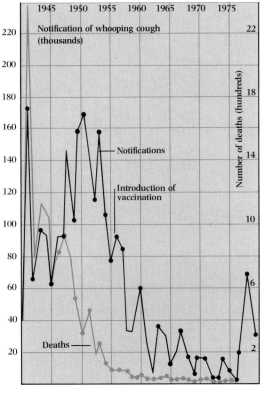

The incidence of whooping cough among children in England and Wales shows a distinct three to four-year cycle, *left*. This alarming disease is caused by a bacterium which is abundant in the sputum of infected children, and is spread by inhaling the drops expelled by the cough. Prior to 1950 the death rate was high, but it was reduced nearly to zero after vaccination was introduced in 1957. The total number of reported cases fell dramatically although the cycle of the disease persisted. Recent doubts about the safety of the vaccine have made many parents expose their children to the hazards of the disease rather than risk any side effects of the vaccine. As a result the incidence of whooping cough showed a sudden increase in late 1977 and 1978.

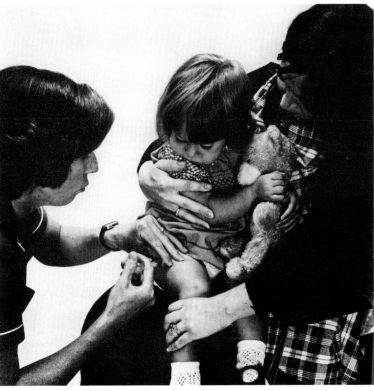

Rhythms of growth

No gardener, who completes the back-breaking task of clearing his vegetable patch only to find his precious plants overrun with weeds again in just a few weeks, could fail to believe that living things are capable of the most spectacular rates of growth. Or think of a human baby with a birthweight of 8 pounds (3.6 kg). Only nine months before, that baby was a living microdot weighing only a fraction of an ounce. Never again is human growth as fast as during these months in the womb. For other animals babyhood is also a time of staggering spurts of growth. Pups of the southern elephant seal, for example, who suckle on high-fat milk made by their mothers from stored blubber, put on weight at the rate of 20 pounds (9 kg) a day. Among plants, climbers are often the fastest growers, and a vigorous passion flower can add 20 feet (6 m) to its stature in a single growing season.

To increase in size, whether quickly or almost imperceptibly like a hardwood tree, living things do not simply blow themselves up like balloons, nor, like growing crystals, do they add on ready-made units to those already in existence. The secret of growth lies in the organization of the bodies of plants and animals into cells, units of life which are usually considerably less than 0.04 inches (0.1 mm) across. Whether an organism is constructed of just one cell or, like a human being, hundreds of millions of them, each cell has limited powers of expansion. For true growth to take place that cell must divide into two, and cells divide in a cyclical and often rhythmic way.

Unlike technological advance, which we accept with hardly a moment's thought, conceptual advances in science take years, if not centuries, to become accepted. Some of the slowness in coming to grips with new concepts arises because new hypotheses run counter to the superficial rules of common sense. If, for example, we cut a pat of butter in half we can go on cutting it into smaller and smaller pieces—but not for ever. The fact that butter is made of fat molecules which, when divided, turn into something else, is a conceptual leap that is in no way aided by common sense. The concept that living things are made up of cells has entered the common currency largely because it has become possible, with the help of sophisticated equipment, to look at and dissect them. Just as butter can only stay as butter if its molecules remain intact, so life can only continue if the integrity of its living boxes, its cells, is preserved. The complex molecules that make up those cells are not in themselves living but, to accomplish the process of growth, the cell has to ensure that all the essential molecules it contains, from its information store to its machinery for making new molecules, can be faithfully reproduced every time the cycle of cell division is completed.

The cycle of cell reproduction which underlies growth processes cannot continue unless, like all other living processes, it has a constant supply of energy. In the final analysis all this energy comes from our radiant provider, the sun, and is used by plants to make the energy-rich molecules which are either used directly by the plants themselves and by herbivorous animals, or indirectly by carnivores. The feeding chains of all the organisms on our planet interlock to form a vast energy cycle in which substances are constantly being trapped and used, processed and then pushed back into the atmosphere, from where they are trapped again and reused. Without this conservative cycle there would be no way in which the earth could support, in the long term, any form of life, least of all the voracious human race, seemingly intent on disrupting the cycle by wasting its energy-giving molecules.

Common sense, that most down to earth of human attributes, is bounded and constrained by the finite abilities of the body organs that help us to perceive the world around us. Our most detailed experience of the physical world is visual, and this vision is a prisoner of the wavelengths of light visible to us—unlike some creatures, such as the bees, we cannot receive and process light at the ultraviolet end of the spectrum. There is also a limit to the size of the objects we can see. Even with the finest optical microscopes we can see nothing smaller than the minimum wavelengths of the light our eye uses—about 0.00004 inches (0.0001 mm). The structure of our retina (the sensitive inner lining of the eye which receives light signals and converts them into nerve messages which are interpreted by the brain) means that only with great difficulty can we see objects less than 0.04 inches (0.1 mm) across with unaided eyes.

It is because of the constraints of our eyesight that emotionally we always regard the growth of living things as a continuous process. Over long time-spans we can see that a plant gets taller and adds new leaves to its collection, or that our children increase in stature, and everything our senses tell us about these natural changes adds up to a story of gradualism. The outward appearance of an elongating shoot is of a smooth, continuous extension, akin to toothpaste being squeezed from a tube.

The underlying process of growth is, however, quite different. Ultimately all sustained growth consists of the production of new living subunits, that is, cells. New cells can only be created by the process of cell division during which a parental cell, after carrying out extensive internal reorganization, divides itself into two. The pair of daughter cells resulting from this process of fission are,

after a time-lag which may vary from half an hour to many days, capable of repeating the fission process. By this essentially discontinuous and cyclical process cells reproduce themselves, new living material is formed and growth occurs. Down at the crucial level of organization, therefore, the growth activities of living things are not smooth expansions. Rather they consist of minute amounts of growth whose sum simply appears to be smooth. The situation perceived by the human eye is the averaged total of thousands or millions of individual cell divisions.

If it is to fit all the circumstances to which our language would apply the word growth, we must expand our detailed explanation of the cellular background to the phenomenon, although most normal examples fit in with the schema of typical cell growth in a straightforward fashion. Take, for instance, the growth of a human being. It is easy to concentrate on the growth processes that take place between birth and adulthood, such as the lengthening of the limbs and trunk, and forget that in some ways the most impressive and intense growth—at least in relative terms—occurs before the great event of birth itself. Each of us was conceived by the fusion of two single sex cells, or gametes, a sperm from our father and an ovum, or egg, from our mother, to form a single generative cell, the zygote, or fertilized egg. From this one individual, effectively invisible cell, a man or woman grew.

The extent of the total growth process is stupendous simply because one cell grows into billions of cells. All of this increase in size is generated by the cycle of cell division. Within each cell division cycle, however, another form of growth must occur. Before each parental cell divides, it must itself grow so that

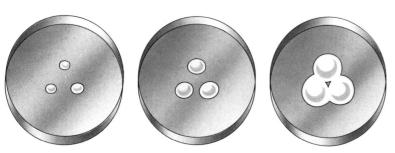

The driving force of nearly all genuine growth is the multiplication of cells, the ultimate units of life. This mechanism can be explicitly revealed by cultivating colonies of bacteria in dishes of the nutrient jelly, agar. Individual bacteria are widely dispersed over the agar surface by making a dilute suspension of the cells in water then spreading this over the agar. The dishes are then incubated and each cell divides many times to become a colony, which is clearly visible, *left*, and whose specific appearance is helpful in identifying the bacteria that produced it.

Stained to show their chromosomes, cells from the root tip of an onion, *below*, are dividing. After the genetic material has been duplicated, daughter chromosomes move to opposite ends of the cell, *left*, before the chromosomes form two new nuclei, *right*.

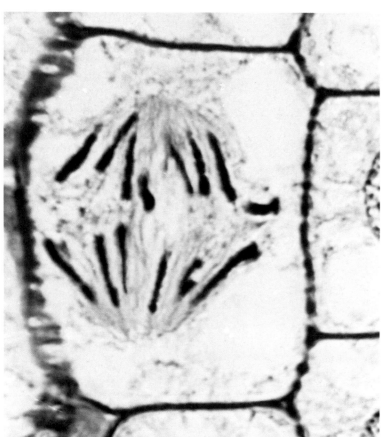

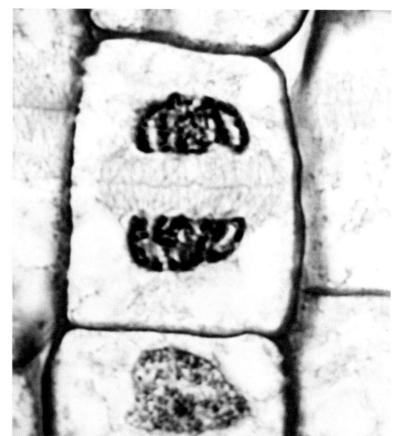

cell size increases. Without this form of expansionist policy each cell would halve in size every time it divides. Growth in cell size, or intracellular expansion, is the result of the production of brand-new complex organic molecules, principally proteins, within the cell, and nearly all dividing cells demonstrate this sort of increase in size.

In the case of one type of cell, specialized to carry out a particular job of work in a creature's life cycle, growth from within becomes enormously exaggerated and finally gives rise to single cells that are easy to see with the naked eye. These cells are eggs. There is no difficulty in seeing a boiled egg sitting in its cup waiting to be eaten at breakfast, and although a bird's egg is a complex structure in which the egg cell is surrounded by nutrients, various membranes and the shell, the ovum it contains is a truly gigantic cell.

The growth process that produces these gargantuan cells involves the laying down of both nutrient reserves, such as yolk, and information-containing molecules which, working together, will ensure that rapid cell divisions can take place after the egg has been fertilized by a sperm. These reserves mean that during the early embryological development of such creatures there is no significant increase in cell size. Instead, the original gigantic cell divides first into two, then four, then into eight cells, and so on until a rudimentary embryo is formed, which consists of hundreds of cells but has the same dimensions as the original fertilized egg.

In the vegetable world, seeds are, for flowering plants, the equivalent of fertilized animal eggs and they, too, are packed with nutrients for use in the first, crucial stages of development. Among plants there are some problems

concerning the definition of growth. The first of these relates to the intriguing fact that a huge proportion of a tree is dead material. The living, cellular, section of a tree is very much smaller than the masses of material made up of cell skeletons. These skeletons are, in fact, cell walls composed of the tough plant carbohydrate cellulose, and it is cellulose, along with its chemical cousins, that makes up the skeletal bulk of a forest. The cellulose and other complex carbohydrates are laid down as secreted material outside the living cells of the plant. Thus a good deal of a tree's growth does not consist of living cells nor does its maintenance involve cell division.

The second growth 'oddity' of the plant kingdom involves cell expansions which do not depend on the standard method of the intracellular growth. When roots or shoots elongate, the new cells produced to sustain this growth undergo a marked expansion process which, although it puffs up the cell dimensions enormously, cannot strictly be thought of as growth. This cell elongation seems to be largely the result of changes which pump extra water into the cells, and these changes are caused by the process of osmosis. As the cell enlarges water is pulled into it as a result of alterations in the concentration of dissolved substances within.

Despite these two exceptions, the cyclical pattern of cell division underlies most biological growth activities. Cells grow internally then divide into two. One cell turns itself into two in a simple form of reproduction. The cyclical and repeating pattern of cellular growth, cellular division, then back to cellular growth is often rhythmical as well as cyclical, since the period of the cycle is reasonably constant in length, although its actual value varies from tissue to ▷

Like a cascade, cell numbers increase when a single cell divides. Cell division involves two separate cyclical processes. One is a cycle of cell enlargement and division, the other a doubling of the genetic material to ensure perfect replicas.

Daughter cell; 1 copy of genetic information (DNA)

Daughter cell after growth; 2 DNA copies present

Nucleus divides; 2 separate DNA copies present

Cell division producing 2 identical daughter cells

As a colony of bacteria grows from a single cell the formation of new DNA and new cells occur in two distinct cycles. If traced out through time, DNA formation appears to progress stepwise, *right*, because DNA is duplicated before the cell substance divides.

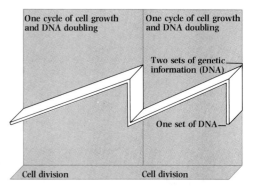

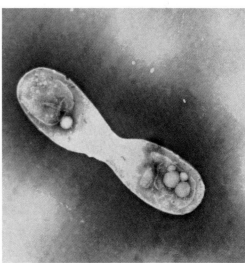

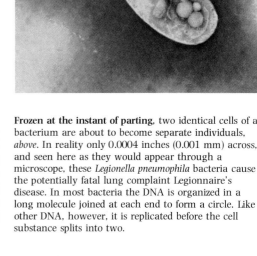

Frozen at the instant of parting, two identical cells of a bacterium are about to become separate individuals, *above*. In reality only 0.0004 inches (0.001 mm) across, and seen here as they would appear through a microscope, these *Legionella pneumophila* bacteria cause the potentially fatal lung complaint Legionnaire's disease. In most bacteria the DNA is organized in a long molecule joined at each end to form a circle. Like other DNA, however, it is replicated before the cell substance splits into two.

▷ tissue and species to species. Fundamental to the cycle is the splitting process. Within this partitioning of material there exists a cycle of change which is more important than the cell-halving process. It is a cycle without which the other processes could not occur and is the division of the material which, in every sense, forms the heart of the cell—the nucleus.

Within each cell of a living organism is a store of genetic information which is transmitted from generation to generation. This store is in the form of the chemical deoxyribosenucleic acid (DNA)—an awesome molecule housed largely in the nucleus—which contains, built into its atomic structure, codes for the construction of the key molecules of life. These molecules are: more DNA, its relative RNA (ribonucleic acid) and proteins, many of which, in the form of biochemical catalysts or enzymes, direct the life of a cell. The DNA information code, organized in the form of genes, is absolutely specific for any organism. Each and every cell in a particular organism bears exactly the same genetic code. If this were not so it would be impossible for the patterns of body shape and behaviour possessed by an organism, and essential to its survival and evolution, to be passed on to and used by future generations.

When a cell divides, the two daughter cells that are the offspring of the process must, therefore, contain an exact copy of the profoundly vital DNA. This biochemical imperative is obeyed in a complex and integrated series of changes which compose the cell cycle. In the cells of higher organisms, which contain separate nuclei within which the hereditary material is organized in long threadlike structures called chromosomes, the unvarying pattern of cyclical change obeys the following rules. The instant after it has been formed by

cell division, a daughter cell contains one complete copy of a unique genetic code sequence in the form of its cluster of chromosomes within the nucleus. Between that state and the daughter cell's next division three phases can be observed. First, there is an apparently quiescent or G_1 stage in which no DNA is produced. During this stage the enzymes needed for DNA copying are formed. Next comes the S, or synthesis, phase in which the exact new copy of the DNA sequences on the chromosomes is produced. The cell now has exactly twice the normal amount of DNA. In another apparently quiet period, the G_2 phase, synthesis of DNA stops while new proteins are made to provide the machinery needed for dividing the two sets of chromosomes by mitosis, the sort of cell division in which two identical daughter cells are created. Now the mechanics of mitosis take place and the cell splits into two, producing a new generation of daughter cells.

In single-celled, simple organisms, such as bacteria, the DNA sequence, although present, is not in the form of multiple chromosomes nor is it neatly packaged in a nucleus. In such life forms all the time between divisions is an S phase. Because DNA synthesis is going on continuously between one cell division and the next, there are no quiescent G stages. But for both higher and simple organisms the machinery of the living growth process is of paramount importance for survival. Except for a few privileged observers, who have seen the cells and chromosomes duplicate and divide by means of sophisticated optical equipment, no one has perceived the deeper processes of growth directly. Although they are carried on in such secrecy, these activities—and their observed results—are essential to understanding biological growth

A tortoise carries a date stamp indelibly marked on its hard shell, or carapace. This shell is divided into plates which give it a characteristic pattern, as shown in the Greek tortoise, *Testudo hermanni*, below. As the long-lived tortoise grows, its hard external skeleton must enlarge to provide space for its expanding body. Because the plates of the tortoise shell grow at different times of year, due to variations in external temperature and food supplies, annual banding patterns *right*, are elaborated in the growing plates.

Variations in colour and texture typify the bands of a tortoise shell, *left*. Until the rings are worn away, later in life, the animal's age can be estimated by counting the bands. For *T. hermanni* the method is reliable up to the age of about 20.

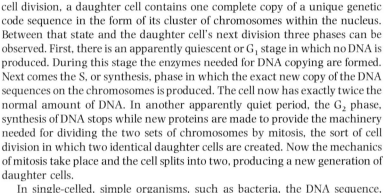

Life between the tides leaves its mark on the shells of many marine invertebrates such as the common cockle, *Cerastoderma edule*, above. The shell grows continuously during the cockle's life and has a complex pattern of bands reflecting annual growth, the tidal cycles and the cockle's inbuilt rhythms of daily life in a fluctuating environment.

The overlapping skin scales of a bony fish may show clear annual growth rings, *right*. These rings can be useful in analyzing the age composition of commercial fish stocks.

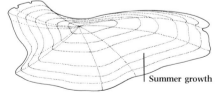

Summer growth

A fish carries its age marked in its ears as well as on its scales. The ear contains a stonelike structure, the otolith, which shows bands corresponding to a rhythmic growth pattern. In temperate species little if any growth takes place during the winter months, and this leads to variations in the rate at which new calcium-containing substance is laid down and, in turn, to the appearance of annual rings.

rhythms. If these rhythms are to be controlled, by the organism itself or by man, then strictures must be applied to the machinery of cell division from without or within.

The task of observing growth, and from these observations distinguishing rhythmical patterns of different frequency, can be difficult, simply because many organisms grow so slowly. For this reason continued experimental monitoring over extensive periods is necessary. In certain circumstances, however, animals and plants carry out their own record-taking for the human observer. These are the organisms that lay down hard skeletal materials internally or externally during part or all of their life-spans. The skeletons may be made of bones, scales, shells, wood or a number of other structures. The fact that such materials are progressively, and more or less continuously, constructed means, luckily for us, that the structures themselves may hold within their physical organization direct evidence of cyclical and rhythmical changes in the way in which they were produced.

At the level of an individual organism and its life-span, such rhythmical growth is conceptually similar to the age-long process of the deposition of sedimentary rocks at the bottom of the sea. In the case of such rocks the presence of different materials and fossils in the sequentially laid down sedimentary strata paints a frozen picture of a sequence of environmental and biological events that occured through those ages. In a similar way the shell of a marine snail might show a patterning that reveals a sequence of shell deposition and tells a clear story of changing shell growth rates tied, perhaps, to rhythms of both the years and the tides.

Among the most commonly recognized examples of such still pictures of rhythmically changing patterns of growth are the rings revealed when, in a forest in a temperate climate, a tree is cut down. Each double band of dark and then light wood, arranged in concentric circles outward from the heartwood to the bark, is a static, ossified image of a tree's annual increase in girth. Such banding patterns can be turned to aesthetic advantage in the beautiful effects of wood veneers, but in forestry, at a more practical scientific level, they are an aid to determining the age structure of a forest. A tree does not have to be felled and thus killed in order to count its annual rings, and so estimate its age. Boring from bark to core will produce a cylinder of wood containing the complete sequence of rings without sacrificing the tree.

Animal analogies of the tree-ring patterns are widespread in most types of invertebrates and in vertebrates which possess any sort of skeletal structures produced over long periods. Most often the practical use to which they are put is that of reasonably accurate ageing. Such knowledge can be of immense commercial significance. Around the world the stocks of many fish are, for instance, in danger of being overfished to the point of near extinction—as has happened in the recent past with the fishing of herring in the North Sea between the United Kingdom, Scandinavia and northern Europe. Only by understanding, in considerable detail, the age structure of a commercially harvested fish population is it possible to determine in advance what fishing strategy should be adopted to ensure a maximal, sustainable yield. Growth rings on fish scales provide sufficiently accurate information about their age to enable this sort of analysis to be undertaken. ▷

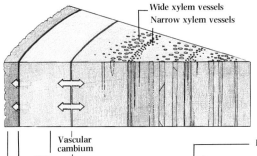

Wide xylem vessels
Narrow xylem vessels

Vascular cambium

Bark
Phloem
Cork cambium

Heartwood

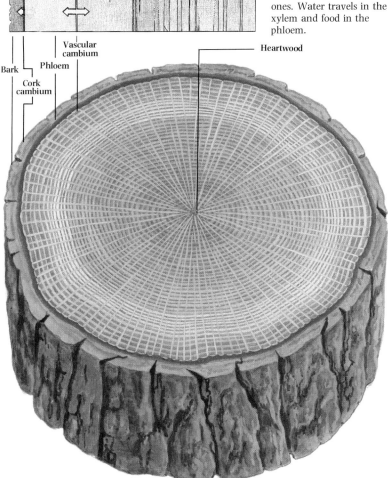

A complete annual ring of a temperate tree contains two kinds of water-carrying tubes, or xylem vessels. The light wood contains many wide tubes, the dark wood fewer, narrower ones. Water travels in the xylem and food in the phloem.

Bristlecone pine
Pinus aristata

The beautiful, complex cross-grain pattern of the wood of a tree trunk, *left*, reflects its cellular composition. Most of the wood consists of the tree's xylem vessels. At the centre of the trunk, in the heartwood, these vessels may be dead or obliterated by the deposition of substances within them. Xylem vessels are made by the division of cells in the vascular cambium, a cylindrical cell sheet around the trunk. The outer surface of this active layer splits off phloem vessels, which transport sugars and all other nutrients down from the leaves to every part of the tree. Outside this is another active layer, the cork cambium, whose cells divide to form the protective bark.

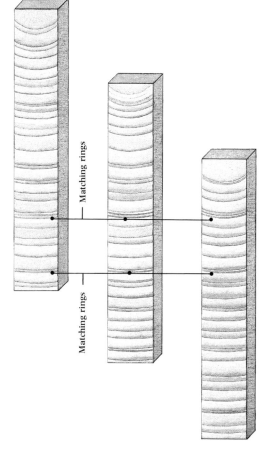

Analyzing growth rings on both dead and living trees can provide much information about past climatic conditions. The rings reflect climate because, for example, more large xylem vessels are made when water is most abundant. In a discrete area, specific conditions will produce identifiable and consistent variations in the rings. Long-lived trees, such as the California bristlecone pine, *left*, have survived for more than 4,600 years, making it possible to produce an overlapping chronology of ring patterns, *below*, extending far back into time.

Matching rings

Matching rings

▷ The work of biologists in studying growth is closely linked with that of organic chemists. The chemicals around whose properties life itself was constructed are the organic or carbon-based molecules, and they come in a mind-boggling range of forms. Furthermore, the molecules are capable of indefinite expansion in all dimensions, so that their final diversity is effectively infinite. Organic molecules occur in the atmospheres and structures of the outer planets of the solar system, and by the use of spectrophotometry have even been identified in interstellar space. To date, however, planet Earth is the only location in the universe in which this plethora of carbon-centred molecules is known to be associated with life. Among this infinite chemical variation a single molecule stands supreme on earth as the most common—its name is cellulose and it is the structural building material of plant cell walls and, therefore, of wood and paper. There is little doubt that this most ordinary of substances occurs in quantities greater than those of any other organic substance, with the result that there are billions and billions of tons of it worldwide.

Cellulose, and ultimately all the other organic building blocks found in earthly ecosystems, are manufactured by plants in the chain of chemical reactions that compose the process of photosynthesis. Animals are essentially parasitic on the plant kingdom when it comes to obtaining their supply of complex organic molecules because, unlike plants, they cannot construct themselves and replace any deficient parts from a collection of non-organic starting reagents. The chemical magic of a plant is sun-powered alchemy. Radiant energy from the sun is, in the first analysis, the power source that drives the energy-requiring machinery of photosynthesis by which carbon dioxide is 'grabbed' from the atmosphere and fixed. In this fixation process the simple carbon dioxide gas, CO_2, made up of one atom of carbon and two of oxygen, is converted into the dramatically more complicated molecules of life, such as sugars. This increase in complexity can only be be paid for with an energy currency of which the sun is the final banker and it provides an extra dividend because oxygen gas is released into the atmosphere as a by-product of photosynthesis.

Not all parts of a plant can carry out photosynthesis. Only those cells that encounter significant intensities of sunlight will be in a position to carry out this process. Because of this fundamental restriction, roots, many woody stems and the central woody tissues of trees cannot do so, which means that their cells depend on the photosynthetic parts of the plant to provide them with the organic molecules they need for growth. All higher plants have a transport system of long, tubular cells or vessels, which together comprise the phloem tissue, by means of which this sort of nutrient translocation is achieved.

The cells that can carry out photosynthesis have an unmistakable brand mark. They are green. This colour, the indelible signature of the plant world, is formed by the pigment chlorophyll, a molecule that has changed the Earth from a dreary planet on which only yeastlike fermentation could take place in the absense of oxygen to the oxygen-rich globe we know today. It is the first in a highly ordered chain of molecules within green plant cells that effectively convert a photon, a package of light energy, into an energetic electron, and then trap some of the energy of several such electrons to drive vital chemical

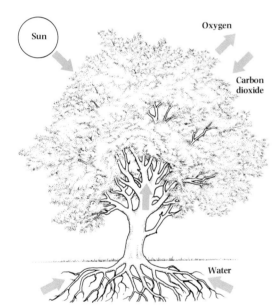

Green plants are the living links between the earth and the sun. The reason for this is that plants carry out the process of photosynthesis in which light energy from the sun is converted into chemical bonds in organic molecules. The chemical complexities of photosynthesis can be reduced to a single equation of change: under the influence of a light-generated energy source, carbon dioxide and a hydrogen-containing molecule combine to make organic molecules, such as sugars, which contain carbon, hydrogen and oxygen, plus water and the remnants of the hydrogen-donating molecule. For normal green plants the hydrogen donor is water (H_2O), so when the hydrogen is removed, oxygen is given off. As well as light, another essential mediator of photosynthesis is the green pigment chlorophyll, which gives most leaves their colour. The exceptions to these rules are some photosynthetic bacteria, which use hydrogen sulphide (H_2S) as their hydrogen source, and produce the element sulphur as a result of their sun-driven chemistry. Much is known about the electronic and chemical changes of photosynthesis. Not all the reactions need light, but the initial light reaction involves the interaction of a quantum of radiant energy from the sun with chlorophyll.

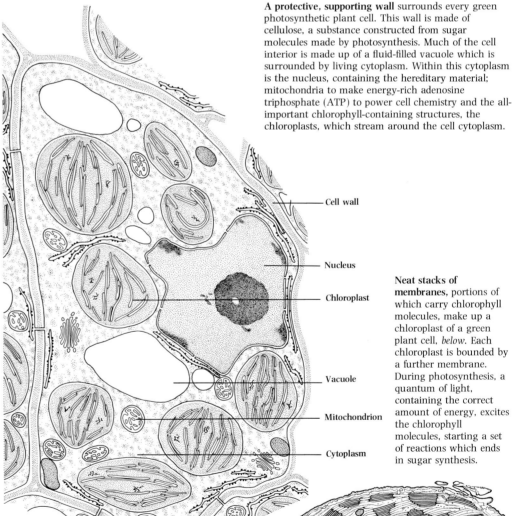

A protective, supporting wall surrounds every green photosynthetic plant cell. This wall is made of cellulose, a substance constructed from sugar molecules made by photosynthesis. Much of the cell interior is made up of a fluid-filled vacuole which is surrounded by living cytoplasm. Within this cytoplasm is the nucleus, containing the hereditary material; mitochondria to make energy-rich adenosine triphosphate (ATP) to power cell chemistry and the all-important chlorophyll-containing structures, the chloroplasts, which stream around the cell cytoplasm.

Neat stacks of membranes, portions of which carry chlorophyll molecules, make up a chloroplast of a green plant cell, *below.* Each chloroplast is bounded by a further membrane. During photosynthesis, a quantum of light, containing the correct amount of energy, excites the chlorophyll molecules, starting a set of reactions which ends in sugar synthesis.

reactions, such as the fixation of atmospheric carbon dioxide gas and the manufacture of cellulose.

Green plant cells, such as those with which leaves are packed, reveal a Chinese puzzle of suntraps within suntraps. Leaves themselves are orientated so that they can intercept the sun's rays most efficiently. Within the leaves green chlorophyll-containing cells are dispersed so as to provide multiple opportunities for green cells to intercept light. Within each cell the chlorophyll and other chemical hardware of photosynthesis are packaged in stacks, and several stacks make up one chloroplast. Monomolecular layers of chlorophyll, bound to membranes in the chloroplasts, are the ultimate targets for the photons of light which left our personal star, travelling at 186,000 miles (297,600 km) per second, eight minutes or so before.

In two vital respects the mechanisms and implications of photosynthesis are cyclical in organization. First, the chemical complexities of carbon fixation itself involve a highly conservative self-replenishing cycle of molecular usage. Some of the key substances in the reaction are not consumed in the process of photosynthetic chemistry but are cyclically reconverted one into another. Second, jumping from the chemical to the more majestic, global scale, photosynthesis is at the heart of a planet-embracing cycle of carbon utilization with the apt name of the carbon cycle.

The carbon cycle of planet Earth involves two states of atomic carbon—organic compounds containing carbon and inorganic, carbon-containing molecules including carbon dioxide. In our world individual atoms of carbon shuttle between these two states. Carbon dioxide is fixed by the processes of photosynthesis and changed into organic compounds. These compounds are ultimately broken down during the process of respiration—either in the plant itself or in an animal that has eaten that plant—which turns the carbon in the respired molecules back into carbon dioxide again. Some forms of carbon, including deep carbonate rock strata and fossil fuels still undisturbed in the ground, are essentially unavailable to life processes in the short term. At another level non-organic carbon stocks in terrestrial and aquatic systems are partially separated from each other. In the air, carbon dioxide is found as a gas, while in the sea, a river or a lake it is present as dissolved carbon dioxide, carbonates or bicarbonates.

In total the absolute quantities of carbon being cycled by the life of our world are staggering. So, too, is the proportion of the available carbon cycled every year. The green plants of the world fix, very approximately, three and a half million billion ounces (a hundred million billion grams) of carbon every year, and pass about a third of it back into the atmosphere and the world's waters as carbon dioxide produced by plant respiration. The remaining two-thirds are eventually returned to the carbon dioxide pool, either as gas or in its dissolved state, as a result of the respiration of bacteria, fungi and animals, including ourselves, who either eat plant food or dine on herbivores who get their energy from plant-fixed organic material. Ignoring the seas and fresh water, it is likely that terrestrial communities or plants, animals and micro-organisms also process as much as 12 per cent of the total carbon dioxide in the atmosphere each year. Thus, on average, every atmospheric molecule of carbon dioxide must take part in the cyclical molecular dance of life every eight years.

All the key molecules from which earthly organisms are constructed are built around atoms of the element carbon. Linked in huge linear, branching or circular configurations, these carbon atoms provide a backbone for the attachment of other atoms. The central role of carbon means that the dynamic pathways, by which its atoms cycle between the living and inanimate components of our world, are vitally important. The most essential non-living forms of carbon are carbon dioxide gas in the atmosphere and dissolved carbon dioxide in the oceans. Together they form a carbon dioxide pool from which living things obtain carbon. These accessible reserves of carbon dioxide (CO_2) are trapped by terrestrial and aquatic plants and changed into sugar molecules by means of photosynthesis. The carbon dioxide is replaced as the plants themselves respire. In addition, animals eat plants and animals eat animals, and each action passes carbon down the chain. Respiration of herbivores and carnivores produces carbon dioxide, as does the decay of dead organisms or their carbon-containing excreta by micro-organisms.

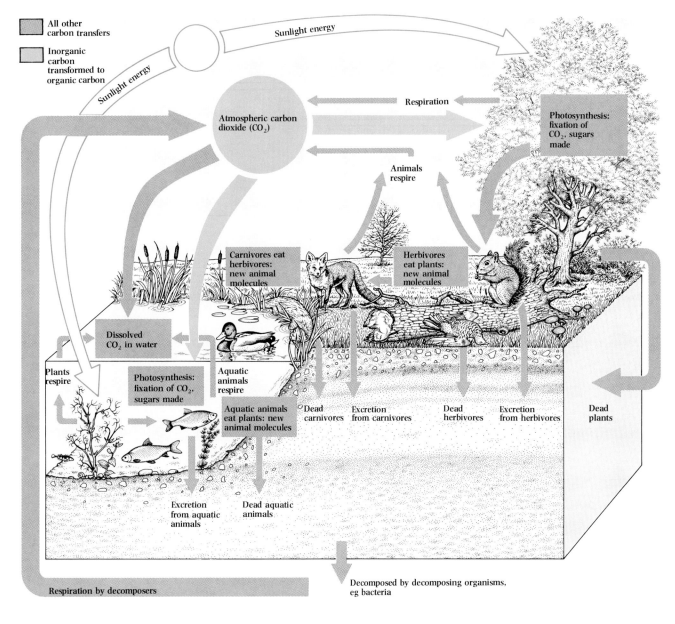

Rhythms of energy

All the activities that make life possible, and which together form the rhythms and cycles of the living world, require energy. The activity of animals is often dramatically obvious, like the sinuously muscular and rhythmic dash of a hunting cheetah in pursuit of an impala, or the laboured beating of the wings of an osprey as it rises from a lake with a trout twitching in its talons. Among plants, activities are usually less overt, none the less the activity patterns are there, from the rapid drooping of the leaves of a sensitive mimosa plant as they are touched by the tongue of a browsing goat, to the gradual blind extension of root tips through the soil in their quest for water or mineral salts.

Inside their bodies, although in a more subtle and covert fashion, animals and plants pulse with change. Food substances are absorbed; cells divide, grow and change their function; secretions are made, transported and even pushed out of the organism's body; nerves transmit electrical signals. At a microscopic level, and in a thousand ways a second, the internal organization of all living creatures teems with change.

All these varied patterns of activity are joined together by many different links, not least of which are the links forged by the connection of all life on planet Earth through a shared evolutionary past. At a fundamental level all the various life processes, from animal locomotion to molecular change inside cells, organs and tissues, need energy from some external source to make them happen. They involve work. They consist of events and phenomena which simply cannot occur without the application of energy.

If an animal is deprived of food for long enough, it will, in the end, not be able to move because it no longer contains any nutrients which can be burned with oxygen in the process of respiration to release the energy necessary for movement. Similarly, if a green plant is kept in the dark for long enough all its activities will grind to a halt because, for a plant, the only significant source of energy is the radiant energy of the sun. A plant can take up a collection of simple chemicals—carbon dioxide, water and mineral salts—and, with the sun's help, turn them into the myriad molecules of life by the process of photosynthesis. But animals must have their nutrients already made into complex organic molecules, such as proteins, sugars and fats, because they cannot build them from the inorganic molecules of the non-living world.

Activity, then, is multiple in form but always displays a basic link with energy supplied from outside. The fact that patterns of activity are so often cyclical or rhythmic in organization is the result of a series of constraints which operate on living things and the ecosystems in which they live, and which favour some processes rather than others. It is difficult to conceive of efficient locomotory systems, for instance, which do not have the cyclical features demonstrated by all running, swimming and flying.

The rhythmicity of locomotion arises because non-cyclical performances are intrinsically 'once only' affairs. A solid fuel rocket burns its fuel and produces movement. When the fuel is gone the movement stops until the rocket is refuelled. Animals and plants cannot function like this. Instead they power their movement by using small, repetitive event patterns that can be sustained indefinitely. Similarly, the constraints of the conservation of matter mean that many of the biochemical processes of life occur in a cyclical fashion. The biochemical cycles provide useful new molecules or energy, but they also generate some or many of their starting materials so that the system is both cyclical and potentially inexhaustible.

More than four billion years ago, when planet Earth was being formed, there was almost certainly no free oxygen gas in our atmosphere. When the first simple life forms began to develop on the earth's surface, probably between four and three billion years ago, the atmosphere was still devoid of this gas that is so crucial today. The absence of oxygen meant that all the first life forms had to make use of energy-obtaining strategies that had no need to 'burn' foods with oxygen. This meant that the cyclical movement of elements from the atmosphere to living things and back again was very different from the one that now exists.

The coming of green plants, and with them the process of photosynthesis, back in the Precambrian mists of prehistory more than 600 million years ago, induced a gigantic lurch in the organic evolution of life on earth. The first single-celled blue-green algae transformed the world. Because these organisms possessed the green pigment chlorophyll, they were able to take in the carbon dioxide already in the atmosphere and, using sunlight as an energy source and water as a donor of hydrogen atoms (water consists of oxygen and hydrogen), could make oxygen gas.

The oxygen gas that changed the world, which oxidized the minerals on the earth's surface and so potently altered the atmosphere and life that came to breathe it, was simply a waste product of plant activity. To begin with this oxygen was of little use to the plants themselves. The blue-green algae used their chlorophyll and the process of photosynthesis to produce organic molecules, such as sugar. However, the polluting waste from this piece of chemical conjuring turned out, inadvertently, to be the generator of a

biological breakthrough. It opened the door to a whole new cycle of molecular change and made possible a completely new way of producing energy for the activities of life.

If you take a tablet of pure glucose and heat it in the air at extremely high temperatures of several hundred degrees, it will eventually burn producing heat, carbon dioxide, water and nothing else. The matter in glucose is completely conserved by being burned with the oxygen in the air, but its atoms are changed from those of the glucose configuration—a ring of carbon atoms each bearing oxygen and hydrogen atoms—to a new configuration represented by the separate molecules of carbon dioxide and water. The heat energy produced by the process comes from the fracture of the chemical bonds that hold the glucose molecule together. But glucose does not give up its bond-energy easily. The great heating that was initially necessary to make the intrinsically stable glucose molecules burn was the energy input necessary to get the oxygen-adding—the oxidation or burning—process going.

The bodies of all animals, and those of plants and micro-organisms, all contain glucose, but it is inconceivable that they could obtain useful energy from this glucose by the burning process described. Such burning would be impossible for two reasons. First, animal, plant and microbial bodies are destroyed by the high temperatures required to make glucose burn spontaneously in oxygen. Second, the sort of energy provided by the uncontrolled burning of glucose, is in the wrong form.

All the forms of life which make use of oxygen—the aerobic organisms—employ a subtle manoeuvre to surmount these two problems. Instead of

Within each green leaf of any plant are many chlorophyll-rich cells which use solar energy for photosynthesis. In these cells carbon dioxide from the air and water drawn up by the roots are combined to produce sugar and oxygen. In the under part of the leaf, a spongy air-filled layer forms a passage for water vapour, which escapes through pores in the leaf underside as the plant transpires.

The opening and closing of each pore is controlled by water pressure in the surrounding 'guard cells'. During the day when the cells produce sugars more water is drawn into the guard cells making them swell and open. At night sugar is changed into starch, pressure falls and the pore closes.

Levels of light and transpiration follow the same rhythm, reaching a peak at noon. The leaf pores usually open to give off water and receive carbon dioxide, only when the sun provides energy for photosynthesis. Cacti, however, invert this rhythm to conserve water in desert conditions. At night the pores open to take in carbon dioxide. By day, pores closed, they photosynthesize using stored carbon dioxide.

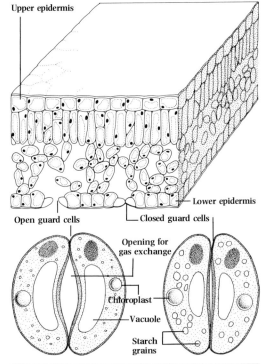

Upper epidermis

Lower epidermis

Open guard cells — Closed guard cells

Opening for gas exchange

Chloroplast

Vacuole

Starch grains

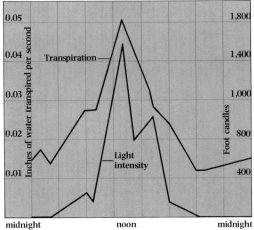

midnight — noon — midnight

burning glucose directly to form carbon dioxide and water, the process takes place as a long sequence of tiny chemical steps, each controlled by specific enzymes. These enzymes, or biological catalysts, work in such a way that glucose can be broken down at physiologically harmless temperatures between 5 and 40 degrees centigrade (41 and 104 degrees F). The enzyme-regulated burning of glucose, which only ever takes place inside the cells of organisms, is thus known as internal respiration. In the same terms, external respiration comprises activities, such as the rhythmical movements of human breathing, which get oxygen into the body of an animal.

As well as taking place at non-lethal temperatures, internal respiration provides a usable form of energy for living things. It is an energy source of an unexpected kind—a complex organic molecule of adenosine triphosphate (ATP) containing three linked chemical groupings, all of which have phosphorus in them. ATP is the bank that issues the energy currency to cells, and it can drive almost any cell or body activity. During the process of internal respiration large amounts of ATP are made, and while it is being 'burned' with oxygen and broken down, one molecule of glucose is responsible for making 38 molecules of ATP.

Without the oxygen waste of the primeval blue-green algae, the energy-grabbing strategy of modern organisms and their countless predecessors would have been impossible. The concentration of oxygen in the atmosphere has slowly climbed since these early times until it now represents about 20 per cent of the gaseous envelope around our planet: a stupendous quantity of gas, all of it generated by living things. Once animals and plants had become biochemi-cally adapted to the immense respiratory possibilities inherent in an oxygen-rich atmosphere, the next steps were those that had the effect of refining the process by which the atmospheric gases are transferred to the cells that need them. Plant cells need carbon dioxide and oxygen, while most animal cells need only oxygen.

Green plants, especially the larger, more complex, higher plants, such as ferns, conifers and the world's wealth of flowering plants, from herbaceous weeds to trees, have evolved a wide range of structures and cyclical processes designed to bring carbon dioxide and oxygen from the atmosphere to their internal cells for the essential processes of photosynthesis and internal respiration. Such plants also show rhythms in the way their water content is controlled. This control is critical because plants tend to lose water via pores, or stomata, in their leaves while they are in the process of trapping carbon dioxide for photosynthesis.

A wide range of animals, both aquatic and terrestrial, have developed respiratory structures to gather atmospheric oxygen for internal respiration. These structures work on the principle that the larger the surface area through which oxygen can diffuse, the more oxygen can be obtained. Gills are the surfaces favoured by water dwelling animals while land livers such as man use lungs. Animals of all kinds also show an elaborate and impressive variety of methods of ventilating these surfaces with oxygen-rich water or air during the process of external respiration.

The ventilating mechanisms used by animals to get oxygen to the tissues that need it for internal respiration are almost always cyclical in nature. ▷

Insects 'breathe' through valve-operated spiracles— tiny holes in the cuticle which open regularly to draw in oxygen from the air and to release carbon dioxide, and close to prevent water loss. An intricate system of air tubes which can sub-divide into minute branches brings oxygen directly to the internal tissues by diffusion. The air flow is controlled by spiracles opening out of phase.

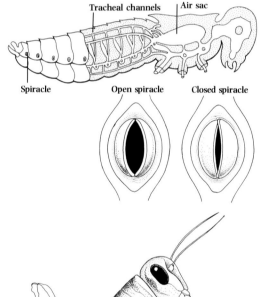

Tracheal channels | Air sac

Spiracle   Open spiracle   Closed spiracle

Many active insects increase the air flow by rhythmically pumping their abdomens. The locust, for example, contracts and relaxes its body in an up and down movement, which alternately squeezes and expands its air sacs, so replacing almost half the air in its body.

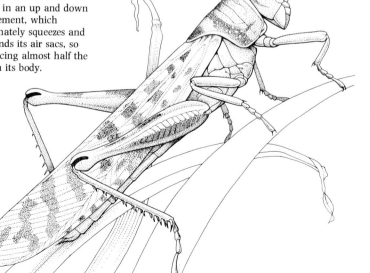

The honey-bee ventilates its body by pumping its air-filled abdomen in and out, accordion-style, in synchrony with the spiracle movements. Inspiration lasts 250 milliseconds during which the thoracic spiracles open for 25 milliseconds. These spiracles then close for one second before a one-second expiration phase when the abdominal spiracles open for 300 milliseconds.

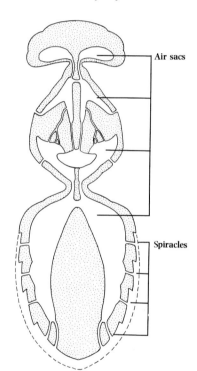

Air sacs

Spiracles

To avoid excessive water loss in dry, hot climates insects keep their spiracles shut for long periods, only opening them for brief bursts to allow carbon dioxide to escape. Before and after each burst, the spiracle valves flutter. The frequency of this cyclic respiration is largely controlled by ambient temperature and the insect's metabolic rate.

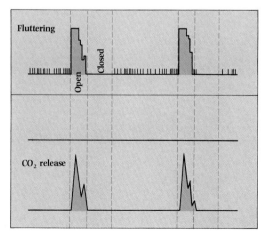

Fluttering

Open   Closed

CO_2 release

▷ Superficially there seems to be little intrinsic reason for using such methods of discontinuous pumping because it is perfectly possible to construct models of continuous-flow systems which would be just as efficient in providing good ventilation for lungs or gills. In fact a few 'greyhounds' of the fish world, bony fish such as mackerel and tunny, do create a continuous flow of water over their gills. These fish keep their mouths slightly open while they are swimming fast. Their swimming speed is so great that it maintains a steady water flow through the mouth and out over the gills, where oxygen is absorbed in exchange for carbon dioxide, which is the waste product of internal respiration. The water exits via the two opercula or flaps, on either side of the body, which cover the gill chambers in a bony fish. Mackerel need so much oxygen that they must swim continuously if the required quantities of water are to pass over their gills. In some species of shark it has been noticed that a mackerel-like continuous-flow strategy for external respiration operates when these predators are swimming, but active cyclical pumping of the water has to begin once swimming stops.

Since this continuous-flow tactic is workable, why is it that almost all other vertebrates, both aquatic and terrestrial, use discontinuous cyclical pumping, like that employed in a pump for inflating bicycle tyres or in any other man-made pumping machine? The answer probably lies in the nature of the equipment that animals have to work with. The basic material at an animal's disposal, for producing movement of either water or air, is muscle. Body muscles are intrinsically cyclical devices. They contract and, while they are contracting, can do work—this is phase one of the cycle. For anything more to

happen the muscle must then relax and elongate back to its resting length; this elongation makes up the second phase of the cycle. This two-phase cycle can be operated indefinitely, but its essentially cyclical nature means that any system of ventilation powered by animal muscles is likely to be cyclical and probably rhythmic. Given these limitations to the design, the range of successful products that has come off the evolutionary drawing-board is soberingly successful. Any commercial research and development team would be justifiably proud of the functional innovation displayed in the product range of respiratory equipment.

The smallest of all the creatures in the animal kingdom do not have to bother about active external respiration. Aquatic animals with a maximum body-length of up to about a quarter of an inch (6 mm) can obtain enough oxygen by diffusion, even at the centres of their tiny bodies. This process does not need an energy source because it is powered by the random motions of all molecules at temperatures above absolute zero. The process works in this way: a minute animal living in water containing dissolved oxygen has internal oxygen levels in its watery tissues that are almost equal to those of the surrounding water. If the creature now uses some internal oxygen in the process of burning glucose to make ATP, a new situation is produced in which the concentration of oxygen outside its body is higher than the concentration inside it. The random physical motions of oxygen molecules in the water automatically act to remedy this state of imbalance and oxygen flows into the animal.

The problem with diffusion, as far as animals are concerned, is that it is only

The muscles around the mouth and mouth cavity of bony fish operate to produce a muscular breathing pump that sends a continuous stream of water over the respiratory surface of the gills. The anterior and posterior cavities move slightly out of phase with each other, which ensures a continuous flow.

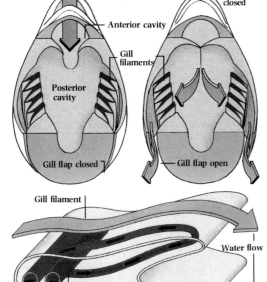

Fish absorb as much as 80 per cent of the oxygen that reaches their gills. This is partly because blood to be oxygenated in the gill runs in the opposite direction to the water flow, allowing the maximum amount of oxygen to pass from the water into the blood.

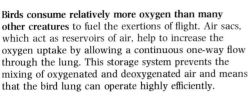

Birds consume relatively more oxygen than many other creatures to fuel the exertions of flight. Air sacs, which act as reservoirs of air, help to increase the oxygen uptake by allowing a continuous one-way flow through the lung. This storage system prevents the mixing of oxygenated and deoxygenated air and means that the bird lung can operate highly efficiently.

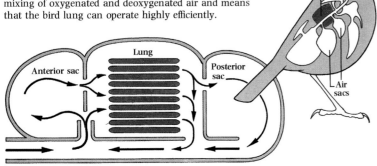

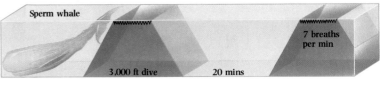

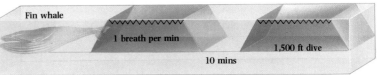

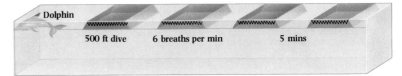

effective over short distances, hence the upper size-limit on the animals that can depend on it as a mechanism for obtaining this metabolically vital gas. Unless they can play a physical trick and cut down the length of the transport path to the tissues so that diffusion can operate efficiently, all animals must supplement the diffusion mechanism by large-scale, usually muscle-powered, mass movements of air or water.

Fish are respiring aquatic animals, that is, they get their oxygen from water. This poses problems, not least because water is about 1,000 times more dense than air so much more muscle power has to be used to shift it about than is needed for moving air. In addition, any given volume of air will contain more oxygen, in absolute terms, than the same volume of water at the same temperature, so the respiratory dice are heavily stacked against water breathers. Fish are, in fact, highly efficient oxygen extractors. By means of a one-way flow system over the gills, powered by multistage muscle pumps in the mouth and gullet (pharynx), elegant gill design and subtle counter-current arrangements for transferring oxygen from the water to the blood within their gills, fish manage to remove oxygen from water more efficiently than man does from air. Some fish can extract 80 per cent of the oxygen in water while man removes only about 25 per cent of oxygen from the air he breathes.

For respiratory surfaces, the parts of animals that actually absorb oxygen (and at the same time get rid of carbon dioxide), design awards go to those which can pack the greatest area into the smallest volume. Absorption is a surface phenomenon—the more surface that can be crammed into a gill, the better it will work. The large surface areas inside the air-breathing lungs of

amphibia, reptiles and mammals are ventilated by tidal flows of air. Air is breathed in by muscle power into a highly subdivided respiratory sac, the lung, then the lung is partially emptied of air during expiration, like the tide rushing out. This in-out cycle is one reason why land dwelling vertebrates cannot attain the oxygen extraction results achieved by fish. The one-way flow arrangement of a fish's gills is intrinsically more efficient than our 'breathing bags' because the gills are constantly in oxygen-rich water, whereas the inner surfaces of our lungs are only in oxygen-rich air during one phase of the breathing cycle. As we breathe out the alveoli, the ultimate microscopic subdivisions of our lungs which ensure a big surface area for oxygen absorption, are exposed to a gas relatively low in oxygen.

For a cockroach, a pike, a sparrow or even an elephant, the ventilating rhythms of external respiration can only take oxygen as far as the respiratory surface. Oxygen travels across this surface by the process of diffusion and so is inside the animal, but it may still be a long way away from the tissues that urgently require oxygen for internal respiration and energy generation. Breathing rhythms can, for example, take air into a giraffe's lungs, and diffusion will pass oxygen across the giraffe's lung membranes, but neither ventilation nor diffusion can carry that oxygen the many feet that separate the giraffe's lungs from its brain.

All large many-celled animals must possess a transport system that can move oxygen—and at the same time nutrients, hormones and a thousand and one other substances—around the body. The problem is analogous to that of a central heating system. Having a boiler that burns fuel to produce usable heat is ▷

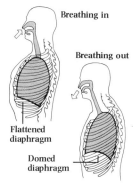

Breathing in

Breathing out

Flattened diaphragm

Domed diaphragm

Man breathes to live in more ways than one, as demonstrated by Dizzy Gillespie, *right*. As air is sucked into the lungs, the rib cage is expanded, partly by a flattening of the diaphragm. Breathing out is a passive process. The rib cage sinks, air is squeezed out of the lungs and the diaphragm relaxes.

Whales and dolphins exhale a spectacular spout of compressed air through a blowhole when they surface for oxygen after a long dive. Their ability to store 40 per cent of their oxygen intake in their muscles, and to exchange 90 per cent of air in their lungs with each breath, allows these mammals to take less frequent though deeper breaths than man. Each species has its own breathing rhythm; the sperm whale breathes a mere 7 times per minute after a deep dive, *left*.

▷not, in itself, enough to warm a house. The boiler must be connected to a circulatory system of pipes to transfer heat from the boiler to every radiator in every room. The blood vessels of vertebrates, and the body spaces filled with the fluid haemolymph that exist in many invertebrates, represent the biological equivalent of the ducting of the heating system. These vessels take oxygen from the gills or lungs and transfer it around the body. The ultimate branches of this tubing system, the capillaries, ramify so extensively through the tissues that there are few cells in the body of a human being, or any other vertebrate, more than a fortieth of an inch (1 mm) away from these capillaries and the blood they contain.

Carrying the central heating analogy one stage further, blood is the working fluid of the body, equivalent to the water in the pipes and radiators. In fish, amphibians, reptiles, birds and mammals the blood in the circulatory system can carry oxygen in two distinct ways. In the simpler of these two forms of oxygen transport, the oxygen is dissolved in the watery background fluid or serum of the blood. Like seawater, rain, or the water in a lake or river, blood can dissolve a certain amount of gaseous oxygen. It is this oxygen, along with other gases, that emerges as bubbles when water is heated.

Merely dissolving oxygen in water is not a particularly efficient method of acquiring oxygen for active vertebrates that have high oxygen requirements. For this reason most vertebrates have specialized cells in their blood containing pigment molecules which combine with oxygen in the most avid fashion. The most common of these pigments is haemoglobin, a complex organic molecule consisting of a protein, globulin, linked to a ring-shaped cage of atoms enclosing a single atom of iron, and it is the red colour of haemoglobin that gives blood its characteristic hue. In the oxygen transport system, haemoglobin is packed inside disc-shaped red blood cells.

The pigment haemoglobin is particularly useful because it has an amazing ability to combine with oxygen to make a new molecular arrangement, oxyhaemoglobin. The formation of oxyhaemoglobin happens most readily when haemoglobin is present in situations of high oxygen concentration, as it is in the lungs of a mammal or the gills of a fish. At the respiratory surfaces of gills and lungs, blood in vessels close to those surfaces picks up oxygen and immediately makes it into oxyhaemoglobin. In the tissues, where oxygen levels are lower, the oxygen contained in the oxyhaemoglobin becomes detached and can move into tissue cells to be used for energy production.

All blood circulatory systems, like all central heating systems, must have a pumping mechanism to maintain the flow around the system's closed network of ducts. In the blood system this pump is the heart and it is in its activity that some of the most clear-cut rhythmic aspects of bodily physiology are manifest. What can be more single-mindedly rhythmic and cyclical than a pulsating heart? The healthy human heart, from its elaboration in the early weeks of foetal life until its eventual failure at the end of a lifespan, is the ultimately efficient machine. It will contract ceaselessly, without pauses for maintenance or repair, for 70 years or more, which adds up to a quarter of a billion pulsations in a lifetime.

All hearts are built essentially to the same design. A portion of the blood vessel tubing is thickened by extra muscles in the walls. The space inside this

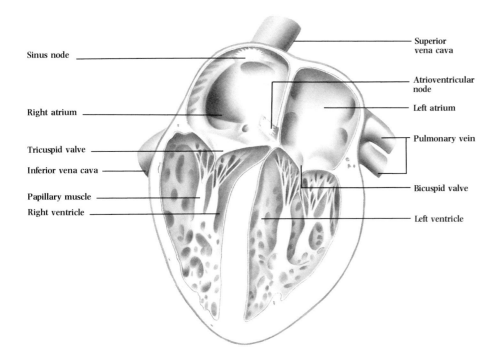

Sinus node

Right atrium

Tricuspid valve

Inferior vena cava

Papillary muscle

Right ventricle

Superior vena cava

Atrioventricular node

Left atrium

Pulmonary vein

Bicuspid valve

Left ventricle

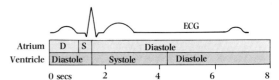

ECG

Atrium	D	S	Diastole		
Ventricle	Diastole		Systole	Diastole	

0 secs   2   4   6   8

The heart's natural pacemaker, the sinus node is a small piece of muscle which initiates electrically induced rhythmic contractions, spreading first to the atria and a split second later to the ventricles. An ECG trace, *above*, measures the electrical charges triggering the contraction (systole) and relaxation (diastole), which produce the double heartbeat sounds as the valves close in rapid succession.

Heart rates of animals	Beats per minute at rest
Grey Whale	9
Elephant	25
Salmon	47
Cockroach	60
Housefly	60
Man	70
Seal	80
Sparrow	500
Shrew	600
Hummingbird	1200

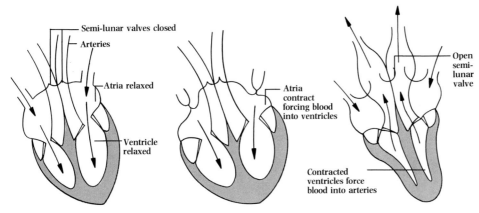

Semi-lunar valves closed

Arteries

Atria relaxed

Ventricle relaxed

Atria contract forcing blood into ventricles

Contracted ventricles force blood into arteries

Open semi-lunar valve

The continuous pumping of the heart ensures a constant oxygen supply to all parts of the body. Arteries carry oxygen-rich blood away from the heart, while veins bring blood back from the organs to the heart to be re-routed through the lungs. A separate circulatory loop feeds blood through the lungs for oxygenation before it returns to the rest of the body via the heart. Both sides of the heart beat in unison to ensure sufficient pressure is maintained. As the atria relax, blood flows from the lungs into the left atrium and from the rest of the body into the right atrium. Meanwhile, with a forceful contraction, the ventricles pump blood out into the arteries, then relax and refill with blood from the atria. A normal human heart beats 70 times a minute at rest, but increases to 200 after a release of adrenalin or heavy exercise.

muscular section of the ducting is functionally separated from the rest of the circulatory system by valves, flaps of tissue that allow flow in the appropriate direction, but restrict any backflow or make it impossible. As the muscular wall contracts, blood pooled within this region is forced out into the circulation via the open exit valve. Then the valve shuts, the wall relaxes and blood is drawn into the heart region through the entry valve, bringing the system back to its original condition, primed for another contraction or beat. During this two-phase sequence blood can only flow in one direction because of the presence of the valves—without them the system would produce no net circulation.

However, even in fish, the most primitive of the vertebrates that we commonly come across, the simplest of pumping mechanisms—the single-chambered heart—is not powerful enough to push the blood round the long circulatory path on its own. In a fish, blood from the heart travels in a single unbroken circuit from the heart, through the gills for oxygenation, and from there to all the other parts of the body before returning to the heart. On its journey the flow of blood is assisted by muscles in the walls of the vessels containing it and by other body muscles alongside them which, by contracting, help to squeeze the blood along. The fish heart has also a low pressure area, or atrium, that receives blood from the tissues, and a high pressure ventricle connected to the atrium that pushes blood round the body, via the gills.

In a mammal such as man a more sophisticated four-chamber pump is the rule. Here the lungs and the rest of the body each have a two-chamber pump of their own. The right side of the heart circulates blood through the lungs, so that it can pick up oxygen and dump its load of carbon dioxide waste, then passes this replenished blood to the left side of the heart from where it is delivered to the rest of the body. The blood coming back from the lungs rich in oxyhaemoglobin is bright cherry red; deoxygenated carbon dioxide-loaded blood returning from the body tissues is a dark purple-red colour. So blue blood is not the prerogative of kings nor of aristocrats—each of us is half blue blooded and half red blooded.

From the left side of the heart oxygenated blood travels outward through narrower and narrower arteries until it reaches the capillaries. Now the oxyhaemoglobin gives up its precious oxygen to tissue cells which use it to burn nutrients, such as glucose, in the controlled combustion of internal respiration which is governed by a series of biological catalysts. In fact, all cells which carry out this life-preserving process do so in specialized areas of their inner substance. These minute organelles, the mitochondria, each four hundred thousandths of an inch (0.001 mm) across, contain internal shelf-like partitions, the cristae, on which are packed in highly precise array the enzymes and other molecules needed, with the vital help of oxygen, to produce ATP. Like microscopic power houses, hundreds or even thousands of them to a cell, the mitochondria burn fragments of glucose to carbon dioxide and water in a self-replenishing chain of reactions named, after its elucidator, the Krebs cycle, which, at the level of molecular interactions, is a cyclical phenomenon.

The carbon dioxide released in the process of ATP formation now diffuses back into the blood in the capillaries. To complete the cycle, this deoxygenated blood flows back to the lungs to replenish its oxygen supply. This oxygen, the gas that respiratory and circulatory systems are designed to deal with, is the intangible, unseen agent that not only allows the mitochondria of an aerobic ▷

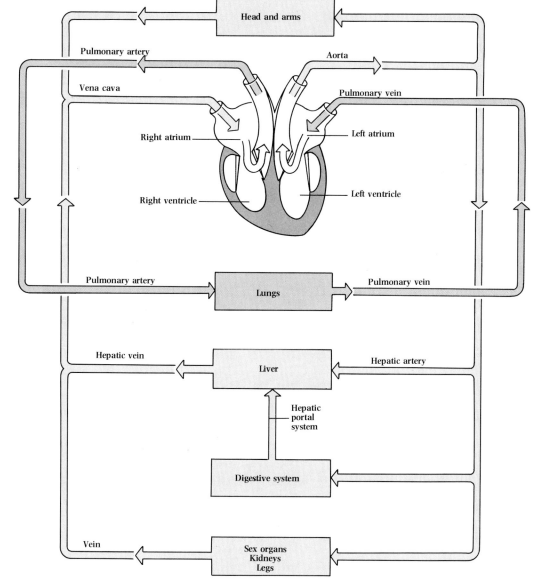

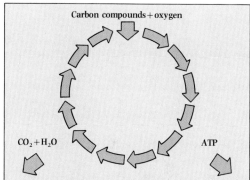

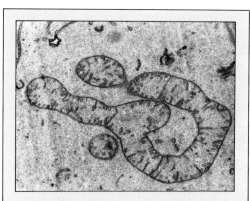

Energy is 'manufactured' within the body cells by minute membranous organelles called mitochondria, *above*. The carbon-containing breakdown products of glucose from other parts of the cell are passed into the mitochondria where a complex cycle of chemical processes, called the Krebs cycle, occurs. Air from the lungs oxidizes the carbon in these substances, producing carbon dioxide. At the same time water and molecules of ATP (adenosine triphosphate) are synthesized. The ATP molecules are an energy store that can be used to drive the energy-requiring processes of the body.

▷cell to produce ATP and so supply its energy needs but also enables a bonfire to burn and a car—or internal combustion—engine to run. But this is only half the story of energy generation. In animals the other half is ultimately food.

Why do we need to eat? The buying and growing of food, its preparation, the ritual of mealtimes, the inclusion of food imagery as a central part of our language are all such unexceptional and ordinary parts of our lives that we rarely stop to question the activity upon which they all centre. At the level of immediate motivation we eat because we feel hungry. Such an explanation can, however, be only the first step in unravelling the biological necessity for food consumption and merely leads us to another question. Why is it that we, and presumably other higher animals, have motivations or 'drives'—such as the internal trigger of hunger that pushes us into finding something to eat once the feeling of satiation created by our last meal has worn off—so firmly built into our patterns of behaviour?

Such patterns of internal behaviour control are necessary because we 'eat to live'—it is only the glutton who is chastised for 'living to eat'. Unlike plants, which can make all the sustenance they need, every animal, from the smallest microscopic creature to the mighty whale, must have food. The reason for this is, that in contrast to their green counterparts of the plant kingdom, even the most complex animals are absolutely incapable of constructing the complex molecules of life from simple inorganic ones.

The only means that an animal has of obtaining energy in the usable form of ATP is by the process of internal respiration in which sugars such as glucose, but also fats, are broken down. The metabolically incompetent animal must,

therefore, obtain its complex organic molecules ready-made. It may be difficult to think of lobster thermidor, ratatouille or a crusty loaf of bread with butter as fulfilling this chemically intimidating role, but that, apart from the pleasure of consumption, is all that any meal is for. The food we eat, thus, has two distinct but overlapping functions. First, it provides us with the organic building blocks originally constructed by plants from inorganic molecules with which we, as animals, can make new cells to replace those that have worn out, or can supplement existing structures to maintain the vast collection of cells we already possess. Second, food acts as fuel for the energy-releasing business of internal respiration to power all the energy-requiring processes and activities which our bodies carry out every minute of our lives.

The two functions of supplying raw materials for building and energy generation overlap because some of the substances that are good for burning as fuel also make excellent building bricks for constructing living cells. To take just two examples: the sugars that are the main starting point for energy production within cells are also essential components in the organization of DNA and RNA, the materials of the genetic code, which all life forms possess. Similarly, fatty acids, which form part of the fats, such as butter and margarine, in our diet can be burned as an energy source in place of sugars. Fats are also of vital importance structurally since they are employed as important raw materials in the construction of the membranes that surround all living cells.

A few of the foods that animals eat are already 'pre-processed' and arrive in the form of the small organic subunits that animals can utilize easily. Honey, for instance, which is made from plant nectars and pollens, contains both simple

Eskimos (the Inuet) are bound to a regular seasonal hunting pattern by the extreme variations in temperature and daylight. In Greenland, seal hunting for meat, oil and fuel begins in November when the sea forms a frozen sheet and seals can be caught coming up for air at their blowholes in the ice. In the summer many Eskimos decamp inland to hunt migratory birds and caribou, to fish river trout and gather autumn berries such as cranberries before returning to their coastal snow houses to prepare for the next sealing season.

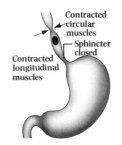
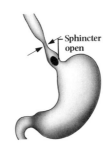

Contracted circular muscles
Sphincter closed
Contracted longitudinal muscles
Sphincter open

Eating involves a number of rhythmic processes which break food down into a digestible form. After each mouthful has been chewed into a manageable ball or bolus it passes into the oesophagus where a wave of muscular contractions pushes it towards the stomach. Circular muscles contract above the bolus forcing it downward and longitudinal muscles shorten the passage ahead, allowing the food to reach the stomach in 10 seconds. Contractions continue inside the stomach every 20 seconds while the food churns round.

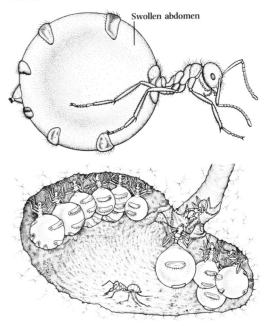

Swollen abdomen

The honeypot ants, *Myrmecocystus*, from the Colorado desert feed almost exclusively upon the sugary juices secreted by galls left on desert scrub oaks by gall wasps. Galls are only available for a few months of the year, so the ants have devised a storage system. They feed up several sterile female workers until they swell to eight times their normal size and leave them hanging from the ceiling of their burrow like a larder of honey sacs. In the dry season the females regurgitate food for the rest of the colony.

sugars and amino acids, the latter being the building blocks from which proteins can be constructed. These small organic substances can be directly absorbed across the gut wall of an animal by a mixture of diffusion—the same process by which oxygen travels into the bodies of minute organisms—and an energy-using system of active transport by which the nutrients are picked up by molecular ferries and given a ride from the space, or lumen, inside the intestinal tube into the bloodstream for transport to where they are needed.

Most of the food we eat is not, however, in such a readily available form. Rather it consists of complex mixtures of enormous molecules of fats, starches, proteins and nucleic acids such as DNA. These substances consist of huge molecules with molecular weights of more than 100,000 or even 1,000,000 compared with glucose with a molecular weight of only 180. These giant molecules cannot easily be absorbed in the state in which they are consumed. So first they have to be broken down into their constituent subunits—starches into simple sugars, fats into fatty acids and glycerol, proteins into amino acids and nucleic acids into building blocks named nucleotides.

It is the process of digestion which fragments these large molecules into their constituent parts. The mechanics of digestion begin when food is chopped up into small pieces by beaks, teeth and tongues; a task supplemented by the churning stomachs of mammals or grinding gizzards, filled with grit, in birds. The result of all these activities is a finely dispersed slurry of particles of the giant food molecules which is moved down the intestine by peristalsis, a kind of rhythmic pushing effected by muscular contraction. In its liquid form food is acted upon by digestive enzymes produced by the wall of the stomach and the

intestine and by the neighbouring pancreas, which passes its secretions to the intestine through a tube. These enzymes do the job of chemical splitting so that the large molecules are converted into smaller ones which can be absorbed across the gut wall.

Just as the provision of oxygen for an organism is built around a number of rhythmic and cyclical processes, so feeding and food handling are often modulated in such a patterned way. Most of these patterns are imposed by, or firmly tied to, external environmental rhythms. So rats, for instance, even those kept in a laboratory with food and water instantly available for 24 hours a day, have a circadian rhythm of nocturnal feeding. During the hours of daylight the rats sleep, feed little, if at all, and produce almost no faeces. At dark they become active, start feeding and release faeces throughout the night until, near dawn, their activity subsides again. This complex pattern of cyclical change, of which feeding is an essential part, has nothing to do with the laboratory rats' food supply, but instead is a manifestation of their natural strategy of life which includes nighttime eating.

The circadian feeding rhythm of these laboratory rats is inextricably tied to the 24-hour cycle of day and night. The same rhythm can be used as a model for any animals that are decidedly nocturnal or diurnal—that is, daytime active—in their habits. Such patterning of general activity will almost always entail a parallel variation in feeding behaviour with nocturnal animals feeding, as one would expect, at night and diurnal ones during the day.

When displayed by predatory animals, such cycles usually imply that there is some restriction on available prey, or that the predator has become ▷

The short-tailed vole, *Microtus agrestis,* displays two different feeding rhythms. Within its longer 24-hour feeding rhythm, which reaches a peak just after sunset, the vole also feeds at 2–4 hour intervals as its stomach is so small.

Many game birds, including the ruffed grouse, *Bonasa umbellus,* of North America, select their diet seasonally according to both availability and preference. They enjoy the abundance of fruits and seeds in the summer months and substitute buds and twigs for them during the winter. The increase in insect consumption during the summer, however, reflects a physiological need, common to vegetarian birds, for protein before the breeding season. Males eat more during courtship; females double their food intake when egg-laying.

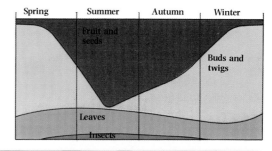

Spring	Summer	Autumn	Winter

Fruit and seeds

Buds and twigs

Leaves

Insects

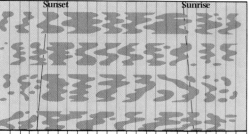

23 Oct
1 Nov
2 Nov
11 Nov
15 Nov
26 Nov
29 Nov
4 Dec

Sunset   Sunrise

1pm 3pm 5pm 7pm 9pm 11pm 1am 3am 5am 7am 9am 11am

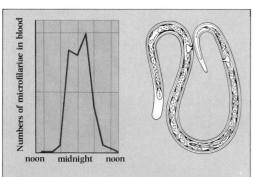

Numbers of microfilariae in blood

noon   midnight   noon

Three rhythms are involved in spreading the disease elephantiasis. Mosquitoes such as *Culex pipiens fatigans* pass larvae of the infecting worm *Wucheria bancrofti* into man's blood when they bite at night. During the day, the larval worms retreat into the victim's lungs, but return to the blood every night, ready to be sucked up by a mosquito, which will infect another man. If the host changes his pattern of sleep and waking, the worms adapt within a week.

▷ specialized to hunt only one or a few types of prey animals. So, for instance, a pied flycatcher, a bird that never flies at night, hunts solely during the hours of daylight for flying insects, and these it seeks and catches with the help of its keen sense of sight. In contrast a horseshoe bat, a totally nocturnal airborne mammal, catches night-flying insects, such as moths, on the wing. The bat is not hampered by the fact that there is no light and that it cannot use its eyes effectively because it uses a high precision system of ultrasonic radar to guide its movements. The bat emits pulses of high frequency sound through the strange horseshoe-shaped folds in its nose.

Other marked feeding patterns of animals have a seasonal or a yearly periodicity. Such rhythms may be tied, for instance, to the annual breeding activity of an animal, a period of migration or the seasonal availability of food. Animals such as birds, which collect food for their offspring confined to the nest, will often gather a quite different selection of food items for their young than they would to sustain themselves. Since this period, during which the young are being fed, normally occurs at a specific time of year, such changes of foraging or predatory behaviour add up, over a period of 12 months, to a reasonably regular annual cycle of food utilization.

Migrating animals commonly need to build up large reserves of fat before embarking on lengthy and arduous migrations during which the opportunities for feeding are likely to be few and far between, and also less successful than they are during the rest of the year. This requirement means that there is an intensive burst of feeding in the period immediately before the migratory journey. Fats are foods favoured by migrating animals, such as birds and butterflies, because, weight for weight, a food-store of fat burned with oxygen will give considerably more energy-providing ATP than the equivalent weight of any other nutrient, such as carbohydrates or proteins. For a flying migrant, payload considerations mean that the weight-to-energy ratio of the food (that is, fuel) it carries is of prime importance.

Naturally enough, feeding behaviour is only one facet of the many activity patterns exhibited by animals which demonstrate a regular rhythmic periodicity or a clearly cyclical organization in their lives. It is probably true to say that whatever energy-requiring activity one consideres in the animal world, whether it is related to large locomotory movements, internal patterns of physiological activity, such as digestion or the rate of the heartbeat, or the performance of mental tasks, it is always possible to demonstrate that in some animals this behaviour exhibits rhythmical properties.

This generalization about animal activity can be shown to be valid for creatures as different in their size, habits and complexity as minute marine plankton and man himself. Marine plankton, made up of microscopic adult organisms and eggs and juvenile, larval stages of larger marine creatures, might easily be regarded as animals which could only be of interest to specialist zoologists with peculiar tastes. In fact the population sizes, distribution and migrations of plankton are of hard economic importance to every nation with a fishing fleet.

Zooplankton, the small planktonic animals in the sea, are absolutely crucial links in the complex web of food chains that exist in the seas of the world. Zooplankton feed on phytoplankton, microscopic free-floating marine plants

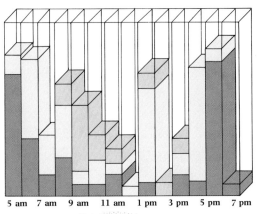

Other activities
Feeding
Preening
Singing

Birds time their daily activity according to their diet, size and predators. The activity of blackbirds, *left*, grows increasingly during the morning. Their feeding rhythm peaks at 5 pm. At dawn and dusk males sing to advertise and defend their territories.

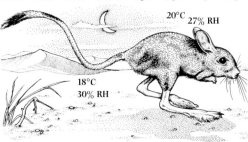

Many desert rodents, including the jerboa, *left*, spend the day in burrows to avoid losing precious moisture in the intense heat. At night, when the fierce drying sun sets and cooler conditions return, the jerboa ventures out, returning to its humid burrow at dawn.

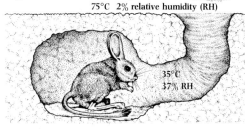

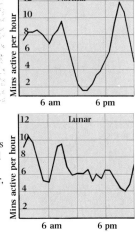

The night-active, red-backed salamander, *Plethodon cinereus, right*, normally follows a distinct pattern of movement. After a burst of activity at 10 am, it spends most of the day sleeping under stones and wood, until it re-emerges in the evening, and reaches a peak of activity at 9 pm. On nights with a full moon, however, this rhythm is subdued and the salamander remains quiet and still, shielding itself from the eyes of nocturnal predators.

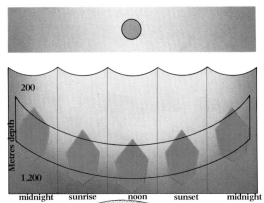

Many light-sensitive plankton species migrate into the ocean depths during the day and move back to the surface at night. Activated by light, temperature and salinity, the plankton swim to varying depths, determined by age, sex and water conditions. Many plankton-feeding fish follow their prey, so adopting the same rhythm.

which comprise the other half of the planktonic equation. Because they are green, the members of the phytoplankton are the primary producers, via the process of photosynthesis, of new organic material in the oceans. The fish that we eat feed on zooplankton or on smaller fish which have themselves grown fat on zooplankton. Much observation has been carried out by fisheries biologists of the daily activity patterns of these tiny, yet vital animals. Many of the zooplankton have been shown to carry out staggeringly extensive vertical migrations in the sea, moving downward during the day then returning to the surface each night in a regular circadian rhythm.

Although they may have the most powerful and versatile brains of any animal on earth, human beings are not excluded from the rhythmical imperatives of life. Much work has been done to examine the way that humans perform certain tasks and solve problems in an attempt to discover circadian, 24-hour rhythms in these abilities. The motivation for carrying out much of this research has been the possibility that work efficiency could be greatly improved if proper account were taken of intrinsic, through-the-day variations in the ability of all men and women—not just a selected few—to carry out specific types of task.

Most studies of this sort have concentrated on the normal, 'wakeful' or daytime phase of the sleep-wake cycle. A wide variety of scorable, or measurable, tasks have been devised by ingenious investigators and have been carried out by willing 'guinea pig' humans. For a remarkably wide range of problem-solving tasks, requiring either small or considerable muscle activity, a similar but distinct diurnal pattern of variation becomes obvious. And this

pattern holds true whether the subjects are tested on their ability to deal and sort playing cards, to draw on a piece of paper the image they see in a mirror or to perform multiplication sums. Within an hour or two of waking in the morning efficiency climbs relatively rapidly, reaching a peak sometime in the middle of the day, although these peaks are differently timed for different tasks. Thereafter, and often more gradually, efficiency declines again as nighttime approaches.

In a general way this pattern of change in performance corresponds to internal physiological rhythms of the human body, such as that of body temperature. When examined in detail, however, the two types of rhythm are rather different because, while efficiency rises rapidly and declines slowly throughout the day for moderately active adults, body temperature rises slowly throughout the day and in two phases. The first of these phases is fast and the second more gradual, but after the end of this second phase the decline in body temperature is rather rapid, so there is no simple and direct correlation between body temperature and levels of dexterity and skill.

The studies carried out on the ability of the human brain to solve problems, such as multiplication sums, were structured around a cycle of human existence that is normally thought of in terms of two phases—daytime wakefulness and nighttime sleep. The results concentrated on circadian patterns, that is, those with a periodicity of about 24 hours. In fact there is some evidence that underlying these relatively long-term cycles is a rhythm of change in human mental activity which has a far shorter period than 24 hours and which is present during both wakefulness and sleep. ▷

Many animals have become nocturnal in response to negative pressure imposed by daylight. Frogs need humidity and are night-active to avoid dehydration. Mice come out at night to escape the many predators hunting by sight. Owls, however, have filled a niche with little competition from other predatory birds.

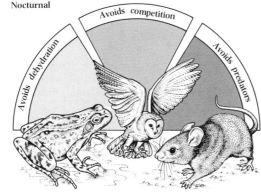

Nocturnal

Diurnal animals are often specifically adapted to take advantage of the daylight hours. Chimpanzees rely on their good eyesight to live their opportunist and inquisitive lives to the full. Hawks use their sharp diurnal vision to identify prey from great heights. Many snakes and other cold-blooded animals need daytime warmth to hunt efficiently.

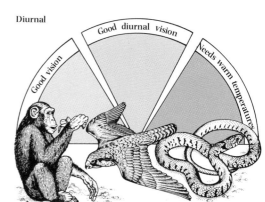

Diurnal

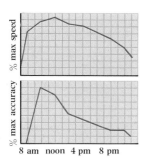

8 am noon 4 pm 8 pm

Human efficiency levels, which fluctuate during the day, often follow the body temperature curve with one peak around midday and a low in the early hours of the morning. Mental performance is generally highest in the morning— multiplication is often most accurate at 10.30 am and most speedy at noon, *above*. Recalling digits also reaches a peak at noon, many important commercial decisions, *right*, may be influenced by the time of day.

▷ The first findings to suggest that these relatively rapid cyclical changes occur in the human body arose from important studies carried out in the 1950s. These studies showed that the sleep phase of the sleep-wakefulness cycle is not a blank, constant, unconscious nothingness. Indeed, in a prescientific way, the human dream experience showed long ago that this could not possibly be the case. Studies on human volunteers hooked up to electroencephalographs (machines which measure the electrical activity of the brain and produce traced recordings, or electroencephalograms (EEGs), of mental activity) as well as the results from a wide variety of other monitoring hardware, showed that sleep is divided into its own cyclical phases which alternate throughout the night in a way that is approximately rhythmic.

Sleep was, thus, found to have two phases which were remarkably different from one another in both behavioural and physiological ways. The first sort of sleep is rapid-eye-movement (REM) sleep because, as its name implies, during this phase rapid and transient eye movements occur beneath our closed lids. The second sort of sleep is non-rapid-eye-movement (NREM) sleep, and for its duration there are none of these paradoxical eye movements. So, rather surprisingly, a normal adult undergoes a complete cycle of REM then NREM sleep every 80 to 120 minutes, which means that, during an ordinary night's sleep of, say, eight or nine hours, we pass through at least four of these cycles.

Although the presence or absence of eye movements are the most obvious manifestations of the two phases of sleep, they are not the only differences between REM and NREM sleep. The pattern of electrical activity of the heart, as measured on an electrocardiogram (ECG), and the heart's rate of beating, the breathing rate and the tone, or tautness, of the muscles all vary consistently between the two types of sleep. So it seems that something extremely basic alters within the body when we switch from one state of sleep to the other.

An intriguing and daring hypothesis has been put forward to account for the profound difference between these two states. This hypothesis suggests that the REM/NREM rhythm involves an alteration in the relative extent to which each of our two cerebral hemispheres controls our mental activities. The uppermost, newest, intelligent part of our brain, the cerebrum, is divided into distinct right and left halves, known from their shape as the two cerebral hemispheres, and between the two are some complicated criss-crossing nerve connections. There is little doubt that in most people the two halves of the cerebrum specialize in different mental tasks. In right-handed people (for handedness is part of this complex story of how the brain works) the right hemisphere is the expert on non-verbal, visual, spatial and artistic aspects of our perception of the world and has little to do with performing processes of logical analysis. The left half of the brain of a right-hander is the hemisphere concerned with written and spoken speech and features of our lives that are analytical and logical, such as mathematical ability.

The idea behind the new hypothesis of the phases of sleep is that during our sleeping hours the two halves of the brain take it in turns to run our mental processes. While the right hemisphere is in charge we pass through a phase of REM sleep, but when the nighttime autopilot sits on the left-hand side of the cerebral cockpit, we go through a period of NREM sleep. The eye movements themselves are an important clue to the veracity of this theory. The eyes are,

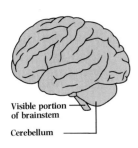

Visible portion of brainstem

Cerebellum

Evidence suggests that sleep may be induced by hormones which pass messages to an area of the brainstem, known as the reticular formation, which acts as the brain's switchboard and controls levels of arousal.

Rapid-eye-movement (REM) sleep, associated with one sort of dreaming, is typified by volleys of spiky brain waves. Breathing and pulse grow irregular, movement increases and body temperature rises.

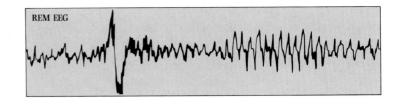

REM EEG

Human sleep is organized into four stages which can each be identified by the extent of electrical activity in the brain, *right*. One complete cycle lasts about 90 minutes and recurs 5 or 6 times each night, *far right*. Stage 1, the light sleep produces small undulations in the brain waves. As the brain reaches increasingly deep stages of sleep the EEG waves become larger and more random until the spindles disappear completely in Stage 4. The quality of sleep changes during the night, becoming lighter and more full of dreams towards morning, as body temperature rises and hormones thought to induce wakefulness become active. Poor sleepers tend to experience deep sleep spasmodically.

Awake

Stage 1

Stage 2

Stage 3

Stage 4

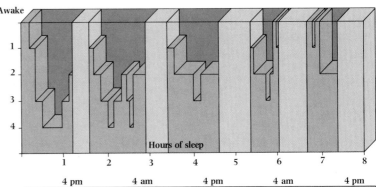

Awake

Hours of sleep

The length of a night's sleep and its internal rhythm varies with age. The most dramatic change is in the proportion of Stage 4 sleep which declines from 18 per cent of the night's sleep as a child, to 3 per cent in middle age, *right*. Stage 4 sleep is thought to fulfil a number of functions including the stimulation of the release of the growth hormone. Active, growing children thus have a far greater need for deep sleep. Tests have proved that children continually deprived of sleep do not grow fast and may become stunted.

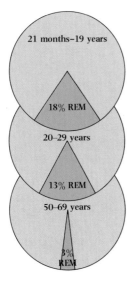

21 months–19 years

18% REM

20–29 years

13% REM

50–69 years

3% REM

Experiments on volunteers in deep caves show that our regular 24-hour sleep-waking cycle is largely coordinated by light cues. Without these external signals, the natural inclination is to allow the sleep pattern to lag slightly each day until it has drifted all the way round the clock. Most people have an innate cycle of about 25 hours which they entrain to the earth's 24-hour rhythm. Body temperature has an independent rhythm and during a period of free-running falls to a low at the onset not the end of sleep.

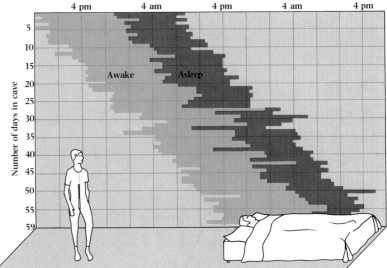

4 pm   4 am   4 pm   4 am   4 pm

Awake   Asleep

Number of days in cave

obviously, most active in the phase of sleep in which visual aspects are favoured, that is, right-sided REM sleep.

For the relationship between brain hemispheres and phases of sleep to be further substantiated, data was needed from people willing to act as subjects in sleep experiments which involved both monitoring and many disturbed nights. When these people were purposely woken up from continuous sleep in either the REM or the NREM phase (as judged by the evidence of EEG tracings), they reported different types of dream. On waking from REM sleep, volunteers described the types of dream associated with specific and clear visual images in which reality appears to be distorted—I dreamed that I was having dinner on top of the Statue of Liberty, for example.

In contrast, subjects surfacing from NREM sleep talked of a less easily remembered form of dream that does not readily fit into the common conception of what dreams are made of. These NREM dreams are less bizarre, less vivid, more thoughtful and rational than those of REM sleep. A dream in which a person is worried about passing an exam could be an example of a NREM dream. In other words, quite apart from the Freudian or other interpretations that might be placed on the content of our dreaming, it displays just the sort of dualism one would expect if our sleeping lives were controlled alternately by the left and right sides of our brains. All vertebrates are like us in having brains split into symmetrical halves about a line drawn from the nose to the top of the spine, and many mammals other than man have been shown to possess REM/NREM cycles of sleep. But what the functional significance of these cycles might be in, say, a cat or a mole is at present unclear.

One intriguing extension of the idea of the REM/NREM cycle has recently occurred to behavioural scientists, namely, that it is possible that the appropriate 100-minute mental rhythm is not confined to the sleeping phases of our lives. There appears to be some evidence that alternating dominance of the right and left sides of our brains also takes place during waking hours as well. Minutely detailed recording of human physiology during the day suggests that there are 90 to 110-minute rhythms in such human attributes as heart rate, the ability to maintain vigilance for a specific but unlikely event, and the fantasy content of our day dreams.

Following up these suggestions, Canadian workers have asked volunteers to carry out two types of task during their periods of wakefulness. The first of these tasks was highly orientated towards visual, non-verbal problem solving, while the second was specifically and closely linked to a written linguistic problem. Interestingly, these workers found significant 90 to 100-minute oscillations in the efficiency with which these two sorts of task are performed. Even more interesting is that while performance in one type of task improved, the other diminished. So it seems that there is a lot of sense in having a break from work when you feel that your performance is flagging, or leaving a problem aside and coming back to it an hour or two later; it could be that by then your brain is back in the correct phase to find the solution rapidly. Thus we are rhythmic in more ways than we know, for it is becoming increasingly clear that the two halves of our brains may rhythmically pass an important part of overall mental control to one another, like a tennis ball being hit back and forth across the net, throughout the hours of both wakefulness and sleep.

Awake

NREM

REM

The world's greatest sleeper appears to be the opossum which sleeps for 19 hours a day, of which 5.7 hours is spent in REM sleep. Turtles experience no REM and it comprises only 1 per cent of sleeping time in birds. Mammals such as moles cats, chimpanzees and man spend about 25 per cent of their sleep in REM while sheep show a mere 3 per cent. Humans, however, sleep for fewer hours than most animals.

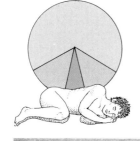 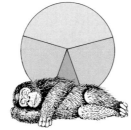 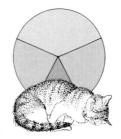  

Electroencephalograms (EEGs), *right*, monitor electrical frequencies in different areas of the brain. They are used to trace some of the internal rhythms of sleep. By identifying individual stages it has been possible to discern the effects of deprivation and thus the function of each phase of sleep. Severe lack of REM sleep can provoke emotional disturbance and even hallucinations, which suggests that it helps to maintain a psychological balance. Sleep loss, however, is normally made up the next night.

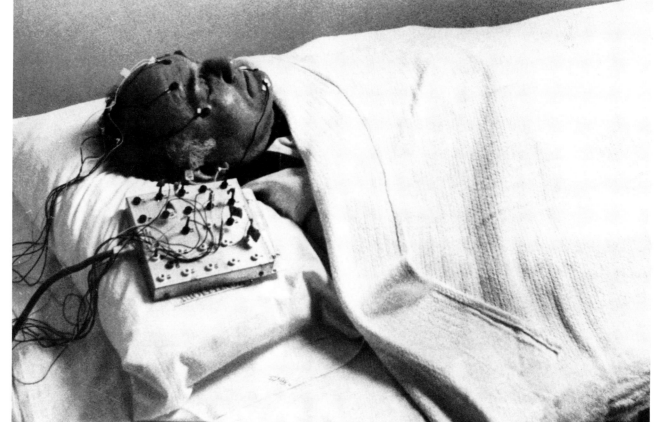

Rhythms of motion

Movement is an animal imperative. In sharp contrast to plants which have a settled existence, nearly all the animals on earth have the power to propel themselves through air, water or even the soil to collect food or find a mate. Because locomotion is a basic animal attribute it is not surprising that it exists in a huge range of patterns. Equally, since animals are dwarfed by the space in which they live, efficient locomotion demands the repeated performance of small cyclical acts, each of which moves the animal only a tiny distance. It is the repeated sameness of these acts that gives locomotion its essentially rhythmical character.

Of all the rhythms of life, those of animal locomotion are among the most obvious and overt. Movement rhythms occur quite fast—sometimes hundreds of times faster than the human eye can follow—and are usually both aesthetically and intrinsically pleasing. What can be more relaxing to the gaze than the unhurried wing-beat of the majestic swan or the regular jumping of a school of porpoises as it leads a ship to harbour? Some of the rhythms of locomotion are better understood than others and they can be explained in the language of physics, but they represent some of the most beautiful examples of functional adaptation in animals.

Over millions of years of evolution the rhythm of locomotion has been moulded according to the template provided by the physical development and needs of each species. The powers of evolution have also made locomotory rhythms extremely economical. Each separate propulsive cycle is as efficient as it can possibly be, and often one cycle leads into the next with some overlap so that the inertial force of the moving body is used to its fullest extent. In maximizing efficiency, counterbalances play a big part. Thus legs and wings, and often the tail or even the neck of an animal move in a particular way to help keep the creature balanced, or to pull it forward so that its centre of gravity falls within the base formed by its limbs. Some seemingly awkward rhythms, for example the swaying gait of the camel, are highly efficient in energy usage and admirably suited to the animal's body structure and its environment.

While some animals use the same pattern of locomotion irrespective of rapidity, others change their gait according to the speed at which they travel. Thus for man, running is little different from walking: the left foot still follows the right and the arms still swing to provide some balance, but a horse has a symmetry of movement when it is walking that is lost once it starts to gallop. Compared with many of his mammalian relatives man has rather an unspecialized means of locomotion and only puts up a mediocre performance. Yet man can run or walk tens of miles simply by getting into his stride and settling down to a steady rhythm. Top-class runners say that rhythm is all-important to success on the track and that any disruption, for instance by a stitch or cramp, can be disastrous.

Man may be able to run, but he certainly cannot fly. His flying machines, although they rely on the rhythmical and mechanical action of engines, do not mimic the flight patterns of birds and are, as a result, far less efficient. Similarly his boats are poor performers compared with their animal equivalents. In translating the rhythm in an engine to a rotating rhythm of an air or water screw, much energy is lost and the shapes of man's machines produce far too much drag. The more we understand about the design of machines, the more we marvel at nature which solved all the problems so elegantly many millions of years ago.

From the languid flapping of a stork to the angry buzz of a hornet, the rhythmic beating of wings gives animals the ability to fly through the air. Man has long marvelled at flight and, over the centuries, has made countless attempts to copy the birds he sees in the sky above him, but always without success. Strapping feathers to his arms, the birdman tries to fly by frantically beating his wings, but to no avail. Only after man discovered how to build successful aircraft did he begin to comprehend the mysteries of bird flight, and to understand how the rhythmic movement of wings allows a bird to climb heavenward with such apparent ease.

For any object—be it an aircraft, a bird or a paper dart—to travel through the air the force of gravity, which tends to pull the object to earth, must be counteracted by an opposite force of at least equal strength. This is lift, and the heavier the object the greater is the lift required to keep it airborne. But to hang motionless in the air is of little practical use to either animal or man and, furthermore, lift cannot normally be generated by a stationary organism or machine. So all fliers need the force of propulsion to drive them forward. The faster they wish to fly, the stronger that propulsive thrust must be. Like all fluids, air has a certain thickness and exerts a resistance on anything passing through it.

Of course air is much thinner than water or, say, molasses, but it is thick enough to stick to a wing and to try to stop it moving forward, so the propulsive force must be at least as great as this hampering force or drag. The three forces of lift, propulsion and drag underlie the physics of flight and every flying object, whether natural or man-made, must obey their rules.

Flying animals differ from man-made machines in that a single structure—the wing—provides them with both the lift and the propulsive force they need to become and to stay airborne. A bird's wing consists essentially of two parts, the inner part from 'elbow' to 'wrist' and the outer section from 'wrist' to 'fingertip'. It is the inner part of the wing that serves as an aerofoil to provide lift. In cross-section it looks like an aircraft wing with a blunt leading edge, a curved upper surface, concave lower surface and a thin trailing edge. As the wing is pushed along, the air striking the leading edge is deflected in two ways. Part of it shoots upward above the wing and part slips underneath it. The air that is deflected upward travels high above the wing and creates a slight suction above the wing's upper surface. To release the suction, the wing tends to be moved upward and this is what creates lift.

The mechanics of flight are complicated by the fact that air does not flow freely over the wing—instead it swirls around in small eddies. The efficiency of flight depends on the angle of attack between the wing and the direction in which the bird is flying. If this angle is too large the eddies become bigger and bigger. As long as the bird is flying fast enough the eddies are dissipated, but if it flies slowly the same eddies destroy the suction and make the lift force disappear in a sudden stall. To overcome this stalling effect at landing and take-off, birds push out a tiny tuft of feathers from the leading edge of the wing which helps to dissipate the eddies.

All the time it is flapping its wings in a repeated cycle of activities, a bird must have lift to counter the force of gravity. It is the outer part of the wing that provides propulsion and it does so in a rhythmical fashion. The two key factors in

Great black-backed gull

Stock dove

Hummingbird

Frequency of wing-beats is inversely proportional to the size of bird. The tiniest hummingbird performs a miraculous 100 beats per second, while a large gull languorously flaps twice per second. Crows and pigeons average 6 to 12 cycles per second.

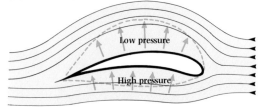

Air travels farther over than under the curved leading edge of the wing. The resulting pocket of low pressure sucks the wing upward, lifting the bird.

Low pressure

High pressure

How hummingbirds beat their wings 100 times a second remains a mystery, as each cycle needs two muscle contractions, each lasting 10 milliseconds. These tiny birds hover motionless as the figure of eight pattern of their wings produces lift on both the up and down strokes.

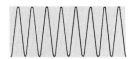

Doves are strong, versatile fliers, adapted to cope with a variety of environmental conditions. In normal flight they flap 12 to 14 times a second, but at high speeds their wing-beats become shallower and less frequent. Racing pigeons can sustain speeds of 100 miles (160 km) per hour.

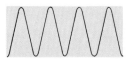

Slow, but powerful, crows cover long distances without rest, flapping their broad wings at a rate of 6 to 8 beats a second. To prevent stalling at low speeds, the crow spreads the tips of the primary feathers, allowing high pressure air to flow through the wing and smooth out turbulence above it.

The gull's flight muscles contract slowly, limiting the wings to two graceful flaps a second. This allows the gull to cruise nonchalantly over the ocean at about 34 miles (55 km) per hour. A huge wing area of over 155 in² (1,000 cm²) enables the gull to glide between periods of flapping flight.

propulsion are the mobile wrist joint, which allows the outermost or primary feathers of the wing to be turned outward so that they will beat almost backward, and the asymmetrical shape of the primary feathers themselves. As the wing is driven through a figure of eight, the wrist flexes back and forth, and in doing so provides maximum contact between the primaries and the air on the down beat, the minimum on the up beat. This is exactly how an oarsman feathers his oars: he rotates them at the same time as he speeds the blades forward for a new stroke. Just as the wing finishes its up stroke the pressure of air on the asymmetrical primaries makes them rotate and, as the next down stroke begins, they are beating backward; this is how propulsion is effected. The rhythm of the bird's wing-beat is synchronized with its rhythm of breathing.

Most birds fly with their bodies held almost horizontal, but one well-known exception to this rule is the tiny hummingbird as it hovers in front of a flower. Hummingbirds have mastered the rhythms of flight so effectively that the precision of their hovering defies the abilities of even the best helicopter pilot. The hummingbird can hold its body almost vertical so that the down and up strokes of its wings become, respectively, forward and backward strokes. The 'hand' part of the hummingbird's wing is relatively much longer than that of other birds. Because it is the 'wrist' that imparts flexibility, the hummingbird wing must be flexible over a greater part of its length than in most other birds. In hovering flight the wing-tips move through a symmetrical figure of eight in each cycle, with lift being provided by both the forward and backward sweeps. The propulsion created in the forward sweep is cancelled out by that created in

the backward sweep. The combination of forces generated over the whole cycle results in an effective upward force alone, but even a tiny change in the power supplied to one or other wing strokes is sufficient for the bird to move its position.

Every beat of the hummingbird's wing is effected by a contraction of the huge wing muscles—they may account for 40 per cent of the bird's body-weight—and the rapid repetition of the cycle, often as frequently as 80 times every second. The contraction of every muscle fibre is triggered off by a separate signal from a nerve, but how the nerve impulses and muscle contractions can work so fast is still a mystery. The frequency of the hummingbird's wing-beat, although remarkable in avian terms, is slow indeed when compared with some insects. The biting midge may beat its wings at the incredible rate of almost 1,000 times a second, but control of this rhythm is quite different from that of the hummingbird. It is impossible for nerve signals to be transmitted at this rate, so what the insect does is to switch its flight mechanism on to 'automatic pilot'.

Insects differ from vertebrate fliers, such as birds, in having skeletons outside their bodies, and their wings are merely extensions of that skeleton. The flexible cuticle of the insect skeleton can be bent by muscular action, with the energy generated by that distortion being dissipated via the wings. Furthermore, once the flight muscles have been switched on they produce their own impulses, so bypassing the higher nerve centres of the brain. The result is a rapid fibrillation that can rise to extraordinary frequencies. Despite its different mechanics, insect flight is aerodynamically essentially similar to bird flight and some insect ▷

Hooded crow

The pigeon's wing-tips trace a figure of eight in each flapping cycle. The wings, fully extended backward, are pulled forward and down by strong breast muscles (1). Half way down, pressure rotates the wing-tips (2) and as the wing swings forward propulsion is generated (3). For the recovery stroke, the wrist and elbow flex (4) and the shoulder rotates (5) to bring the wing up and back for the next downstroke (6).

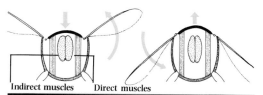

Indirect muscles   Direct muscles

The wings of tiny insects vibrate 1,000 times a second. Such frequencies are beyond nervous and muscular control and are powered by pulsations in the body of the insect. A contraction of the indirect flight muscles distorts the skeleton, forcing the wings down. The skeleton then clicks back into place, pushing the wings up. Once switched on, this mechanism runs by its own momentum, boosted by an occasional nervous impulse.

Dragonflies oscillate their two pairs of wings 25 times a second under the control of direct muscles attached to the wing-base —the system used by most large-bodied insects. Dragonflies perform sudden bursts of rapid wing-beats if chasing prey.

The tiny, delicate thrip is designed to drift rather than fly. Before launching into a zig-zagging flight, the thrip combs its finely feathered wings with its hind legs, which also provide the initial lift-off. Once airborne, the wings fan out and flutter rhythmically, just long enough for the wind to whisk the thrip high into the air for many miles.

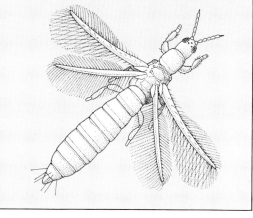

▷ species have become specialized to ways of life requiring fast flight, slow flight and hovering. Honey-bees have evolved an extremely specialized launching technique. As they are unable to create lift for long on still days, they gather in small groups to fan their pairs of wings and build up a communal 'draught', strong enough to propel them into the air.

While many small animals make use of flight rhythms to get from place to place, comparatively few use jumping rhythms. Many small insects, such as fleas and springtails, jump to escape the attention of predators, but there is nothing cyclical or rhythmical about a single jump. Movement by rhythmical jumping or, more correctly, ricochetal locomotion, has been perfected by a number of mammals, most notably the kangaroos, and a few amphibians. But because jumping is only efficient when the body is long the common frog is about the smallest jumper, and even this creature has evolved certain anatomical adaptations to increase its body-length. There are no insects that use true richochetal locomotion and in man, for whom jumping rates low among his physical skills, the only sort of jumping that comes anywhere near it is the 'hop, step and jump', or triple jump, of athletic competition.

The least efficient sort of hopping is that employed by a frog. Each jump cycle has a clear beginning and end and does not lead naturally into the next. From a sitting position, with its head held up at an angle of about 45 degrees, a single backward kick of the hind legs launches the frog's body into the air. The front limbs are held pointing backward in a streamlined fashion, and at the peak of its jump the frog is aerodynamically quite stable. As it starts its descent the forelimbs are brought forward and extended, for they will be the first part of the

animal to make contact with the ground. The hind limbs come forward, too, so that once all four limbs are down, the frog can kick off on another cycle. An advantage of this type of hopping is that each jump can be made in a different direction from the previous one, so confusing a would-be predator.

Since all the power for the frog's jump comes from a single kick of its hind legs, it is essential for them to be in contact with the ground for as long as possible. The pelvic or hip girdle of the frog has become adapted to meet this requirement by an elongation of the bone on to which the thigh bones are attached. Furthermore, a strengthening strut, the urostyle, runs the length of the pelvic girdle and helps to translate the backward push into the considerable amount of kinetic energy necessary to propel the frog through the air. Just as the frog's pelvic girdle is adapted to jumping, so its shoulder or pectoral girdle is strengthened to absorb the shock of landing.

Compared with the ungainly hopping of a frog, the leaping of the kangaroo is a poetry of flowing grace. With seemingly effortless ease an adult red kangaroo in full flight can cover 14 feet (4.2 m) in a single jump, and can do so repeatedly for miles on end. Kangaroos only use true ricochetal locomotion when they are travelling at speeds above about 20 miles (32 km) an hour. When they are moving more slowly than this the short forelimbs are put on the ground while the massive hind limbs are slid outside them. The thick tail seems an unwelcome encumbrance during such slow movement, but when the kangaroo gets into its rhythmical stride, the tail comes into its own. At the start of a single jump cycle the tail is held out behind the body, then strong thigh muscles contract—just as in the frog—which tends to straighten the legs and

Long muscular legs unfold to launch the frog into the air at an angle of 45°. As the frog's toes spring off the ground, the body is already fully stretched, the extended backbone providing extra length. At the top of the jump, the frog draws up its hind limbs and brings the shock-absorbing forelimbs forward to cushion the impact of landing. It lands with a splat, readjusts the splayed hind legs for the next jump and the cycle begins again. Unlike the kangaroo, the frog can not use the momentum from the previous jump but treats each leap individually, the better to dart unpredictably away from a predator.

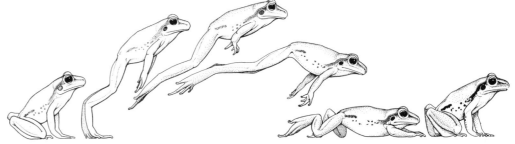

Small desert rodents find ricochetal jumping an essential adaptation to the desert environment, where scarcity of food makes speedy and long-distance travel necessary for survival. This two-legged hop not only enables the kangaroo rat and jerboa to bounce quickly across long stretches of sand but burns up much less energy than four-legged running. In these creatures a long tail acts as both a stabilizing counterbalance and a rudder. The tuft of fur at the tip provides directional control and a lash of the tail, mid-air, effects a change of direction. On landing, the jerboa's tail arches to make an air brake and prevents the animal toppling over. Larger species of jerboa can spring as high as 8 ft (2.4 m) and the legs of the tiniest species appear to vibrate, for their cycle of movement occurs at such high frequencies. The short forelegs, which are important for gathering food, are rarely used for running except at slow speeds.

The kangaroo rat hops on both legs simultaneously. Its toes spread for take-off, then gather and flex slightly as they swing forward for landing.

When jumping, the jerboa lands on one hind leg, then pushes forward taking one step before springing upward from the other hind leg.

lift the kangaroo forward and upward. As the kangaroo lands its body becomes more nearly horizontal. Because the tail assumes the same position relative to the body throughout the cycle, it is raised up, well clear of the ground. Acting as the perfect counterpoise, the tail now swings downward, so lifting the body up and forcing the centre of gravity rearward and the kangaroo is now ready for its next jump.

Speeds of up to 35 miles (55 km) per hour have been recorded for male kangaroos, while the female can achieve 40 miles (65 km) per hour for short bursts. When pursued and in danger a female kangaroo may jettison her young from the pouch to increase her speed. The pouch relaxes and the young drops, somersaulting along the ground for some way before ricocheting away at high speed. A kangaroo increases its speed by lengthening its stride rather than by hopping more frequently, so there is no marked change in rhythm with increasing speed as there is when a man changes from a jogging pace to a sprint. The duration of each hop thus remains constant over a wide range of speeds, although the time on the ground decreases and the time off the ground increases. The frequency of hops of each individual is determined by the length of the animal's leg.

In order to achieve the longest possible jump kangaroos, like frogs, have long legs and greatly elongated toes, but they do not have elongated pelvic girdles. Where kangaroos score over frogs is that they can use the elasticity of their leg muscles and tendons as springs which can be stretched and compressed. At the start of hopping, considerable energy is required to get the animal upward and off the ground. On landing, muscles which had contracted to initiate the previous take-off are stretched, and the elastic recoil of these stretched muscles helps to provide some of the energy for the next cycle. So by contracting the muscles slightly before landing, their stretching releases some tension which, together with the stringy Achilles tendons and the pendulum-like motion of the tail, propels the kangaroo into the next cycle. Recent studies into the physiology of kangaroo locomotion have shown that this action of storing energy is highly efficient and requires far less energy for a given speed than would be required by a running species. To be able to travel so cheaply is a great asset to the kangaroo since it may have to go 20 miles just for a drink of water. This efficiency may also explain why the kangaroo survived in great numbers while its quadripedal counterparts, such as the marsupial lions, *Thylacole carniflex*, and the Tasmanian wolf, *Thylacinus cynocephalus*, became extinct when man and his dogs became hunters in Australasia.

It is interesting to note that among the other mammals that have taken up hopping many are desert dwelling species, living in a world of sparse food and even sparser water. The jerboa or desert rat shares many of the kangaroo's traits—long, strong hind legs with tough, elastic tendons and a long, counterbalancing tail. During its rhythmical hopping the jerboa's tail normally bobs up and down, but if the animal wants to change direction it need only lash its tail in the opposite direction to that in which it wishes to go, because a tuft of fur at the tail-tip acts like a rudder or sail to impart directional stability. The springhaas of southern Africa is about the size of a jackrabbit or European hare, and can hop tirelessly for immense distances of 5 miles (8 km) or more. As it does so, the heavy tail rises and falls, ever compensating for the rhythmic fore-and-▷

Rhythmic grace and speed characterize the movement of the red kangaroo, *Megaleia rufa*. In one bound it can cover 27 ft (8.1 m), reach a height of 10 ft (30 m) and move at 40 miles (65 km) per hour. In the air, the heavy tail counter-balances the forward inclination of the body.

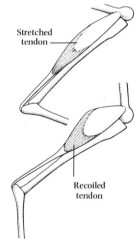

Stretched tendon

Recoiled tendon

At speeds over 11 miles (18 km) per hour, a kangaroo uses much less energy than a horse. The secret lies in the long Achilles tendon which runs from the calf muscle to the heel. On landing, the tendon tenses slightly, then stretches under the weight of the kangaroo. It then recoils by elastic contraction, providing the catapulting impulse for the next jump.

▷ aft movement of the centre of gravity. To stop the head from bobbing so much that it might cause damage to the brain, the neck is made inflexible by having bones (vertebrae) that are short and partially fused together.

Hopping is a very refined method of travel that almost certainly arose from the more usual four-legged or quadripedal running. But even running on four feet is restricted to comparatively few animals because most of the species that inhabit planet Earth have more than four legs, and some of them many more. Slow-moving invertebrates, such as millipedes and centipedes and even marine lugworms, have a large number of short limbs. To move forward, waves of locomotory activity pass forward from the rear to the front of the animal. Each limb is placed on the ground fractionally before the one in front, and also lifted from the ground fractionally before it. The result is a rippling type of leg movement, a metachronal rhythm.

A similar effect is seen in microscopic single-celled organisms such as *Paramecium* which live in stagnant water. These animals are covered with hairlike projections, or cilia. As the animal moves each cilium stiffens up and is beaten backward, to be followed by the neighbour in front of it, by the one before that, and so on. Like a whip being dragged up for another lash, the cilium then rises, again followed immediately and in succession by the neighbour in front of it. The ripple of movement of the cilia is like that of a field of corn blown by a gust of wind.

The golden rule of terrestrial locomotion is that if a creature has many legs and wants to travel fast, it must reduce the number of legs used in moving so that only a small number of limbs stays in contact with the ground. Insects have only six legs, and are capable of much higher land speeds than millipedes and centipedes. A cockroach, for example, can move more than a hundred times faster than a millipede, reaching a speed of 3 miles (4.8 km) per hour compared with the millipede's 0.02 miles (0.032 km) per hour. Typically, the rhythmic movement of an insect, such as a cockroach, is effected by an alteration of the triangular support, from the front and back legs on one side and the middle leg on the other, to the middle leg on the first side and the front and back on the other. The triangle formed in this action allows the legs previously used to recover and be ready for the next cycle. At high speeds the pattern changes, but it does not do so in a uniform fashion. Some species become quadripedal, simply ignoring the rear pair of legs, while others, such as the mantis, which are quadripedal at normal speeds, take to a six-legged gait. Spiders normally travel on all eight legs, which are well splayed out to prevent their tips from becoming entangled with one another, and in this way they can achieve a speed of 1.1 miles (1.76 km) per hour. Insects lying on their backs will continue to display rhythmic rather than random patterns of movement.

Like the wheels of a train all running on the same track, the limb-tips of millipedes must follow in the footsteps of their predecessors, because the limbs are all the same length. The variable limb-length of fast-moving invertebrates, such as insects, spiders and crabs, allows each limb to make contact with a different piece of ground from that of its neighbour. To get from place to place all the land-living vertebrates, the animals with backbones, have either four or two limbs. Compared with the many-legged invertebrates, and making due allowance for their size, the locomotory ability of these animals is far superior.

The sea gooseberry, like many other aquatic organisms, moves by waving its eight bands of tiny hairlike cilia in a rapid succession of ripples. Each cilium swings rigidly forward against the water a split second before the next, to produce a straight power stroke, and bends limply back like a whip for the recovery stroke to minimize the resistance to the water, *far right*. Each metachronal wave passes down the band a fraction of a second before the next. Viewed from above, the individual cilium of the sea gooseberry traces a clockwise circle, although in some other organisms, the cilia travel anti-clockwise.

The running millipede appears to travel on a gently moving wave which is caused by the action of the legs. In most species this wave passes forward over the body as each leg lifts, a split second before the leg in front. During the propulsive phase the legs spread to give a broad push-off and bunch together when swinging forward for the next stroke. Each pair of legs moves simultaneously and at any one time there will be as many legs on as off the ground, although each limb will be $\frac{1}{8}$ to $\frac{1}{12}$ of a step out of phase with the next. On average, 38 waves a minute pass along the body.

Long flexible legs allow the spider to scuttle along at a top speed of 1.1 miles (1.76 km) per hour. As the four pairs of legs are wide-spanning, each needs to be a slightly different length to prevent the spider tripping. Although the second and fourth legs are often synchronized, the spider's rather disorganized gait sometimes falls into a wavelike sequence. To speed up, the spider increases the frequency of each step, without altering its pattern of movement. A primitive device controls the spider's every stride: the legs can only be stretched by a sudden increase in hydraulic pressure of blood, although muscles bend them.

Beetles, cockroaches and all six-legged insects walk as if on a tripod and rhythmically alternate their three-legged triangle of support. The middle leg on one side is always synchronized with the two outer legs on the other side. To maintain the rhythm, the periods on and off the ground are equal and each limb is half a complete cycle out of phase with its opposite number, to prevent entanglement. This triangular gait makes the insect veer slightly from side to side with each step. An insect's speed is largely controlled by temperature: in hot surroundings, the muscles contract faster.

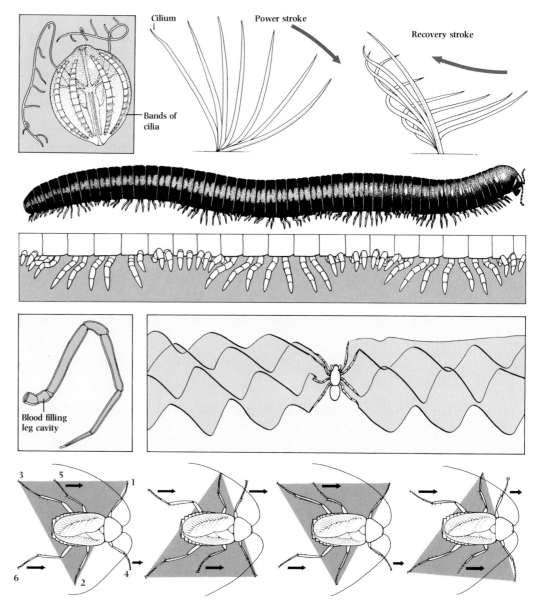

The fundamental rhythm of movement for a terrestrial species, such as the horse, is essentially symmetrical—each leg is raised from the ground in turn and for the same amount of time. The pattern of hoof-fall is front left, back right, front right, back left. When viewed from above the horse's left side mirrors its right.

On increasing its speed from a walk to a trot, the horse has fewer hoofs in contact with the ground at any one time, so the pattern of the rhythm changes. Typically, the left hind and right front hoofs are in ground contact together, and are raised together, as the left front and right hind hoof are placed on the ground. But the symmetry of this gait is destroyed when the trot breaks into a gallop. Starting with the left hind foot striking the ground, the right hind, left front and right front follow in rapid succession, launching the horse upward and forward. All the feet are off the ground for perhaps four or five yards (3.6 to 4.5 m). This period of suspension, during which the horse is airborne, effectively increases the length of the stride, and with it the speed of movement.

The rhythmic pounding of a galloping horse's hoofs occurs more frequently than the rhythmic silent footfall of the cheetah because, during the suspension period, the feet of the horse are gathered up underneath its body. During the cheetah's suspension period its legs are stretched out to front and rear, and the animal's immensely flexible vertebral column curves up and down as the cheetah stretches for the longest stride. Rabbits and stoats are among the animals that employ a slightly different sort of rhythm, the half bound, in which both front feet touch the ground in unison, after which the rear feet are both put down together.

When humans ride terrestrial mammals, they have to flex and relax their legs in the stirrups in unison with the rise and fall of the animal's back and the constant shifting of the position of its centre of gravity. Yet even trained jockeys find difficulty in riding camels because, like giraffes, hyenas and a few other species with sloping backs, camels move both legs on one side forward, followed by both legs on the other side. A giraffe gallops in an ungainly fashion—its massive size precludes it from leaving the ground for more than the briefest moment—with the long neck swinging forward and backward twice in each stride, pulling the centre of gravity forward in concert with the forward movements of the legs. Because of its huge size, the maximum speed that a giraffe can achieve is little over 34 miles (55 km) per hour.

Using only two legs for walking and running—as man does—is a minority means of locomotion in the animal kingdom, but it is particularly fascinating because it is the human gait. Raising the body up from the ground and on to two legs means that the period of suspension during the running cycle may be more prolonged, so helping to achieve a high speed. However this is not the only critical factor, for fleetness of foot also depends on the flexibility of the spine, the length of stride and the power to weight ratio.

Among the lizards there is clear evidence that running on two legs and high speed are linked. Almost 40 species of lizards are quadripedal when moving slowly, but rear on to their hind legs, tuck in their forelimbs and run fast on two legs when danger threatens. The collared lizard, for instance, reaches speeds of 24 miles (38.6 km) an hour in this way. Unlike the legs of birds and mammals, which are positioned directly under the body, those of lizards project at right ▷

The horse's slow rhythmic walk follows a regular, symmetrical pattern of raising hind and forelegs on opposite sides. At normal speeds three legs are in contact with the ground at once. This rhythm is evident to the rider who bobs up and down twice in every cycle.

As a horse moves from a walk to a trot, the pattern and frequency of movement changes. Diagonally opposite legs move together, changing between swing and support phases, so that two legs are always on the ground. The head lifts to improve balance and stability.

For most of the galloping cycle, the horse is in mid-air. As it lands on one foot, the ligaments stretch to take the impact, *below*, before the other three feet follow asymmetrically, in rapid succession. A rider must move in unison with his horse to keep his seat.

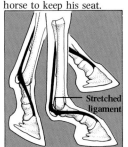

Stretched ligament

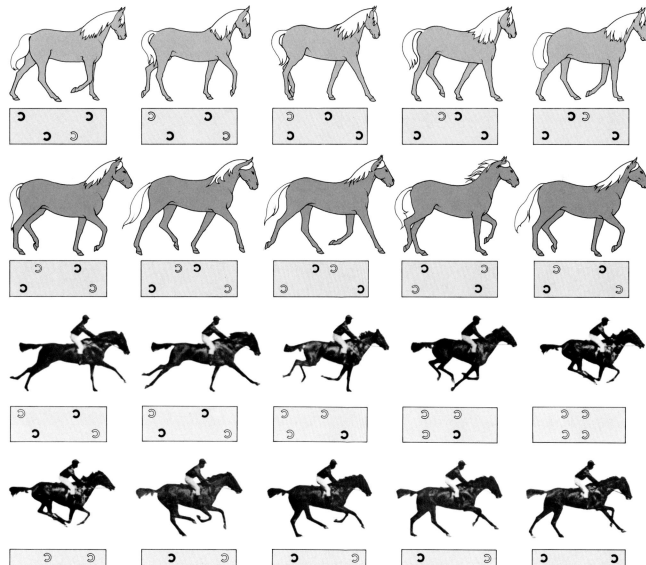

▷ angles from the body and this gives them a rhythmical, rolling gait. During each cycle of limb movement the long axis of the body pitches up and down, being more steeply sloped during the early stages of the retraction of a hind limb and more nearly horizontal during the later stages. At the same time the body yaws from side to side—like a rowing boat riding out the wake of a ship. Both these characteristics of bipedal lizard locomotion result from the forward propulsive thrust being delivered by the foot some distance away from the centre of gravity. Bipedal lizards have long, heavy tails which bob up and down once during each cycle of movement of a hind limb and act as counterbalances. However, lizards are often found with damaged tails, so it seems that having a counterbalance is not vital to running on two legs. It is also interesting that the body-weight of these lizards is so widely distributed during running that some of them can travel several yards across water without risk of sinking and drowning.

Among the birds, ostriches, emus and cassowaries are all adapted for bipedal running, as was the moa of New Zealand which became extinct about 300 years ago. The ostrich is the fastest of all the running birds alive today and is reliably credited with speeds of 50 miles (80 km) an hour and more. In proportion to the size of its body, an ostrich's legs are extremely long, giving it a massive stride, and hingelike knees allow each foot to be raised far off the ground. Ostriches also have long necks which counterbalance the slight rolling motion produced by lifting first one, then the other leg. They appear to run effortlessly, almost as if they were on wheels. This is because several compensatory movements, built into the cycle of leg action, cut down

inefficient pitching and yawing to a minimum. Apart from the counterbalancing role of the long neck, the feet are long and pliable. As in a horse's foot there are strong elastic ligaments, which take the strain of landing and give some spring to the next stride, but while a horse breaks into an asymmetrical rhythm as it accelerates from a trot to a gallop, an ostrich merely increases the frequency of its symmetrical running cycle.

Walking upright on two legs, man is not nearly as well adapted for fast movement as an ostrich. Yet man can perform a great many other actions completely beyond the powers of the ostrich. It is the rhythm of man's walking that keeps him from falling on his face. As man walks, each stride starts when the calf muscles relax and the body sways forward under the influence of gravity. This sway moves the centre of gravity forward and outside the base provided by the feet. If nothing were done to correct this, the body would fall, so man moves one foot forward to widen his base and capture the moving centre of gravity. As he does so, the pelvis rotates and many muscles interplay to keep him on balance. Now the rear leg provides propulsion, using muscular energy which is transmitted first to the ball of the foot and then to the big toe, so terminating the 'stance' phase of the walking cycle. As the rear leg starts to move forward, so the 'swing' phase starts. Facilitated by the leg bending at hip, knee and ankle, the foot clears the ground and is straightened again immediately before footfall. The ankle stays bent, so the first contact with the ground is made by the heel. This action concludes the swing phase and starts a new stance phase, during which the point of contact between the foot and the ground moves inexorably from the heel to the tip of the big toe.

The ostrich, the world's fastest two-legged animal, can maintain and exceed speeds of 50 miles (80 km) per hour, over twice man's top speed. Adapted for speed, rather than variety as man is, the ostrich has powerful muscular legs, which are long in proportion to the body, and flex acutely at the knee to allow immense strides of over 12 ft (3.5 m). Strong elastic ligaments in the foot lend a spring to the ostrich's step and cushion the blow of landing. Evolution has reduced this pliable foot to two toes so strengthening and lightening it. To improve stability and counteract the swaying of the body, the slim neck swings back and forth and sideways. This 8 ft (2.4 m) tall bird may plunge its head in the sand, but it never takes flight.

	Maximum speed	Frequency of movement cycles
Millipede	0.02mph (0.03km/h)	38 per min
Giant tortoise	0.18mph (0.29km/h)	30 per min
Spider	1.1mph (1.76km/h)	10 per sec
Eel	2.2mph (3.52km/h)	1.7 per sec
Cockroach	3.13mph (5km/h)	20 per sec
Man	27mph (43km/h) for sprint.	6/7 per sec
Man	22mph (37km/h) for 100m (Olympic speed)	6 per sec
Cat	30mph (48km/h)	3.4 per sec
Giraffe	35mph (56km/h)	1 per sec
Greyhound	36mph (58km/h)	3 per sec
Jackrabbit	40mph (64km/h)	1.5 per sec
Horse	43mph (69km/h)	2.5 per sec
Red fox	45mph (72km/h)	2.6 per sec
Ostrich	50mph (80km/h)	2.5 per sec
Antelope	60mph (96km/h)	2.4 per sec
Cheetah	70mph (112km/h)	3.5 per sec

Throughout a stride, the human pelvis does not remain in quite the same plane, simply because the action of the rear leg in the stance phase drives it upward slightly—once for the action of the left leg and once for the right. Two people walking alongside one another, who wish to talk, find the rhythmical bobbing up and down irritating, which explains why they usually fall into step to synchronize their pelvic movements. The pelvis is not merely a passive partner in the walking cycle; it, too, undergoes rhythmic movements. During the stance phase, the contraction of the gluteus minimus and gluteus medius muscles underlying the buttocks tends to tilt the pelvis and stabilize it relative to the rear leg. At the same time it rotates slightly, so increasing the stride. This rotation, which is exaggerated in the mincing walk of theatre comedy, is not the same in both sexes. Because the proportions of the female pelvis are different from those of the male to allow for childbearing, a woman's hip cannot be thrown as far forward as a man's, so that for a given stride length a woman must rotate her pelvis through a greater angle. Fashion designers have exploited this by encouraging women to wear high heels, which further exaggerate the pelvic tilt and angle. However, styles of walking are much influenced by both individual anatomy and personality.

Accompanying the leg movements of human locomotion is a rhythmic pumping of the arms, which is much more pronounced in running than in walking. In walking, the right arm swings forward as the left leg starts its swing phase. As this leg enters its stance phase, the right arm travels back and the left swings forward, so that the four limbs move exactly the same as those of a walking horse. As the frequency of movement cycles increases, and man breaks into a run, the arm action becomes more prominent. This is partly to correct the yawing motion the rotating hips impart to the whole body, and partly to offset the imbalance produced by the body's being clear of the ground for part of each stride. Otherwise the rhythm of running is essentially the same as that of walking, although the heel is not placed so firmly on the ground at the end of each stride, and the body is bent forward so that the centre of gravity is pushed just that little bit farther from the base formed by the feet. For a fast sprint, these motions become more exaggerated: the body leans forward at a 25 degree angle, only the ball of the foot touches down, the knees are kicked higher and the arms pump aggressively through a wider angle and are held closer to the body for more efficient thrust.

The rhythmicity of human movement develops gradually. The hesitant movements and shaky coordination of young children improves as they grow and as the nervous system, which controls balance, timing and muscular control, becomes properly tuned.

Man's method of movement is well suited to a terrestrial existence, but is ill adapted for performance in water. Humans have to learn how to swim—but at least they can do it. The great apes, man's closest living animal relatives, simply do not possess the necessary motor skills for learning to swim, which means that apes can be kept safely in zoos behind a small water-filled moat. In the water man is very ungainly, for he must weld together the kicking of his legs and a sweeping, or rotation, of his arms to force his unstreamlined body through a medium some 800 times more dense than air.

Truly aquatic animals have a great degree of side to side or lateral flexibility ▷

Finding his rhythm becomes all-important to a top-class runner and this rhythm is largely dictated by the body's anatomical proportions. Swinging the arms in concert with the opposite leg helps to maintain rhythm and restores the balance, disrupted by lack of body support. As a human breaks into a run, the body leans forward to draw the centre of gravity forward, but then pulls upright again to maintain stability.

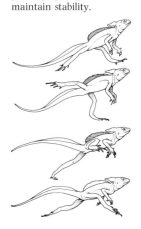

Some lizards are able to run on two legs. Their enlarged hind limbs are attached to the sides rather than the base of the body so they swing in wide circular arcs. The resulting sideways roll is partly corrected by the tail. The earnest-looking basilicus (left) can even sprint over water.

▷(an attribute severely limited in man and extending only as far up as his hips) which provides them with extremely smooth and rhythmic activity of movement. All fish swim by making use of lateral undulations of their bodies. Those like eels, which are long and thin, rely on both wavelike undulations of their bodies and the lateral beating of a flattened tail fin to provide propulsion. Others, such as the trout, herring and tuna, adopt a combination of body undulation and fin action, while some highly specialized forms, which have abandoned fast swimming in favour of camouflage or some other device, use only fin action.

As an eel swims along, a wave of contractions of the muscles at the side of its body spreads backward down the fish, twisting the body first to the left and then to the right. In an adult eel there may be as many as three bends in the body at one time, but in younger, shorter individuals there may be only one or two. Each bend serves as a brake to stop the bent part of the body from slipping backward through the water as the part in front of it straightens up and forges forward. The forward propulsive force comes from the rhythmic lateral sweeping of the flattened tail, as well as from the forward thrusting of the body as the waves of undulations pass along it. If you watch an eel swimming just above the bottom of a muddy stream, you see a regular pattern of small whirls of water which seem almost to represent solid posts against which the undulations push.

Most fishes swim by flexing their tails rhythmically from side to side. This lateral flexure may be amplified by a rhythmic contraction of the body muscles which drive the base of the tail. As it moves from side to side, the tail fin bends

because of the pressure of the water upon it, in rather the same way as the primary flight feathers on a bird's wing rotate as they move through the air. But water is much 'stickier' than air and so the lateral force of the tail fin is translated into a forward movement of the body to which it is attached. Directional stability is provided by other fins, which can be raised or lowered like brakes to keep the fish on the course it wishes to take. The speed of the fish depends on the dynamic shape of the body as well as the frequency and amplitude of the tail-beat.

Seahorses, boxfish and porcupine fish all have rigid skeletons that allow no lateral flexure of the tail base. Swimming is accomplished solely by the beating of a fin. This fin is not dragged from side to side, rather a ripple of waves runs down it from top to bottom. In the dorsal fin of the seahorse as many as seven waves may be in train at any one moment, with each wave lasting a tenth of a second. The many planes of pressure this series of waves produces means that the tiniest modifications allow for precise changes in position.

Seasnakes have a rhythmicity of action which closely parallels that of eels. Being air-breathing vertebrates, seasnakes often swim with their noses breaking the surface, and in calm water the bends in their bodies—the points of lateral thrust—can be seen as a tell-tale series of ripples which fan out at right angles from their point of origin. In swimming through water seasnakes are making the same movements as their terrestrial cousins do on dry land. In water, which has a much higher resistance to movement, a bend in the seasnake's body provides sufficient braking power to stop it from being pushed backward, but on dry land a simple bend is not enough. Most land snakes live in areas of dense vegetation and push against clumps of plants in order to achieve

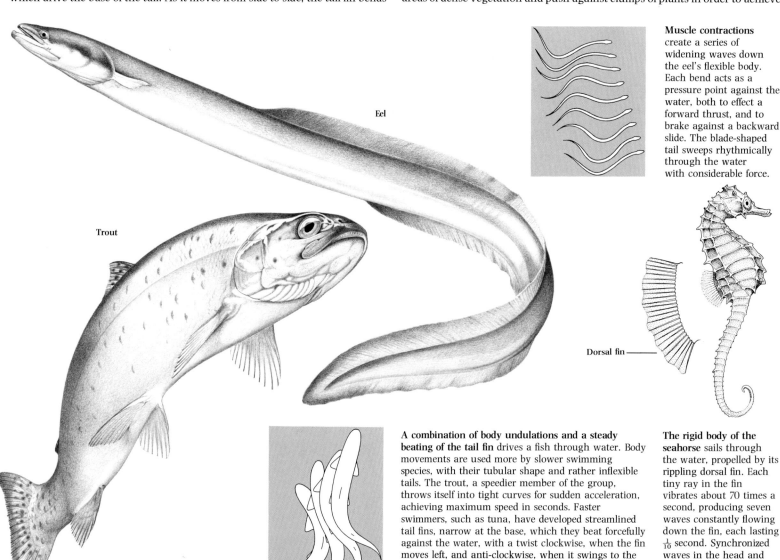

Eel

Trout

Dorsal fin

Muscle contractions create a series of widening waves down the eel's flexible body. Each bend acts as a pressure point against the water, both to effect a forward thrust, and to brake against a backward slide. The blade-shaped tail sweeps rhythmically through the water with considerable force.

A combination of body undulations and a steady beating of the tail fin drives a fish through water. Body movements are used more by slower swimming species, with their tubular shape and rather inflexible tails. The trout, a speedier member of the group, throws itself into tight curves for sudden acceleration, achieving maximum speed in seconds. Faster swimmers, such as tuna, have developed streamlined tail fins, narrow at the base, which they beat forcefully against the water, with a twist clockwise, when the fin moves left, and anti-clockwise, when it swings to the right. The frequency and span of the tail gradually increase with speed in these fast species, just as the undulations become more exaggerated in slower fish such as the herring.

The rigid body of the seahorse sails through the water, propelled by its rippling dorsal fin. Each tiny ray in the fin vibrates about 70 times a second, producing seven waves constantly flowing down the fin, each lasting $\frac{1}{10}$ second. Synchronized waves in the head and anal fins help increase the power and precision of these slow yet agile creatures.

forward thrust.

To exploit all the features of their environment as fully as possible, snakes can change their locomotory rhythms. If a snake enters a tunnel of any kind, such as might be found in a pile of boulders, it can switch from a simple undulatory rhythm to a concertina one. In this sort of movement a region at the front of the snake is thrown into two loops which press outward at their crests on to the wall of the tunnel. The rear of the snake is brought up by muscular contraction, then some new loops form at the rear. Thus held, the front relaxes, elongates and forms a new anchor.

Some snakes—notably vipers—are able to travel in a straight line, without the need for undulatory or concertina rhythms, although they do so only extremely slowly. These snakes have well-developed musculature and heavy scales on the underside of their bodies. At any one time there may be two waves of contraction passing down the body, but two short patches of the underneath, or ventral wall, of the body are firmly fixed to the ribs and act rather like athletes' starting blocks. Once the section of the snake in front of the patches is fully stretched, they relax and another two points immediately behind them take over.

Desert dwelling snakes, such as the rattlers, have to live in an environment often composed almost exclusively of sand. These snakes use the basic undulatory rhythm, but the whole body moves sideways. The snake only makes contact with the sand at two points, but because the whole body-weight bears down on just two points, enough static friction is developed to prevent these points from slipping. Against these effective anchors the body muscles can exert a backward pressure, and so move the snake forward. This rhythmic 'sidewinding' leaves characteristic tracks in the sand, like parallel strokes inclined backward at an angle of about 30 degrees to the forward axis of the snake.

Because water is many times thicker than air, its extra resistance can be exploited in many ways and there are even more rhythms of locomotion associated with water than with air. Jet propulsion, for instance, is unknown in terrestrial animals but relatively common for aquatic species. Many types of clams and all octopuses use jet propulsion to escape their predators, but this is seldom rhythmical. Among the jellyfish and their relatives, however, rhythmic pulsing is well developed. Even the most sedentary of polyps send out tiny jellyfish barely visible to the naked eye to carry their reproductive cells into the upper layers of the water. Like their larger, more familiar, relatives, these jellyfish rhythmically pulsate their domes and so gain both height and sideways movement.

Although the musculature of jellyfish is primitive, each transparent mass that composes a jellyfish is supplied with a strong circular or coronal muscle running round the edge of its bell, and several radial muscles like the spokes of an umbrella, starting from the umbrella point and radiating out to the periphery. A single pulse starts with a slight shortening of the radial muscles as a result of impulses generated by nerve cells in the animal's sense organs, and is followed immediately by a strong contraction of the coronal muscles. This causes the dome to narrow and suck water in through an aperture of decreasing size, before a rapid contraction of the radial muscles thrusts the water out with ▷

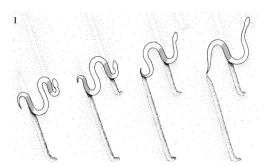
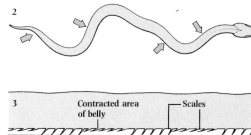
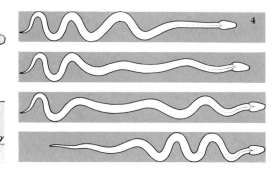

Snakes have adopted four different undulatory patterns, each using different push-off points, to suit a variety of habitats. The desert sidewinder leaves 'tramline' tracks in the sand (1). The more common laterally bending snakes thrust sideways against plants and stones from alternate sides of the body (2). Rattlesnakes and vipers creep slowly forward in a straight line, using contracted areas of the belly for 'feet', as muscles haul the body forward inside the skin (3). When confined in a crevice a snake sends tight loops, concertina-style, down the body to press against the boulders, squeezing itself forward (4).

The sidewinders of North America have developed a highly efficient means of 'running' across the slippery desert sands. The snake is constantly poised on two anchoring push-off points with the rest of the body arched between. As the head reaches forward, the body follows and the snake lands on two new push-off points, resulting in a rapid sideways 'hobble'. The parallel tracks left in the sand by the contact points lie at a 45° angle to the forward direction of the snake.

▷considerable force. An adult individual of the common European jellyfish, *Aurelia*, blows out about 0.04 pints (22 ml) of water with each contraction. Slow, rhythmic pulsing of the bell can continue for long periods and carry the jellyfish many miles. As a general rule, the rate of muscle pulsation in jellyfish is higher in younger individuals, but it also increases in frequency with a rise in temperature of the sea water in which the animal lives.

Many aquatic species row themselves through the water. At its simplest this sort of action is similar to human rowing in boats with oars, but at its most elaborate becomes almost a sort of flight. Water beetles propel themselves by making beatlike movements of their middle and, sometimes, their hind legs. Starting with the 'oars' at the end of a stroke, the cycle of movement is as follows: the legs are rotated slightly to present a thinner profile and minimum surface area to the water. The legs are then drawn forward as far as they will go, rotated again so that the area of contact with the water is at a maximum, and then they are forced backward. Stiff bristles fringing the extremity of each leg now open up to provide an even greater area of contact with the water. At the end of the propulsive stroke the bristles fold down so that they do not hinder recovery.

A similar sort of rhythmical rowing is seen in the backswimmers and water boatmen, small bugs with elongated hind legs which pull rhythmically so drawing the creatures through the water. In water boatmen the body is flat and the back slightly concave. The creature's flat, paddlelike legs are employed to scull the animal over the water surface, while the long, slim middle legs are used for clinging to vegetation while the water boatman feeds. Once the animal releases its hold on these food materials it floats to the surface because its body

is lighter than water, and it will often shoot off through the water surface and take flight. Backswimmers are similar to water boatmen, living and swimming very close to the surface of the water. The big difference is that, as their name suggests, backswimmers swim on their convex, keeled backs. Backswimmers store bubbles of air under their wings in two channels formed by rows of hairs, and force themselves down into the water by rowing with their hind legs which, like those of water beetles, have hairy fringes for greater propulsion.

Whirligig beetles are also aptly named because they swim in circles on the water surface, although they move in straight lines when moving beneath the surface and when they dive. Both sorts of movement are made possible by rowing action of the four flat swimming legs (the other two foremost legs are much smaller), but in circular swimming the legs on one side of the body pull harder and more frequently than those on the other, rather in the same way that a novice rower with two oars fails to coordinate the actions of his two arms.

Turtles move through the water pulling rhythmically on their paired, bladelike forelimbs. These limbs are feathered during their recovery strokes and—like all oars, natural and man-made—are twisted at the start of the power stroke to present a broad, flat surface to the water. Although both the turtle's flippers beat together, the passage of the turtle through the water is not jerky because its streamlined shape allows its inertial mass to move fast enough between strokes to maintain its speed. Careful analysis of slow motion film of a turtle swimming reveals that the movement of the flippers shows some of

Rowing requires perfect synchrony for speed. A Cambridge boat crew, *below right*, mimics the techniques of the 'back-swimmer', *top*, presenting the broad side of the oar blades for the strong power stroke and the narrow side for the recovery, to minimize air resistance. Long oars produce a wide stroke in relation to the movement at the rowlock.

Gentle rhythmic pulsation propels the jellyfish through water. Spokelike muscles in the body dome contract, then a rapid contraction of a second set of muscles causes the dome to narrow and water to be sucked in. Further contraction of the spokelike muscles forces water out, driving the jellyfish upward.

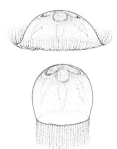

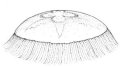

the characteristics of birds' wings. On the down stroke the tip trails behind slightly, flicking backward, and producing forward propulsive thrust at the end of the power stroke and again at the start of the recovery stroke.

Even more like flying are the wing-beats of the rays, huge fish which can travel swiftly through the water with the grace and ease of a stork in the air. A large manta ray will beat its 'wings' once every three seconds—about one-third of the frequency of a stork's wing-beat. As the ray's huge wings beat downward, lift is produced exactly as in birds, with an area of low pressure being created above each wing, but it does not need to be as strong, since the relative weight of the ray in the dense water is low. A wave of ripples travels around the edge of the wing and serves to trim it to the particular manoeuvre being undertaken. During the recovery stroke the wing bends, allowing the springy wing-tip to travel back to its starting position without creating a downward force on the fish. Of all the rhythmic patterns of locomotion seen in aquatic animals, there can be few as aesthetically pleasing to the human eye as the graceful beating of the ray's wing.

Dolphins are streamlined, water dwelling mammals that have long fascinated man. Highly successful swimmers, dolphins can reach maximum speeds of about 25 miles (40 km) an hour. Most of the dolphin's normal swimming movement is brought about by the beating of its tail fin, powered by immensely strong body muscles. The tail fin, which is attached to the dolphin's body in a horizontal plane, beats up and down with a slight horizontal twist. Rhythmical swimming in the dolphin is made more efficient by adaptations of the animal's skin. When the dolphin suddenly increases or decreases its swimming speed, the skin of the lower half of the body wrinkles and these folds act to dissipate eddies of water round the body and so reduce drag. Fluid trapped within the wrinkles of skin is squeezed from high pressure areas of the body to low pressure areas, so that the water exerts a minimum of resistance on the dolphin. This pattern of skin wrinkles differs in males and females, but when individuals of either sex are moving at high speed, the waves move toward the rear of the body, each folding cycle lasting about two seconds.

As well as being excellent swimmers, dolphins and their close relatives make spectacular rhythmic jumping movements, which propel them out of the water. In such movement, the body muscles contract against the large vertical and horizontal extensions from the bones of the vertebral column, so that a wave of muscular contraction passes rapidly down the body. After the initial contraction, the rest of the jump cycle is essentially a passive process, until the next wave of contractions again forces the dolphin's body up above the surface of the ocean.

Sadly, man has exploited the low-drag, high-efficiency swimming abilities of dolphins and their close relations in two ways. For commercial purposes he has established dolphinariums in which animals are trained to perform before an audience. The body design of dolphins and porpoises has also attracted the attentions of military scientists. A number of experiments has been carried out in order to discover whether hull shapes or fluid-containing surface layers based on dolphin designs could be usefully applied to atomic submarines to improve their efficiency and reduce their drag as they surge powerfully through the world's oceans.

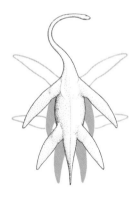

Plesiosaurs were agile paddlers of the prehistoric world. Although their 'oars' were rigid and could not be raised above shoulder level, they could flick them to make sharp turns and paddle backward. Such agility was useful when searching for fish which it would snap up by plunging its head into the water.

The loggerhead turtle, *Caretta caretta,* glides through the water by sculling both its elongated front flippers simultaneously. The smooth front edges of the flippers slice cleanly forward through the water before twisting to sweep the water back, and so propel the turtle forward. The flexible flipper-tip acts as a sensitive steering system. Some species have four flippers and may move diagonal pairs together.

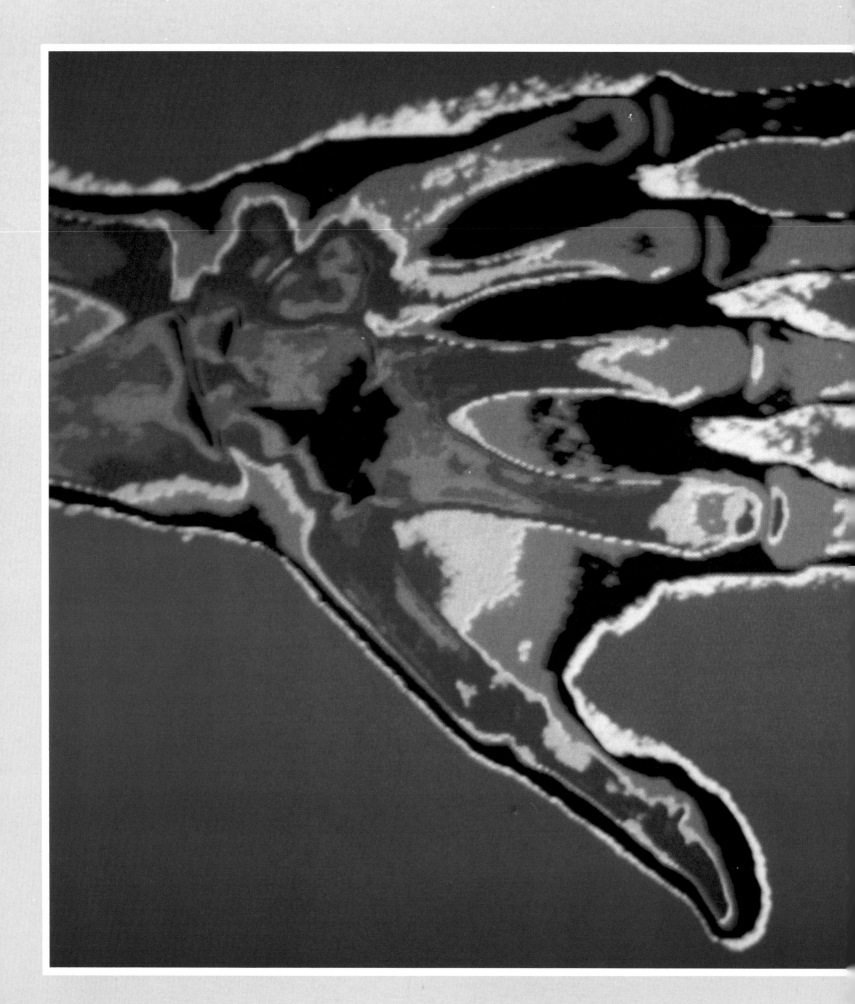

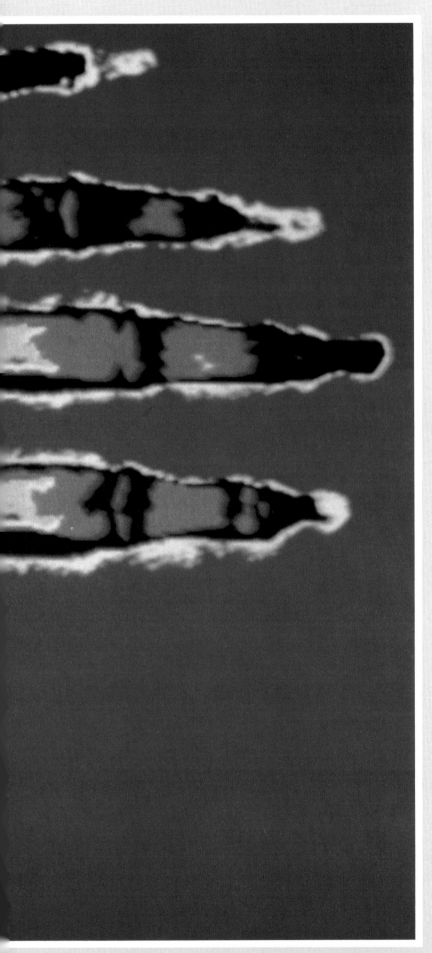

Rhythms of health and disease

By understanding something about our body rhythms we can all take steps to improve the way we do our jobs and make the decisions vital to our day-to-day well-being. Just judging whether you are at your best at crack of dawn or after dinner in the evening, and timing your work to fit in with this assessment, is a simple way to make a significant difference to the quality and quantity of your work. Knowledge of the rhythmic nature of human life is also beginning to have important implications in the diagnosis and treatment of disease. During every 24-hour period, for example, the body's internal chemistry changes, so making the symptoms of some illnesses show up more clearly at some times of day than others and, at the same time, altering the way in which the body will respond to treatment with drugs.

The human body is a remarkably rhythmic piece of machinery. Without any conscious control on our part, our hearts beat about 70 times a minute while we breathe in and out between 12 and 15 times. More subtle than these minute-by-minute rhythms are our daily or circadian rhythms, of which the most obvious by far is the rhythm of sleeping and waking. Other circadian rhythms involve biological functions which, at first sight, seem to be going on at a constant rate. The human pulse is a good example, for it does not stay at 70 beats a minute all the time. Over every 24 hours it shows a distinct variation, slowing down at night when the body is normally at rest and speeding up during the day. This rhythm continues even when the body is deprived of the natural daily change in illumination signalled by dusk and dawn, but not on a strictly 24-hour basis. Thus the heart rate rhythm of someone incarcerated in a pitch dark bunker or cave, and the rhythm of sleeping and waking, are found to have periods of between 25 and 27 hours, not 24 as you might expect, so that the body gradually gets out of phase with the time of day outside the cave, on the surface. The reason for this is that the biological clock, which drives the rhythm, normally synchronizes itself with the outside world but, when deprived of all time checks, free-runs and assumes its own slightly longer natural period.

Since the first orbital manned space flights of the 1960s scientists have taken great interest in the way in which human physiology changes under conditions of weightlessness and a day that is 90 minutes rather than 24 hours long. Indirect monitoring of the blood pressure of astronauts, for example, has shown that it follows a normal, approximately 24-hour rhythm in space, despite the drastically altered daylength.

In keeping the body healthy, and treating the diseases to which it succumbs, it is the approximately daily circadian rhythms that have so far proved most relevant, although the body does have medically significant monthly rhythms such as the female menstrual cycle. Many of the circadian rhythms which have been discovered show such a small variation between their highest and their lowest points that they can only be detected by meticulous and frequent measurements throughout each 24 hours, and may require the help of computers to reveal their underlying rhythmicity. But some rhythms in body functions have a larger variation and so are much easier to detect and study. Such rhythms include the pulse rate, blood pressure and body temperature, the production and release of hormones from some glands, the rate at which the kidneys produce urine and the brain's alertness. Knowing that the brain is least alert in the early hours of the morning, for example, gives the police a potential advantage over weary terrorists holding hostages.

The more we look for rhythms in our lives, the more we find. But despite an immense amount of study the true site and mechanism of the biological clock is still a mystery. Of all the many biological components from which the human machine is constructed, no one organ, tissue or group of cells can be singled out as the ultimate biological clock that drives all the body's 24-hour or circadian rhythms. Whether there is just a single clock or several, the human brain probably houses a master clock, which seems to synchronize itself with the outside world. This overall control ensures that our bodies are usually most active when we are awake and that internal body functions operate more slowly at night when we are asleep. The chief synchronizer of the master biological clock is probably light from the sun, which is received by our eyes, then transmitted to the brain by the optic nerves for interpretation. Other signals, such as the temperature of our surroundings, noise, and contact with our fellow human beings, which are received by other sense organs and sent to the brain for analysis, also act as synchronizers of human circadian rhythms.

For prehistoric man, who presumably had to hunt by day and hide from predators at night, a biological clock that stimulated him by day and encouraged him to hide and sleep when it was dark was a valuable aid to survival. Using artificial lighting, modern man has managed to change his environment so that, unlike his ancestors, he is no longer bound by the confines of the natural day; but the ordering of our activity in time is apparently so vital that the biological clock still persists. Experiments show that the clock is so strong that it is impossible to stop its natural rhythm for any length of time.

If the biological clock is in the human brain, it may control rhythms

throughout the body directly via the nervous system, of which it is the ultimate coordinator, or indirectly by influencing the manufacture and release of hormones, chemical messengers which are carried in the blood to target organs where they exert their effects. A master clock in the brain may simply coordinate the timing of several subsidiary clocks in other parts of the body, or every individual body cell may have its own microscopic clock. It has been shown, for example, that animal hearts will retain their heart rate rhythm even when removed and kept alive in nutrient fluids—and even individual heart-muscle cells maintain a circadian rhythm, beating fast to correspond with daytime and moving slowly to correspond with nighttime.

The biological clock in the brain may regulate rhythms by operating through the unconscious or autonomic division of the nervous system. Without our knowing it, the autonomic nervous system sustains and controls all the essential functions that keep us alive. It keeps our hearts beating and the blood flowing round our arteries and veins. It makes the lungs pump air in and out and regulates the many aspects of the digestive process. It is also the autonomic nervous system that is operating when you get goosepimples from the cold or sweat because you are too hot.

The autonomic nervous system is itself divided into two parts whose actions tend to be contrary to one another, so that between them they keep the body 'on balance'. The two parts of the system act to control the flow of physiological traffic in the body rather like red and green traffic signals. When the 'green light' is showing, this means that all is well and the body can calmly get on with its work. The parasympathetic part of the autonomic nervous system is more

The machinery driving body rhythms includes both nerves and hormones. Information about the changes occurring in the course of the natural day is perceived by our sense organs and carried to the brain where it may act to synchronize a master clock. This clock drives rhythms in organs such as the heart and lungs. The system of control works through the unconscious or autonomic nervous system, either directly through nerves or indirectly via the release of chemicals. It also works through the controlling effect of the pituitary gland on hormone production. The autonomic nervous system is divided into two parts, the sympathetic and the parasympathetic. Nerves of the sympathetic system travel from the brain to the spinal cord and out between the bones of the spine to connect with relay stations, strips of nerve tissue (sympathetic ganglia), lying alongside the spine. Further links are made by different sets of nerves supplying individual organs. Most of the actions of the parasympathetic nerves are fulfilled by means of instructions carried in the vagus nerve supplying the lungs and abdomen.

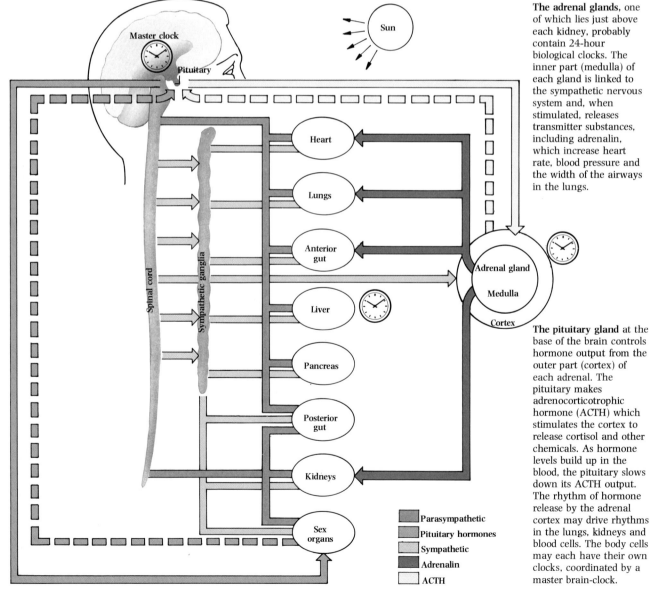

Parasympathetic
Pituitary hormones
Symphathetic
Adrenalin
ACTH

The adrenal glands, one of which lies just above each kidney, probably contain 24-hour biological clocks. The inner part (medulla) of each gland is linked to the sympathetic nervous system and, when stimulated, releases transmitter substances, including adrenalin, which increase heart rate, blood pressure and the width of the airways in the lungs.

The pituitary gland at the base of the brain controls hormone output from the outer part (cortex) of each adrenal. The pituitary makes adrenocorticotrophic hormone (ACTH) which stimulates the cortex to release cortisol and other chemicals. As hormone levels build up in the blood, the pituitary slows down its ACTH output. The rhythm of hormone release by the adrenal cortex may drive rhythms in the lungs, kidneys and blood cells. The body cells may each have their own clocks, coordinated by a master brain-clock.

active at these times. It slows the heart rate and, at the same time, encourages the breakdown and absorption of food. The parasympathetic predominates during the night when the body is most relaxed.

Imagine you are asleep in bed and suddenly you are awakened by the knowledge that there is a stranger in your room. In an instant you are wide awake. Your heart is pounding, but you are ready to jump up and fight off the intruder or call for help. In such a situation it is the second part of the autonomic nervous system—the sympathetic section—that is working overtime. As a result of its stimulation of the body's internal organs, not only do heart rate and blood pressure increase but the hairs stand on end, just as they do on the neck of a frightened animal, and the pupils dilate. To help prepare the body for immediate action blood rushes to the muscles and is diverted away from those parts of the body, such as the intestines, whose working has temporarily become inessential. The feelings that your heart is in your mouth, that you are sick with terror, are rooted to the spot or want to run for your life are all brought about through the action of the sympathetic nervous system.

In the course of a normal day events are rarely as extreme as this, but the sympathetic and parasympathetic systems act as fine balancers. As you relax watching television the parasympathetic predominates—until something horrific comes on to the screen then the sympathetic has the upper hand again. It has been suggested that the circadian rhythms of the heart and lungs may be driven by the brain, the body's ultimate traffic controller, simply by varying the balance between these two parts of the autonomic system.

The job of preparing the body for action in times of stress is not the sole prerogative of the nerves that make up the sympathetic section of the autonomic nervous system. Hormones, too, are involved and these are released by the inner region (medulla) of the adrenal glands. These small glands are located over the kidneys and are supplied with nerve branches from the sympathetic system. It is the action of the adrenals that makes the body respond so instantly to danger that it can be fast asleep one minute and awake and ready for a fight the next. The secret ingredient that makes immediate action possible is the hormone adrenalin. As it pours out into the blood, adrenalin is carried to the internal organs and mimics the action of the sympathetic system, speeding up the heart, increasing the blood pressure, dilating the pupils and so on. Without this powerful adrenalin-motivated response the body would not be able to react quickly enough to save itself from danger—the midnight intruder would have burgled you or knocked you out before you were even properly awake. Adrenalin is not only involved in preserving the body from danger, but it also helps the body achieve the peak of performance. It is adrenalin that spurs on the marathon runner to reach the tape when he feels he cannot take another stride.

The body's output of adrenalin and other related hormones from the adrenal glands is not usually constant. In ordinary, everyday circumstances it, too, has a circadian rhythm and is lower during the night than during the day. Because of this rhythm, adrenalin and other adrenal hormones may act as a kind of local biological clock to control pulse rate and other body rhythms, either alone or by acting as relay stations from a master clock in the brain. Two circadian rhythms that seem to involve adrenalin can easily be shown. When your body ▷

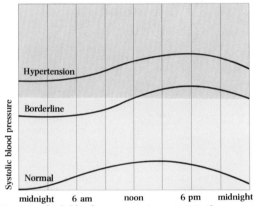

Systolic blood pressure

Hypertension

Borderline

Normal

midnight   6 am   noon   6 pm   midnight

Detecting high blood pressure is important in the prevention of heart disease. Although there is a small rhythmical variation in blood pressure over 24 hours, most people are clearly normal or have high blood pressure. In borderline cases an evening reading may give an erroneous impression of high blood pressure.

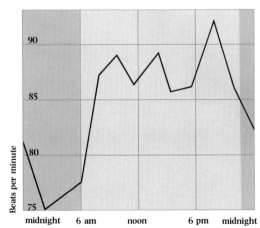

Beats per minute

90

85

80

75

midnight   6 am   noon   6 pm   midnight

The heart beats slowest at night and fastest in the afternoon, even if you lie in bed all day. The heart rate rhythm, *above*, may be controlled by a clock in the pacemaker, by the rhythms in the output of adrenalin and similar hormones from the adrenal gland, or by body temperature rhythms.

Taking the blood pressure gives a doctor valuable information about the state of the heart and arteries. As blood is pumped from the heart it presses on the elastic walls of the arteries. The resistance of the arteries to the flow of blood can be measured by finding out how high a column of mercury the pressure in the artery can support. To measure blood pressure, the brachial artery of the forearm is compressed by pumping air into a cuff. Pressure in the cuff is recorded by joining it to a mercury pressure gauge, or manometer. When the pressure in the cuff is gradually lowered, blood spurts through the artery as soon as the pressure in the cuff is the same as the maximum pressure the heartbeat can generate. This is the systolic blood pressure which produces a shock wave and a knocking sound that can be heard through a stethoscope. The knocking disappears when cuff-pressure falls far enough for the blood to flow continuously. The measurement at this moment, when the heart is resting between beats, is the diastolic pressure. The rhythm of blood pressure is most noticeable for the systolic measurement—the higher of the two.

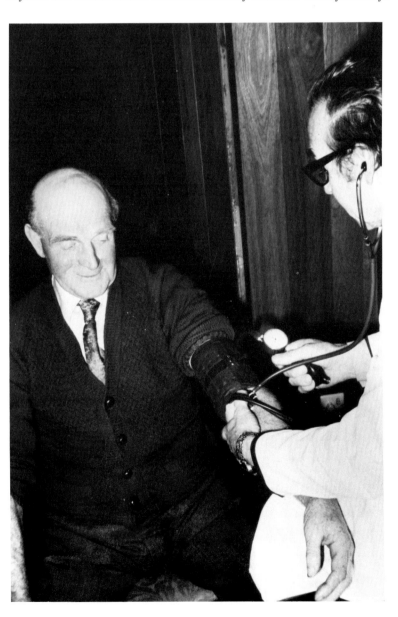

▷ is resting—that is, not being made to perform vigorous exercise—the heart beats slowly after you wake up in the morning and rises stepwise during the day. The total rise in pulse rate is often more than 10 beats a minute. In a similar way your blood pressure is lowest first thing in the morning and rises as the day goes on. Both rhythms are probably controlled, at least in part, by the adrenalin rhythm, which is at its lowest point in early morning, and both have some medical importance because changes in pulse rate and blood pressure may be early indicators of disease.

The outer parts of each adrenal gland secrete other hormones known as corticosteroids, which are made and used in artificial forms as cortisone and other steroid drugs. These corticosteroids are powerful chemicals and produce a huge range of vital effects in the body. They control the rate at which the body's metabolic reactions take place, help balance its mineral content, influence blood pressure and kidney action. To add to that, they have the ability to damp down the process of inflammation, which explains why the artificial versions are so valuable in treating rheumatoid arthritis, a disease in which the joints become inflamed, and why they help in preventing the rejection of organ transplants.

Control of corticosteroid output from the adrenals is not via the nervous system, as it is with adrenalin, but is governed by the pituitary gland, a pea-shaped projection on the underside of the brain. The cells of the pituitary have a remarkable inbuilt measuring machinery to 'test' the blood for the presence of one of the most important of the corticosteroid hormones, namely, cortisol. If the amount of cortisol in the blood is too low this triggers off the alarm bells in the pituitary. On 'hearing' these bells, the pituitary cells respond and set to work to make and send out a hormone to the adrenals so that cortisol output is stepped up. Laboratory measurements of cortisol levels in the blood throughout the day reveal that it has a marked circadian rhythm. This rhythm may be driven by the brain, which has a strong influence on the actions of the pituitary gland, or by a local clock situated in the adrenal gland itself.

The cortisol rhythm is so powerful that its effects overflow into other body systems. One of these is the blood system itself which contains two sorts of cells, the red cells, which carry oxygen round the body, and the white cells, which are involved in the body's natural defence against disease and play a part in allergic reactions. As the level of cortisol in the blood rises it makes the blood inhospitable to a type of white cell called an eosinophil, which is important in allergic reactions, and their numbers fall. So the circadian rhythm of eosinophil cell numbers is driven by the cortisol rhythm but is the mirror image of it. This is not altogether surprising because it has been shown that the rhythmical activity of the brain and of the heart, lungs and circulation are very closely related. What is more, most of the organs involved in these reactions are influenced by both the autonomic nervous system and the hormones released by the adrenal glands.

Many sorts of mental activity show circadian rhythms but the most obvious is the rhythm of sleep and wakefulness. Keeping the body awake depends on the activity of a system of nerve fibres joining the cerebral hemispheres of the brain and the brain stem, which contain centres which act as motors to drive breathing, the beat of the heart and the blood pressure. The rhythm of sleep and

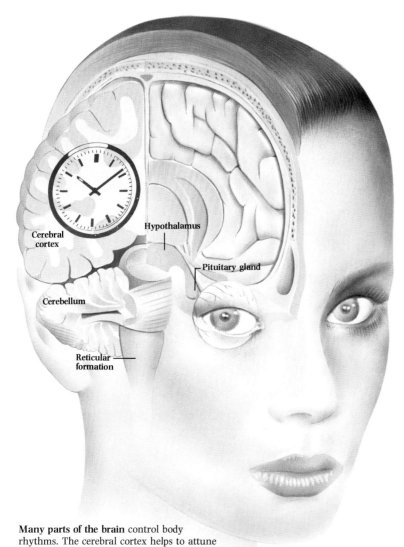

Many parts of the brain control body rhythms. The cerebral cortex helps to attune the biological clock to the environment. The hypothalamus controls temperature, with which many mental rhythms are linked. The pituitary controls hormonal rhythms, the reticular formation the rhythm of sleep and waking.

Cerebral cortex

Hypothalamus

Pituitary gland

Cerebellum

Reticular formation

The ability to work well varies during the course of the day. 'Larks' do best in the morning but tire early in the evening. 'Owls' struggle to get going in the morning but liven up later. Both larks and owls experience a 'post-lunch dip', a sharp fall in performance around 1 pm, even if they do not eat.

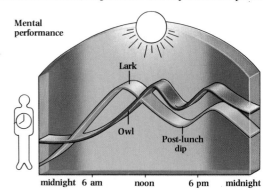

Mental performance

Lark

Owl

Post-lunch dip

midnight   6 am   noon   6 pm   midnight

Measurements of the amount of alcohol in the blood, after regular, hourly intake of equal quantities, reveal a circadian rhythm in the rate at which alcohol disappears from the blood. Alcohol is metabolized most slowly from 2 to 7 am, when it depresses mental ability most. For least effect drink alcohol in the early evening.

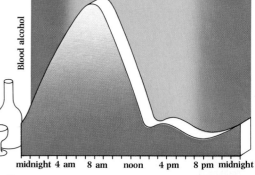

Blood alcohol

midnight   4 am   8 am   noon   4 pm   8 pm   midnight

Rhythmic fluctuations in body temperature seem to be closely related to mental performance. The rhythm is easily shown by carefully taking the temperature inside the mouth at hourly intervals. This graph shows average results for a group of normal people over several days. Body temperature is lowest just before waking, rises to a peak around 9 pm, then falls rapidly. Values vary by 0.5°C (1°F) at most.

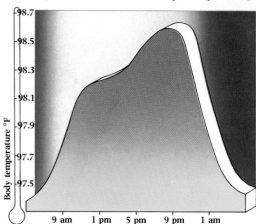

Body temperature °F

98.7
98.5
98.3
98.1
97.9
97.7
97.5

9 am   1 pm   5 pm   9 pm   1 am

wakefulness develops shortly after birth. To begin with, babies spend about eight hours awake out of every 24 but these hours are equally divided between day and night. By the time babies are two or three months old, the adult pattern is emerging, with most sleep taking place during the night. The adult rhythm depends on external stimuli to keep it synchronized with the outside world. Without such stimuli it free-runs and develops its own period of about 25 hours. Fortunately for modern man the rhythm of sleep and wakefulness adapts rapidly to new circumstances so that some of us can do shift work reasonably efficiently, although during the night we can never achieve the standards reached during the day. If, however, you are forced to stay awake without sleep the rhythm persists, which explains why people deprived of sleep find it hard to stay awake during the night, but in the morning find their fatigue is less marked although it returns more strongly the next night.

Mental performance rhythms are closely linked to the body temperature rhythm. Body temperature varies by about 0.5 degrees centigrade (1 degree F) over 24 hours, with lowest values in the early morning. This rhythm is not simply the result of increased heat production from chemical reactions, meals and exercise during the day, since it persists in people confined to permanent rest in bed and given meals equally spaced out over the 24 hours. Body temperature is controlled by a centre in the hypothalamus, the part of the brain that connects with the pituitary gland. The hypothalamus regulates sweating and the dilation and constriction of blood vessels in the skin, and, by stimulation of the pituitary gland, also influences the secretion by the thyroid gland of the hormone thyroxine, a chemical which speeds up the rate at which

cells consume oxygen, metabolize glucose and produce heat. Does this mean that the body's master clock is in or close to the hypothalamus? The answer is that we do not know, but a variation in the sensitivity of the temperature regulating centre may well be responsible for the temperature rhythm, so this rhythm could be a close guide to the time that is actually being shown by the 'hands' of the master clock.

As well as affecting our performance, circadian rhythms affect our moods. Most of us can classify ourselves either as 'larks', who wake early, work well during the morning but need to go to bed early, or 'owls', who find it hard to face the day but, once they get going, can work efficiently into the small hours. Some psychologists have gone further, claiming that larks are introverts, owls extroverts. These relationships—and the effect of the temperature rhythm—have been substantiated by a study on a group of sailors, who were classified by psychological testing and whose temperature rhythms were recorded. Introverts showed an earlier temperature rise in the morning but extrovert temperatures fell more slowly in the evening. Irrespective of psychological type, we all undergo a 'post-lunch dip'—a slight fall in mental efficiency around 1 pm. The cause of this phenomenon, which is seen in several other rhythms, is not known, but it is not due simply to eating lunch or to the effects of alcohol, since it persists even if you miss lunch and stay perfectly sober.

Rhythms of mood and body temperature, and the rhythm of sleep and wakefulness are all closely interrelated. Generally we are most anxious and depressed when we wake up in the morning, but become more cheerful as the day wears on. The sensation of fatigue, and performance in tests of mental ▷

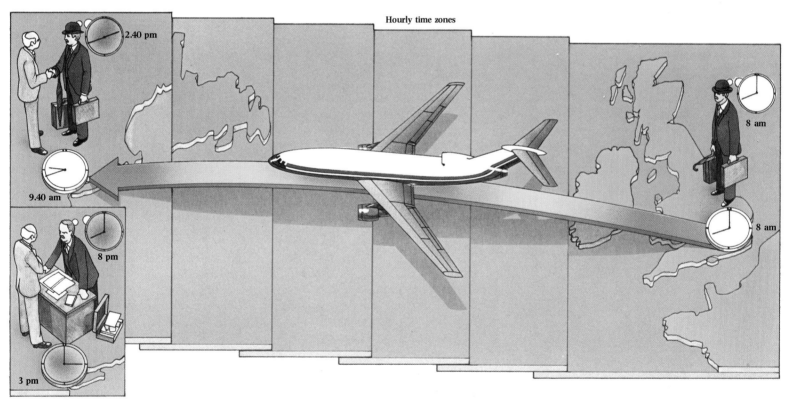

Hourly time zones

A British businessman leaves London for New York at 8 am, He travels 3,500 miles across five time zones in 6 hours 40 minutes and arrives at 9.40 am New York time, but his body clock registers 2 pm London time. Arriving at his American office at 1 pm he is in time for a working lunch and the serious business is done at 3 pm, when his American colleagues are at the end of their 'post-lunch dip' but he is at his mental peak. At 8 pm local time they go out to celebrate. The Englishman whose 'clock' is at 1 am finds that the wine goes to his head, but the Americans are at their peak for alcohol tolerance. It takes three days or more for all his body clocks to be reset to a new local time, right. On day 1, all the clocks tell London time, on day 2 they are all awry. By day 3 only the kidneys have not adapted completely.

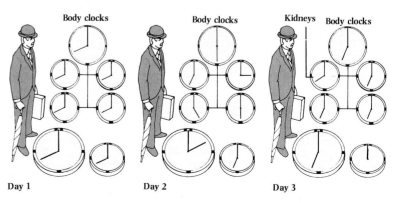

▷agility, such as mental arithmetic, speed of dealing playing cards or reaction times in simulated motor vehicles, can be closely linked with the temperature rhythm, with the poorest responses occurring in the early morning when body temperature is lowest.

To help complete our knowledge of mental performance rhythms we can also consider the effects of alcohol. A group of enthusiastic volunteers has shown that alcohol is broken down and used more slowly, and thus has its maximum effect in clouding mental activity, between 2 and 7 am. The last drinks 'for the road' at the end of a late party can, therefore, be particularly dangerous as they will be metabolized more slowly.

Mental performance rhythms have many practical applications. It is useful to identify yourself as a lark or an owl and plan your day accordingly. If you are a lark, get up early and relax in the evening. If you are an owl, put in a hard stint of work around midnight. It is worth remembering the existence of the post-lunch dip, particularly if you are often involved in 'working' lunches. To gain maximum psychological advantage, tackle the important business before lunch, not after. Similarly, anyone involved in strenuous mental activity will probably benefit from a diversion for an hour or two after lunch.

The rhythm of sleep and wakefulness is vital to our lives. Because it exists, it is inevitable that night shift workers are less efficient than their counterparts on the day shift. The implications of the rhythm are particularly important for those involved in emergency services, for their own safety and that of those they treat. If at all possible, for example, surgeons should avoid performing operations at night as they cannot expect to perform as well as during the day.

This said, some people do adapt to night shift work better than others, and there does seem to be a 'learning' effect so that shift work gets easier the more you do it. To give the body time to adapt, it is sensible to go on night shift for a minimum of a week, because rapid changes of shift are most disruptive to the body clock.

Air travel does not allow time for the clock to adapt and jet lag is at least partly explained as fatigue experienced during the day as the clock remembers that it is bedtime back home. The addition of intercontinental ballistic missiles to the arsenals of the superpowers has given circadian rhythms a military significance. There are some areas of the USA and USSR which are 12 hours apart, although one could attack the other in a few minutes. The aggressor could, therefore, usefully time his offensive so that it occurs during the middle of the night for his enemy; for although his defences are manned around the clock, decision-making cannot be as good at night.

Circadian rhythms are significant to many sorts of disease but particularly to the diseases that affect the adrenal glands and to the problems of using cortisone and other steroid drugs to treat chronic diseases, such as rheumatoid arthritis, over a long period. The key to understanding the problems to which the adrenal glands are prone lies in the way the output of natural corticosteroid hormones is controlled. The mechanism of control is a feedback system which works like this: first the hypothalamus in the brain makes a chemical which travels to the neighbouring pituitary gland. This chemical tells the pituitary to make a hormone which acts as a kind of intermediary between the hypothalamus and the adrenal. The hormone, which has the rather daunting

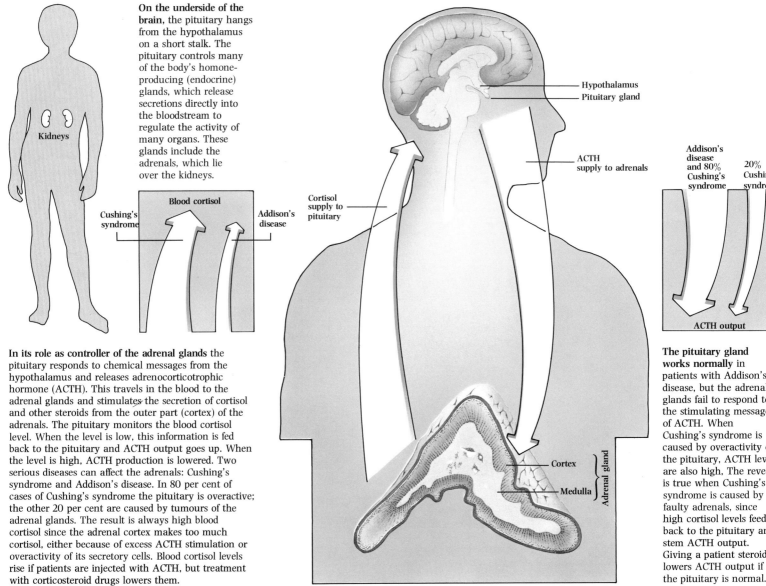

On the underside of the brain, the pituitary hangs from the hypothalamus on a short stalk. The pituitary controls many of the body's homone-producing (endocrine) glands, which release secretions directly into the bloodstream to regulate the activity of many organs. These glands include the adrenals, which lie over the kidneys.

Kidneys

Blood cortisol

Cushing's syndrome

Addison's disease

Cortisol supply to pituitary

Hypothalamus
Pituitary gland

ACTH supply to adrenals

Addison's disease and 80% Cushing's syndrome

20% Cushing's syndrome

ACTH output

Cortex
Medulla
Adrenal gland

In its role as controller of the adrenal glands the pituitary responds to chemical messages from the hypothalamus and releases adrenocorticotrophic hormone (ACTH). This travels in the blood to the adrenal glands and stimulates the secretion of cortisol and other steroids from the outer part (cortex) of the adrenals. The pituitary monitors the blood cortisol level. When the level is low, this information is fed back to the pituitary and ACTH output goes up. When the level is high, ACTH production is lowered. Two serious diseases can affect the adrenals: Cushing's syndrome and Addison's disease. In 80 per cent of cases of Cushing's syndrome the pituitary is overactive; the other 20 per cent are caused by tumours of the adrenal glands. The result is always high blood cortisol since the adrenal cortex makes too much cortisol, either because of excess ACTH stimulation or overactivity of its secretory cells. Blood cortisol levels rise if patients are injected with ACTH, but treatment with corticosteroid drugs lowers them.

The pituitary gland works normally in patients with Addison's disease, but the adrenal glands fail to respond to the stimulating messages of ACTH. When Cushing's syndrome is caused by overactivity of the pituitary, ACTH levels are also high, The reverse is true when Cushing's syndrome is caused by faulty adrenals, since high cortisol levels feed back to the pituitary and stem ACTH output. Giving a patient steroids lowers ACTH output if the pituitary is normal.

name adrenocorticotrophic hormone or ACTH for short, reaches the adrenal cortex through the bloodstream and orders it to begin making corticosteroids, including cortisol. All the time this is going on the amount of cortisol in the blood is monitored by the pituitary. If there is too much cortisol, ACTH production is turned down, but if there is not enough, then output is stepped up.

The circadian rhythm of cortisol production by the adrenal glands can be discovered by measuring the amount of cortisol in the blood at intervals over a complete 24-hour day, or by measuring the concentration of chemicals generated by the destruction of used hormones passed out of the body in the urine. The second of these lags slightly behind the first simply because of the time it takes for the kidneys to do their job of making and releasing urine. What these measurements show is that cortisol levels are lowest at midnight and in the early hours of the morning, after which they rise to a peak around 9 am.

The clock that drives this rhythm of cortisol production by the adrenals is probably in the glands themselves, since it has been shown that adrenal glands removed from hamsters and kept working in nutrient fluid continue to show rhythmic cortisol production. But the fact that the pituitary allows the rhythm to take place also implies that there must be a rhythm in the sensitivity of the pituitary to the presence of cortisol in the blood. As a result of this rhythm the message of maximum cortisol level at 9 am falls on deaf ears, yet at midnight the same cortisol level reaches a receptive audience and ACTH output falls considerably. This feedback mechanism can be tested by disrupting it with drugs; and in this way doctors can differentiate between faults in the pituitary and faults in the adrenals. If a tablet of one such drug is given to a healthy

person at 11 pm in the evening, it will provide the pituitary with exactly the same message and elicit the same response as a natural increase in the cortisol level. As a result ACTH output from the pituitary will go down and will, in turn, result in a lowering of cortisol output by the adrenals.

Understanding the cortisol rhythm is crucial to laboratory tests used to diagnose diseases of the adrenals, particularly Cushing's syndrome and Addison's disease. Cushing's syndrome, discovered by a Boston surgeon Harvey Cushing in 1932, results from excessive production of corticosteroids by the adrenals. In 20 per cent of cases it is caused by the presence of an adrenal tumour—a growth of abnormal cells which may be benign or malignant—while the remaining 80 per cent are due to tumours in the pituitary or to overactivity of the hypothalamus.

The signs and symptoms of Cushing's syndrome result from the presence of too much corticosteroid. Typically, the amount of fat under the skin increases, making the face become moon shaped, and a 'buffalo hump' of fat appears over the lower part of the neck and upper back. The trunk is obese, too, but the limb-muscles waste away, producing an appearance like a lemon on sticks. The skin is thin and bruises easily. Bones—particularly those of the spine—may weaken and fracture, and patients may become psychotic, or diabetic, or develop high blood pressure. To add insult to injury, men can become impotent and bald, while women grow excess body hair, lose their head hair and suffer from irregularities in their menstrual cycle.

From this horrific description you would think that such an unpleasant disease could easily be recognized, but in its early stages it can be hard to ▷

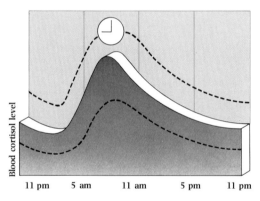

The normal daily cortisol output from the adrenals has a circadian rhythm. Peak output, at about 9 am, *left*, is more than twice as high as at midnight, but actual levels vary widely, as shown by the dotted lines. Because of the cortisol rhythm, cortisol levels must be measured at 9 am and midnight at least if disease is suspected.

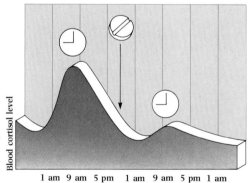

The integrity of the feedback loop between adrenals and pituitary can be tested with the steroid drug named dexamethasone. Given to a normal person at 11 pm the drug will lead to low cortisol levels at 9 am next day, *left*, showing that the pituitary has lowered its ACTH output and that the adrenals have made less cortisol.

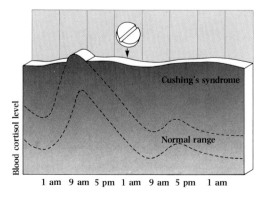

In Cushing's syndrome, cortisol levels are too high but the 9 am reading, *left*, may be within normal limits. Levels at midnight are, however, much higher than they should be. In borderline cases the drug dexamethasone is given to test the adrenals. If these glands are at fault this steroid will fail to suppress cortisol levels.

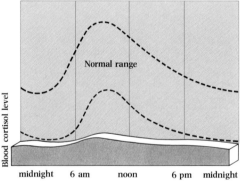

The cortisol rhythm is depressed in Addison's disease. Cortisol levels are low, but at midnight may be near normal in borderline cases. The rhythm is important to diagnosis, since a measurement at 9 am will reveal the low level. When the adrenals fail, the pituitary tries to stimulate them by making more ACTH.

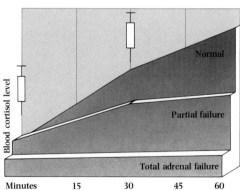

To confirm the diagnosis of Addison's disease, the hormone ACTH is injected and blood cortisol levels measured 30 and 60 minutes afterwards. Normal adrenal glands respond by making more cortisol than usual, but total adrenal failure produces no response. Partial adrenal failure or pituitary disease produces an intermediate response.

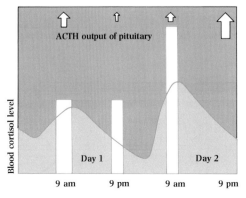

Prolonged treatment with steroids can reduce ACTH output and make the adrenals shrink. The drugs are usually given at 9 am and 9 pm (day 1). But at 9 pm the pituitary is most sensitive to steroids so ACTH output falls, switching off the adrenals and allowing them to degenerate. If all the steroids are given at 9 am (day 2) this effect is reduced.

▷diagnose. In such cases, measurement of cortisol levels is essential, but because of the circadian rhythm in cortisol output, a single high reading alone is not enough to make a diagnosis. At a minimum, measurements are needed at 9 am and at midnight. Only when both results are found to be above normal, and the rhythm abolished, is the diagnosis certain.

Addison's disease gets its name from the London physician Thomas Addison, who died in 1860, and is the reverse of Cushing's syndrome since it involves reduced corticosteroid production. Most cases are due to disease of the adrenal glands, but a few result from pituitary abnormalities. Patients with Addison's disease become very weak, lose weight and may have a craving for salt. They may vomit and become dehydrated. Their blood pressure falls and this can cut down the blood supply to the brain and make them feel faint. In the majority of cases, due to adrenal gland disease, the pituitary, in an attempt to correct the imbalance, becomes overactive and it also releases too much of another hormone which triggers over-production of the brown substance melanin—the pigment of a normal sun tan—in the skin, resulting in a darkening of the skin which is most marked in skin creases. Disruption of the normal 24-hour circadian rhythm of cortisol output is typical of Addison's disease and this fact is a valuable aid in diagnosis; in addition, cortisol levels remain consistently low throughout the 24 hours.

The adrenal rhythm is particularly relevant in the use of cortisone and other corticosteroid drugs in treating asthma, rheumatoid arthritis and other chronic diseases, and to the administration of similar drugs following transplant surgery. In all these conditions corticosteroids are used because they reduce

inflammation, or in the case of transplants, suppress the response which makes the body reject foreign tissue. Unfortunately, prolonged use of corticosteroids often brings about a host of problems, among them the abnormal fat distribution and collapse of the bones of the spine typical of Cushing's syndrome. The adrenal glands shrink and cannot respond to a stressful situation, such as a surgical operation, by producing more cortisol. This explains why steroid drugs can be dangerous if their doses are not kept to a minimum. Using the natural circadian rhythm, doctors can manipulate the drug dosage so it has the best effect. If all the day's steroid treatment is given in one dose at about 9 am, when the pituitary is least responsive to the drug, it will produce least interference with the feedback system. Theory and practice have certainly proved to coincide in the treatment of rheumatoid arthritis and it is hoped they will do so for other diseases.

In asthma there is a circadian rhythm so strong that even the sufferer is acutely aware of it. Even in the second and third centuries AD ancient physicians knew that asthma attacks are most common at night. In the seventeenth century it was thought that they were due to the body overheating at night, and only at the end of the last century did the discovery of an association between allergy and asthma offer a possible new explanation for these nocturnal attacks. Patients were believed to be reacting to feathers in their pillows or the house dust mite that lives in mattresses—agents to which asthma sufferers were less exposed during the day. It has also been suggested that the horizontal sleeping posture is the cause, but all these theories can be disproved by simple experiments. Nocturnal asthma occurs in patients without

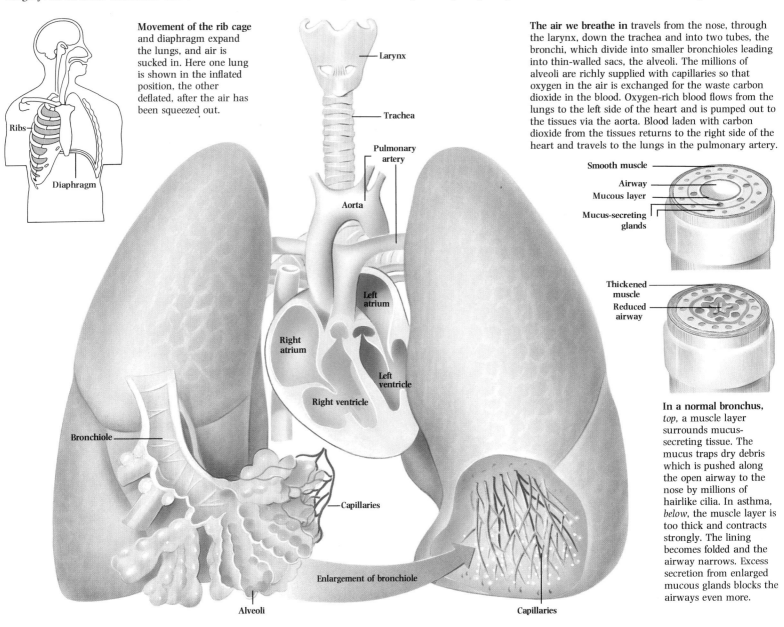

Movement of the rib cage and diaphragm expand the lungs, and air is sucked in. Here one lung is shown in the inflated position, the other deflated, after the air has been squeezed out.

Ribs

Diaphragm

Larynx

Trachea

Pulmonary artery

Aorta

Left atrium

Right atrium

Left ventricle

Right ventricle

Bronchiole

Capillaries

Enlargement of bronchiole

Alveoli

Capillaries

The air we breathe in travels from the nose, through the larynx, down the trachea and into two tubes, the bronchi, which divide into smaller bronchioles leading into thin-walled sacs, the alveoli. The millions of alveoli are richly supplied with capillaries so that oxygen in the air is exchanged for the waste carbon dioxide in the blood. Oxygen-rich blood flows from the lungs to the left side of the heart and is pumped out to the tissues via the aorta. Blood laden with carbon dioxide from the tissues returns to the right side of the heart and travels to the lungs in the pulmonary artery.

Smooth muscle

Airway

Mucous layer

Mucus-secreting glands

Thickened muscle

Reduced airway

In a normal bronchus, *top*, a muscle layer surrounds mucus-secreting tissue. The mucus traps dry debris which is pushed along the open airway to the nose by millions of hairlike cilia. In asthma, *below*, the muscle layer is too thick and contracts strongly. The lining becomes folded and the airway narrows. Excess secretion from enlarged mucous glands blocks the airways even more.

allergies and is not affected by posture, since it occurs just as often if patients spend the night sitting up in a chair.

The answer to the prevalence of asthma attacks at night lies in the true nature of the disease which has only recently been understood. In asthma the airways that lead air in and out of the lungs are abnormally sensitive. These tubes or bronchi are surrounded by coats of muscle which, when stimulated, contract too much, narrowing the airways almost to blocking point. In addition asthmatics produce excess amounts of mucus, a fluid which traps particles of dirt inhaled into the lungs so that they can be carried back up the airways in a constant stream which is regularly coughed up out of the lungs. In asthmatics this mucus is abnormally thick and sticky, so restricting the air flow even more. Even in people without asthma, a circadian rhythm occurs in the width of the bronchi. The tubes are at their widest between 4 and 6 pm, and at their narrowest on waking in the morning. The same rhythm is found in asthmatics, but its amplitude is much greater.

This circadian change in the width of the bronchi can be measured as the peak flow rhythm, the peak flow being the maximum rate of air flow a person can produce by blowing out as hard as he can. When the peak flow rate falls it means that the airways are narrowed, just as water comes out of taps more slowly if the pipes are furred. The mechanism driving the peak flow rhythm is still a mystery, but there are several likely explanations. The airways are dilated as a result of nerve messages from the sympathetic section of the autonomic nervous system—part of the set of reactions preparing the body to deal with stress—to help more air to get to the lungs. On the reverse side of the coin, the airways are constricted by nerve messages arriving via the parasympathetic nerves, and it may be that the rhythm results from the alternate activity of the two parts of the system, the balance being determined by a clock in the brain.

Another possibility is that the asthma rhythm results from the circadian rhythm in cortisol secretion by the adrenal glands. The reason behind this suggestion is that cortisone and similar drugs tend to widen the bronchi in asthmatics—an action described as bronchodilation—and help to cut down excess mucus production. It could be that the low nighttime level in natural cortisol output causes nocturnal asthma, but experiments suggest that this is not the case. The most plausible explanation is that, because adrenalin acts to widen the bronchi, the rhythm results from the rhythm in adrenalin production. The fall in adrenalin output during the night would allow the asthmatic airways to narrow and so bring on an attack.

The peak flow rhythm has become a useful tool for the treatment of asthma and many patients are learning to record their own peak flow rhythm at home. This enables asthma sufferers to recognize the early stages of an attack. If they notice severe falls in peak flow at night then this is a signal that their asthma has become unstable and that they need some treatment. The same practice is carried out in hospital if an asthmatic has to be admitted following a bad attack. Peak flow readings rise as the attack dies down, but large falls in peak flow at night are a sign that a relapse may be imminent. Thus the circadian rhythm provides a constant test of the sensitivity of the asthmatic airways—increased sensitivity means the risk of an attack and is heralded by severe nocturnal falls in peak flow rate. Sometimes asthma is difficult to diagnose because it only ▷

To assess the width of their airways, patients are asked to blow as hard as they can into an instrument which measures the maximum, or peak, flow of air out of the lungs, *left*. In asthma patients, regular measurements of peak flow reveal a circadian rhythm with lowest levels in the early morning and at night, *below*, which may be associated with asthma attacks. A similar pattern exists for people without asthma, but with much less variation between highest and lowest values. Nocturnal asthma seems to be an exaggeration of the normal circadian rhythm in the width of the airways.

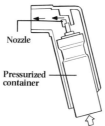

Nozzle

Pressurized container

Routine treatment of asthma is with a drug released as a droplet spray from a pressurized aerosol, *left*, which is then inhaled, or with tablets. The drug works by imitating the effects of adrenalin and similar substances naturally produced by the inner, medulla region of the adrenal gland, which act to widen, or dilate, the airways of the lungs. Steroid drugs can also be useful in treating asthma because they dilate the airways, reduce excess mucus secretion and may also increase the sensitivity of the airways to other drugs.

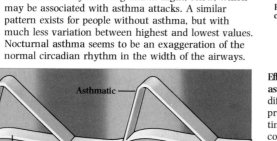

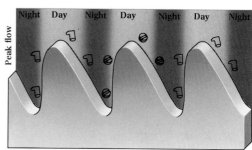

Effective treatment of asthma with drugs is difficult because of the practical problems of timing treatment to correspond with lowest levels in peak flow. Often patients wake up in the night with asthma attacks and need to take extra doses.

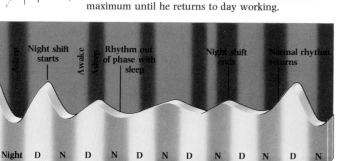

Shift work can be beneficial to an asthma sufferer. On the day shift peak flow is lowest at night, but when a worker goes on to the night shift his sleep-waking cycle is 12 hours out of phase with the normal pattern, *below*. This disrupts the clock that drives the peak flow rhythm and for a while the difference between the highest and lowest readings is drastically reduced, the minimum being when the asthmatic wakes at the start of the 'day'. The rhythm becomes synchronized with sleep and waking, but its amplitude does not reach a maximum until he returns to day working.

In hospital, asthma patients may be given drugs to dilate their airways continually via a drip into a vein in the arm. With constant administration, the nighttime falls in peak flow tend to be reduced, but they are not abolished.

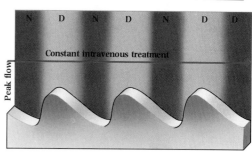

Constant intravenous treatment

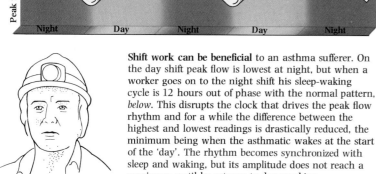

For an active asthmatic not in hospital, the best compromise between theory and practice is to take regular drug doses with an aerosol during the day and at night to take long-acting tablets which gradually release the drug into the blood. Nocturnal asthma attacks can thus be reduced.

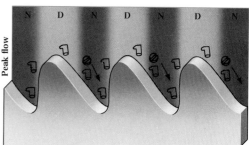

▷causes symptoms at night and, because the lungs appear normal during the day, it may be mistaken for some types of heart disease, but a few days of recording the peak flow rhythm will usually clinch the diagnosis.

Like other circadian rhythms the peak flow rhythm responds to shift work. On changing to night shifts, the rhythm is disrupted for a day or two and its amplitude falls—a rare example of a beneficial effect of shift work. Patients with bad nocturnal asthma often find that their problem improves with a rapidly changing pattern of shift work, because the biological clock is disrupted and cannot stimulate the sensitive asthmatic airways. Because asthma is always worse during sleep and on first waking, if asthmatics go on to a night shift their asthma will become worse during the day.

The rhythm of asthma also affects its treatment. The drugs now given as tablets or pressurized aerosols for inhaling mimic the action of the sympathetic nervous system, to provide maximum stimulation of the asthmatic's defective receptors and relieve the spasm of the muscles surrounding the bronchi. Although effective, these drugs can be taken easily only during waking hours, not at night when they are needed most. To overcome this problem, slow-acting tablets taken at bedtime have proved helpful.

Interest is increasing in the potential importance of circadian rhythms in the treatment of other diseases as well, both by drugs and with the surgeon's knife. If they are to work, drugs must be taken into the body and transported to the parts of the body on which they are destined to have a useful effect. At the same time, some of the drug will travel to and be broken down (metabolized) by the liver. Usually this liver action diminishes the drug's effect, but the liver actually

converts some drugs into more potent forms. Drugs are eventually filtered by the kidneys into the urine, usually after the liver has metabolized them. All these processes may be susceptible to circadian rhythms, and may well be driven by individual clocks.

Most drugs are taken by swallowing, inhaling or injection, or are administered as suppositories. Because it is inconvenient to wake up to take them, treatment is usually at regular intervals during waking hours. In some cases, however, it is crucial for a doctor to consider whether this is the best thing. In diabetes, for example, insulin injections are used to replace deficient natural production by the pancreas. Diabetics have to mimic the natural rhythm of insulin output by anticipating the meals they will eat and controlling the quantity, quality and timing of their meals. Most find it best to take a big dose of insulin in the morning to cope with breakfast and lunch and a smaller evening dose to deal with dinner. Excessive insulin doses at night can be dangerous, for they may make the blood sugar level fall too low while the diabetic is asleep. Similarly patients with Addison's disease, who need to take corticosteroids, achieve the best results if they mimic the natural rhythm of the adrenal glands by taking most or all of their tablets at about 9 am.

Investigation of the liver and the kidneys reveals that both organs show circadian rhythms, so it is reasonable to expect a circadian rhythm in the way they deal with drugs. And if the rhythm is known, then drugs can be administered to maximum effect. Drugs broken down by the liver are best given at periods of low liver activity, but those that are converted to more active substances are most effective if administered when the liver is most active. In

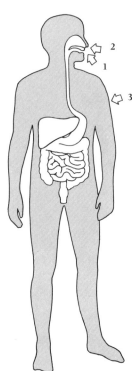

Several biological clocks may operate when the body deals with drugs. However they are given, all drugs are carried in the blood to the liver. Here, many drugs are broken down into inactive compounds or occasionally into more active ones, and liver cells show circadian rhythms in their activity. The end products of liver action are excreted by the kidneys which show circadian activity rhythms. The amount of a drug dose escaping the attention of the liver and kidneys is free to act on the end organ where it is needed; this organ may have a clock making it most sensitive to drugs at a certain time of day.

The most common method of drug administration is by mouth (1), but drugs taken in this way must be able to resist destruction by stomach acid and be easily absorbed through the intestine wall. The lungs have a rich blood supply, so some drugs can be inhaled (2). In emergencies, and for drugs destroyed in the stomach, an injection is made direct into the blood (3).

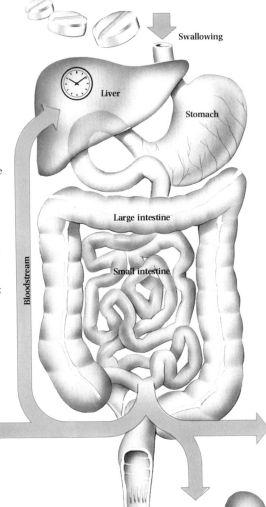

The activity of a drug may show a circadian rhythm. Theophylline, taken by asthmatics to widen their airways and thus increase peak flow, is eliminated most quickly from the blood by the kidneys when given at night. This means that the drug's 'half life', the time it takes for the amount of a drug in the blood to reach half its highest level, is shortest at night when the drug is most needed by the patient to combat asthma attacks.

The breakdown of a drug by the body can be assessed by its stay in the blood. After injection, repeated blood samples are taken to measure the drug's concentration. This rises slowly as the drug leaves the stomach, then climbs to a peak. The level falls as the drug is broken down by the liver and/or excreted by the kidneys. The longer it takes for the level to fall to half its peak, the longer a drug's effects on the body.

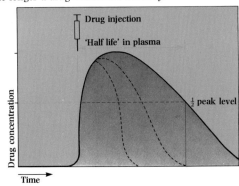

the case of drugs that are got rid of by the kidneys without being metabolized, the timing of treatment should reflect the period of lowest excretion. The organs that react to the drugs may also show a circadian rhythm. The bronchi of the asthmatic are a prime example of this. Because the bronchi are least sensitive at night it is logical to give larger drug doses in the evening.

Studying the timing of drug action has generated results with great significance to the treatment of allergic diseases, such as hay fever and nettle rash, which is also known as hives or urticaria. In both of them substances, such as grass pollen and house dust (collectively known as allergens), to which patients are sensitive, react with antibodies. As a result of this link-up, the chemical histamine is released from most cells, making small blood vessels leak fluid to produce the red skin and the raised weals of nettle rash and the blocked or runny nose and streaming eyes of hay fever. The degree of a person's sensitivity to an allergen can be discovered by injecting a small quantity of it into his skin and measuring the size of the weal that results. If this is done at different times it is found that sensitivity is greatest at night and in the early morning.

Antihistamines are drugs that block the action of histamine, although they cannot stop its production. If antihistamines are given at various times of day, before skin tests with allergens are performed, a rhythm in the degree of protection that they give is evident. Antihistamines have their most prolonged action when taken at about 7 am. At 7 pm they do well at protection but not for as long. One serious side effect of antihistamines, particularly for drivers, is that they cause drowsiness but, applying the results of these experiments, smaller doses should be taken in the mornings, which reduces daytime drowsiness with little loss of protection.

Exciting prospects for the drug treatment of cancer are being based on body rhythms. Studies of artificially induced leukemia, a form of blood cancer, in mice show that there is a circadian rhythm in response to the drugs used to treat the disease. At about 6 pm these drugs destroy the rapidly dividing tumour cells most efficiently. Unfortunately the same drugs also destroy some normal cells, which limits their usefulness in treatment, and can make human patients and laboratory animals very ill. The tests on mice show, however, that the poisonous effects of the drugs on normal cells are greatest in the morning and so treatment at 6 pm is best all round. It is not yet known if similar rhythms exist in human cancer but, if they do, then careful timing could improve current treatments considerably.

Circadian rhythms could also hold the key to better results in transplant surgery. The greatest obstacle in this branch of medicine is rejection of the transplant, due to the activity of certain immune cells specialized to eliminate foreign materials from the body. These cells seem to operate in circadian fashion. The transplant is thus best carried out when the immune rhythm is at its low point, when the ability to start up the rejection reaction will be weakest. Corticosteroids and other drugs are used to suppress the immune reaction, often at the high price of troublesome and unpleasant side effects. Greater attention to the immune rhythm could make these drugs work better—in smaller doses and with fewer adverse reactions on the body—simply by giving them at the most appropriate time.

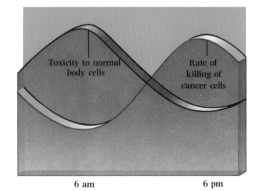

The time of day at which drugs are given may become important in treating cancer. Experiments on mice with leukaemia show that the cancer cells are most susceptible to drugs at about 6 pm. Conversely, this is the time when normal cells are least prone to being poisoned by drugs. If the same sort of rhythm occurs in man this may aid treatment.

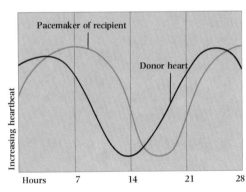

In human heart transplant surgery, *far right*, a small part of the recipient's old heart is left behind. This includes the pacemaker, a piece of tissue which controls the heart beat. The rhythms of donor and recipient pacemakers are slightly out of phase, showing that heart rate rhythm is probably driven by a clock in the pacemaker.

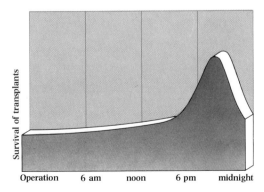

Time of transplantation may be crucial to the success of kidney transplants. A study on rats showed that the kidneys survived much longer if operations were performed at 8 pm. One explanation for this is that the operations coincided with the time at which the circadian rhythm in the immune response was minimal.

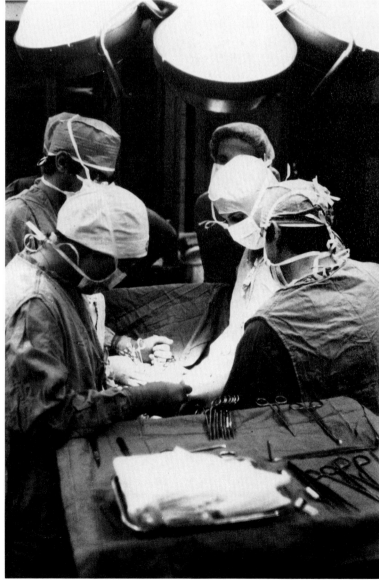

Rhythms of fate

Man is, at heart, a conservative creature. He may desire change in his life and environment but, when confronted with any drastic alteration, he panics. This inner terror is most real and most potent when change adopts the habit of disaster, leaving him with the feeling that his destiny is no longer in his own hands. It is man's sensation of inadequacy when confronted with the horrors of disease or drought, with earthquake and flood, with the death of those he loves, and with his own death, that has led him to try to explain—and if possible prevent—these happenings by summoning up the existence of forces that are not immediately obvious or measurable.

Many of the forces that man has come to believe in as the explanation and source of treatment for the unpleasant events of life are rhythmical or cyclical in nature, and all are controversial. Biorhythms, for example, are thought to operate cyclically within our bodies and brains to control our physical, emotional and intellectual performance, while the technique of biofeedback aims to give us power over the rhythmical activities of our brains by providing us with the machinery to record our innermost physiology. The medical art of acupuncture is based on the ancient belief that the energy that imbues us with life circulates round the body once every 24 hours. Both daily and yearly cycles are the essence of astrology, another of man's most ancient arts, and it is to the heavens that we look when we try to use the cycles of dark spots on the sun to explain the cyclical pattern of earthly weather.

Belief in these cycles and rhythms implies a large measure of faith because they cannot necessarily be measured scientifically. Such faith in rhythmic and cyclical forces is almost as old as man himself, since man has always been prey to death and destruction by natural disaster. The attitude we adopt today may be that nothing more precise could have been expected from prescientific man and that modern man should know better. We may view these forces with scepticism mingled, perhaps, with a sneaking feeling that there is a grain of truth in them somewhere, or we may take the view that the ancients were right after all, and that the trappings of modern life have blinded us to the potency of these rhythmical forces and that there are yet more to be discovered.

So who is correct? Are our lives determined by the cyclical movements of the heavens as interpreted by astrologers, or are we reacting to our inner biorhythms determined, like astrological influences, by our day of birth? Do natural disasters depend on the waxing and waning of sunspots? Can cancer be cured by acupuncture, which alters the cyclical flow of body energy, and can our rhythmical, electrical brain waves be controlled to rid us of anxiety and stress? The answer to most of these questions must be that we do not know. But because some of these rhythmical forces have been part of human belief for so many centuries, and because they are, for many, an ever-increasing part of modern medicine and psychology, they are well worth examining in depth.

In the study of rhythmical or cyclical events, but particularly those that seem to have a mystical component, statistical evidence is often employed to prove or disprove a theory. Every time we are told by the men and women of the media that there is a trend this way or that, we are reaping the results of statistical methods. The problem with such statistics is their precise interpretation. A sceptic would say that anything can be proved or disproved by statistics, so when we are trying to work out in our minds whether mysterious rhythmic and cyclical events are real, and are presented with statistical evidence, we must be as wary of the figures as of the phenomena.

To every owner of a human body, the machine that is the making of the man is the source and object of infinite speculation. But of all the phenomena displayed by the body, it is those that are rhythmical which prove to be among the most fascinating. We can all recognize with consummate ease the circadian, 24-hour cycle of sleeping and waking and could, if we wished, measure the rhythmic ups and downs of our body temperature or heart rate over any period of time, but it is impossible to convert our emotions into hard scientific data. Yet life does, indubitably, have its highs and lows, from the elation of success to the depths of failure; from the days when all seems right with our own personal world to those on which nothing goes well. Between these extremes are all those mediocre days, part good, part bad, on which there is little to record in the diary of events. So do these black, white and grey days fit into any sort of cyclical or rhythmic pattern and, if so, is it possible to predict them and to order our lives accordingly? Many people believe that the answer to this question is 'yes', and their faith has become known as the biorhythm theory.

Biorhythm theory began in the 1890s when a Viennese professor of psychology, Hermann Swoboda, noticed that the behaviour of his patients seemed to show a rhythmic pattern. This observation spurred him to make a meticulous study of his patients' case notes and from this study he arrived at the conclusion that behaviour fell into cycles with two distinct periods of 23 and 28 days. Putting his conclusions to practical effect, Swoboda devised a sort of slide rule from which patients could predict their 'critical' days. The 23 and 28-day cycles of behaviour were also 'discovered' quite independently by a German doctor, Wilhelm Fliess, at about the same time. Fliess was not a psychologist but

an ear, nose and throat specialist. Fliess based his study on examinations of the mucus-secreting cells that line the human nose, and his biorhythm theory depended on the periodicity of cell replacement. For Fliess, the significance of these cells was that they were supposedly male and female.

To the cycles of 23 and 28 days a German engineer, Alfred Teltscher, added another figure of 33 days, so the system of biorhythms now has three separate elements—a physical (P) cycle of 23 days, an emotional or sensitivity (S) cycle of 28 days and an intellectual (I) cycle of 33 days. These innate cycles begin, according to biorhythm theory, at the moment of birth and continue until death. The biorhythm hypothesis states that each cycle is divided into two equal phases and that a person's performance is best during the upper or positive half of the cycle and worst during the negative or lower half of the cycle. Our weakest, most vulnerable or 'critical' days are those on which a cycle crosses the zero line between positive and negative. If the theory is true then we should expect accidents and poor health on critical days in the physical cycle, depressions and bad personal relationships on critical days in the sensitivity cycle and poor judgment and learning on critical days in the intellectual cycle.

The total interpretation of these rhythms depends on the fact that they do not have the same periodicity. Because the cycles cross the zero line at different intervals there is only one day in about every six months on which all three cycles are critical together. This difference in periodicity would also explain, if the biorhythm theory is true, why we have so many ordinary days in our lives and so few extraordinary ones, because for most of the time the cycles are either moving away from or toward the zero line.

Physical

Positive phase

Negative phase

Emotional

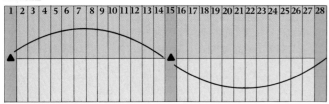

Intellectual

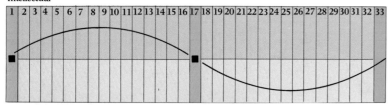

From the moment of birth and throughout life, three basic rhythms are believed to affect us. A 23-day physical rhythm supposedly governs our strength and endurance; a 28-day rhythm governs our sensitivity, moods and relationships; a 33-day intellectual rhythm controls our powers of judgement and thought. Each rhythm alternates regularly from a positive to a negative phase, becoming critical on the one day when it crosses over from one phase to another. A rhythm's phase on a particular day can inform us about ourselves. For example, when the physical rhythm is positive, strength and endurance are at a peak, so physical exertion can be attempted with confidence.

High positive days, low negative days and potentially dangerous critical days can all be predicted by using a biorhythm calculator, such as the Biomate, *left*. The serrated gear wheel at the bottom of the calculator turns the four dials, which show the date, and readings for each of the three biorhythms. To discover your biorhythms for a particular day in any calendar year, you must move the dial until the date of your birthday is aligned with the vertical axis line; the gear wheel is then disengaged. The settings for the biorhythm dials are found in the count list provided with the calculator, each biorhythm having a special setting according to how old you will be in that year. Each dial is turned, using a finger, until the right setting from the count list is lined up with the axis line. At this point the gear is re-engaged and the calculator will show your biorhythms for any day of the year. This biomate is set so that all three biorhythms are critical, which is a rare occurrence.

Racing along a highway near New Orleans, Jayne Mansfield, the film star, made a driving error, crashed into a truck in the opposite lane and died. Her biorhythms on that day, 29 June 1967, showed that she was physically critical a day after she had been emotionally critical, and a day before she was due to be intellectually critical. The fact that she died on the day when her biorhythms were so critical seems to support biorhythm theory. The record-breaking athlete, Sebastian Coe, won the final of the 1,500 metres at the Moscow Olympics on 1 August 1980. His biorhythms showed that he was emotionally critical, that he was physically low three days after being critical, and that he was intellectually low and approaching critical. According to the state of his biorhythms, Coe should not have won such a gruelling race, but the fact that he did win suggests that biorhythm theory is wrong. There are many similar examples that either support or contradict biorhythm theory.

The idea of a birthday-based biorhythm is one that has achieved considerable popularity, and it is now possible to buy a biorhythm calculator to work out your critical days so that you can order your life accordingly. The makers of one such calculator suggest, for example, that you should try to avoid driving whenever S and P criticals coincide and take particular care with business ventures on critical I and S days. Critical days in the P rhythm are those on which you are likely to be badly affected by drinking alcohol, while critical S days are bad for relationships. As you would expect, sporting performance peaks when the P rhythm is in its positive phase, but intellectual achievement is greatest when I is on the up. In business a combination of I and S in the positive phase augurs well for important deals, and the same duo helps give a gambler a lucky streak. People in creative jobs are advised to 'expect extraordinary ideas' on critical I days and to 'take notes for future use'.

All this sounds plausible indeed, and in Japan some companies even take steps to prevent workers from using machinery or driving trains on critical P days. But do biorhythms really exist? The statistical evidence is, at best, unconvincing, and much of it is based on samples too small to give significant results. In statistical studies of biorhythms it is found that the bigger the sample the less it conforms to the theory. In one test of 112,000 car drivers who had been involved in accidents—and for whom the biorhythms were worked out—there was no correlation between critical days on any of the three cycles and the dates on which the accidents occurred. According to one biorhythm protagonist, Arnold Palmer was in positive phase in all three rhythms when he won the British Open Golf Championship in 1962, and lost the American PGA title two weeks later when low in all three. However, a detailed analysis of Palmer's achievements between 1955 and 1971 showed no significant correlation between his performance and his biorhythms.

In biological terms, one of the chief arguments against biorhythms is their incredible regularity. No known physiological rhythms are ever so well ordered and there is no evidence that biorhythms, when measured in subjects protected from all outside environmental influences, take up their own periodicity, or free-run. Another powerful objection to the biorhythm theory concerns the fact that all three cycles begin together on the day a person is born. On this day—which is undoubtedly stressful—all three cycles are set at 'critical' and, if this is so, you would expect many more babies to die at birth than in fact do so. And even after this, we all get through our triple criticals by some means. There is probably no truth in the biorhythm theory, but it might be an interesting exercise in self-analysis to try constructing your own chart, based on your physical, emotional and intellectual performance each day. Even if biorhythms do not work, belief in them may possibly serve the useful purpose of making us more self-aware and bringing us back to reality occasionally.

The improvement of mental and bodily well-being is also the aim of the technique of biofeedback which has a much more scientific background. There are countless processes continuously going on within our bodies of which we are normally totally unaware. The purpose of biofeedback is to make a person aware of these processes, many of which are rhythmic and cyclical, and to teach him to exercise conscious control over them.

The part of the body most intimately involved in biofeedback is the brain, the ▷

▷ final arbiter of all human actions, and it is the rhythms of the brain that were used in the first biofeedback experiments carried out by Dr Joe Kamiya in San Francisco in the 1960s. The human brain is a ferment of electrochemical activity, and the electrical pulses it generates are the physical clues to the work of its cells in sensory processing and the memorizing of information. The rhythms of the brain are recorded as electroencephalogram (EEG) traces, and produced by sticking a forest of electrodes to the scalp.

By means of EEG recordings it has been discovered that four distinct sorts of electrical rhythms take place in the brain. The first of these brain waves to be identified were the alpha rhythms, with a frequency of between 9 and 12 cycles per second, but sometimes between 8 and 13 cycles per second. The alpha rhythm is usually associated with a state of relaxed wakefulness. People undergoing a pronounced period of alpha activity describe their mental state as peaceful and sometimes mention a floating sensation. When your brain is producing alpha rhythms it is receptive, but you are not concentrating on any particular thought or activity.

Theta waves are about half the frequency of alpha waves, from 3.5 to 6.5 cycles per second, and rarely make up more than about 5 per cent of the total brain wave pattern. Theta waves are recorded most often when a person is drowsy or dreaming. Delta waves are even slower, with a frequency of 0.5 to 3 cycles a second, and are almost exclusively confined to the deepest stages of sleep. Nearly all the other electrical activity of the brain is classified as beta and its frequency ranges from 13 up to about 40 cycles a second. Beta is associated with alert behaviour and concentrated mental activity.

As so often in science, things are never clear and simple. Brain wave patterns are, in reality, a combination of the rhythmic and the irregular, and change so much from moment to moment that it is impossible to say with any certainty what a 'normal' brain wave pattern looks like. Despite this, the techniques of biofeedback can help a person to achieve a desired state of brain activity. The key to biofeedback, which has been applied with some success to the control of other body functions, such as blood pressure, muscle tension, skin temperature, sweating and heart rate, is to display to the subject monitored read-outs of the physiological function he wishes to control. Through a system of trial and error he then works at achieving the mental or physical state he is seeking, and can see the results of his efforts displayed before his eyes. The final accolade of biofeedback is awarded if he can achieve his goal without the machine. The technique has definitely shown encouraging results for people suffering from anxiety, who have few alpha brain waves.

Brain wave biofeedback has become a popular aid to meditation. Although it may not be true that experts of yoga and Zen have more slow alpha waves in their brain rhythms than the rest of us, mastering the art of relaxation is a worthwhile aim—with or without biofeedback—in the hurly burly of modern life. Mind control through transcendental meditation is claimed to increase the number of slow alpha and theta waves in the brain wave patterns. Research on the physiological effects of such meditation suggest that it does not result in an altered state of awareness, but in a mental state akin to sleep. This does not necessarily diminish its usefulness as an effective weapon against stress.

Like any machine, the human body is not perfect, and there can be no one

Yin and **yang** are the two opposing energies found in all things. They are both antagonistic and, at the same time, complementary. For example, the night is yin, the day is yang. The acupuncturist aims to restore the harmony between the yin and yang of the body.

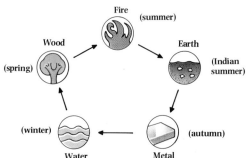

The five Chinese elements encompass all natural phenomena and together form the creative cycle of the seasons. Wood burns to create fire, which leaves behind the ashes of earth; from earth metals are mined, which become molten like water when heated; plants need water to grow and to make wood. These elements also qualify the 12 organs of the body through which the life energy, *chi*, courses every 24 hours. Starting at 3 am *chi* flows through the lungs and activates them. At 5 am it enters the colon and continues on its course, activating each organ in turn until it completes one cycle of this Chinese clock. Acupuncturists use this flow of *chi* in their diagnoses and treatments, and also the relationships of opposite organs on the clock: too much energy in the heart, for example, may upset the gall bladder and vice versa.

When a Zen Buddhist is meditating his body is relaxed, but his mind is alert. His brain is generating alpha, beta and theta rhythms and his skin shows a high electrical resistance. Without knowing it he is practising biofeedback, because he is learning about some function of his body or mind and controlling that function. Research into many forms of meditation has helped biofeedback therapists correlate the various states of awareness with the four principal brain rhythms: alpha, beta, theta and delta. As a result, people can now learn to relax their bodies and minds by finding out, with the aid of a therapist, how to develop their alpha and theta rhythms.

INCREASING YIN, WEAK YANG

3 am
1 am
5 am
11 pm
Liver
Lungs
Gall bladder
Colon
7 am
Triple warmer
Stomach
9 pm
9 am
Pericardium
Spleen
7 pm
Kidneys
Small intestine
Heart
11 am
Bladder
5 pm
1 pm
3 pm

STRONG YANG, WEAK YIN

STRONG YIN, WEAK YANG

who can genuinely claim never to have had a day's illness in his life. The practice of medicine is thus as old as man himself and one of its most ancient branches is the Chinese art of acupuncture, which originated at least 5,000 years ago, but has only recently become popular in the West. All Oriental medicine is based on the twin towers of yin and yang. These are not tangible entities but tendencies in the movement of energy—yin is a tendency toward expansion, while yang is a tendency toward contraction. The theory is that all the rhythmical aspects of human physiology come about because of the constant flux that exists between these two extremes, pulling first one way, then the other. Thus yin and yang occur in a state of dynamic tension and interaction, but the end result is a state of perfect bodily harmony.

In the living world the forces of yin and yang, which underlie all aspects of Oriental life, are thought to operate by means of vibrating energy or *chi* which permeates all creatures. When a human body is healthy, the rhythm with which *chi* pulses is in harmony with the *chi* vibrations of the environment. Within the human body, life-giving *chi* energy circulates in specific systems, the *ching* or meridians. There are 12 regular meridians, each supplying particular body organs, and six 'extraordinary' meridians. Energy is thought to flow through these six emergency channels only when the normal 12 cannot handle the excess flow of energy generated by disease or disorder of any organ. Strategic points, located along all the meridians, and numbering about 1,000 in all, are the control valves that regulate energy flow and are the points for acupuncture treatment; anatomically the most important strands of the nervous system run along the meridians.

Acupuncture has a complex system of laws, but the one that stands out most clearly as a biological rhythm states that the *chi* circulates round the body once every 24 hours. Each of the 12 regular meridians—and hence their associated organs—is thought to have a two-hour period of maximum energy flow and a two-hour period of minimum flow. Using this knowledge, and their centuries of accumulated experience of when the disorders of any body system are likely to show the most acute symptoms, ancient Chinese medical practitioners drew up tables, still used today, of the most favourable times in each 24 hours for the treatment of disease by inserting needles into the appropriate point on a meridian to redistribute the excess energy.

Added to this clock is the knowledge of the opposite yin and yang properties of the various body organs. In acupuncture organs are treated in pairs, one member of each pair being a hollow, yang, organ such as the stomach, the other a dense, blood-filled yin organ such as the spleen. Thus if the kidney (yin) is stimulated the result is an improvement in the large intestine (yang). The treatment is best carried out in the evening because this is yin time and so coincides with the yin organ being stimulated. According to the acupuncture law of the five elements, the organs served by the 12 meridians are in six pairs and five of the pairs are equivalent to the elements wood, fire, earth, metal and water. The sixth pair are also 'fire' organs. Treatment is based on the interaction of the elements—for example, water (the kidney) quenches fire (the heart), so that if the kidney is sedated then the heart is stimulated.

No longer dismissed as 'quackery', but not heralded as a cure-all either, the practice, if not the theoretical background, of acupuncture is becoming ▷

The art of acupuncture is knowing where, in relation to which disease, to pierce the skin with needles. Each acupuncture point relates to a specific imbalance of yin and yang energies, which is diagnosed by the subtle assessment of the qualities of the pulse and the colour of the tongue and skin.

In acupuncture, the life energy that activates the 12 organs of the body is created by breathing and eating. It passes from one organ to the next along channels called meridians which, being neither arteries nor nerves, appear to have no place in the anatomy of the body. Each of the 12 main organs has one meridian associated with it, and each meridian is present on both the left and right sides of the body. Imbalances of yin and yang cause too much or too little energy to flow, thereby creating the symptoms specific to a disease. By carefully selecting the appropriate points, of which there are about 1,000, an acupuncturist can restore the harmony of the body's yin and yang.

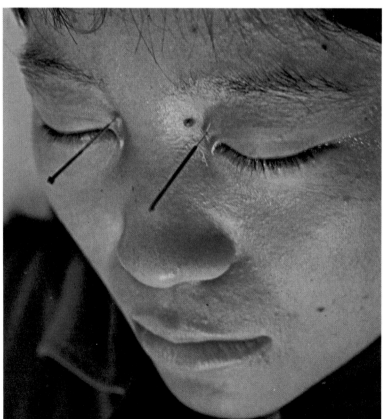

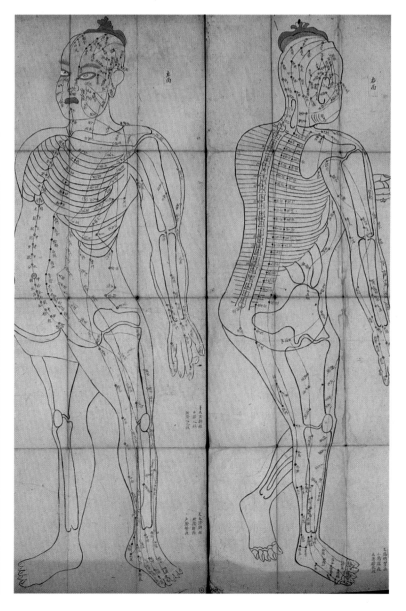

▷ accepted by many Western medical practitioners, particularly for the relief of chronic pain, for asthma and the treatment of stress-related disease. A few thin needles are inserted into points along particular meridians and either twisted or connected to electrical stimulators. The most recent research, carried out in both China and the West, suggests that acupuncture may work to relieve pain because it stimulates the pituitary gland at the base of the brain to release endorphins, natural morphinelike chemicals which numb pain.

Older even than acupuncture is the practice of astrology which, with astronomy, is one of the most ancient branches of science. There can be few of us who have never read our horoscope in the daily newspaper or do not know under which 'star sign' we were born; but what is the true basis of astrology and can it really foretell the future? Astrology has its basis in the cyclical movements that can be observed in the heavens. As for predicting tomorrow, any serious student would say that astrology is concerned more with defining trends than forecasting specific events, and that the horoscopes printed in newspapers and magazines are mere commercialization.

The deductions of astrology are based on the positions of the stars, sun, moon and planets in relation to the Earth. No astrologer would tell you that the Earth is the centre of the universe, this is just a convention that has persisted since ancient times. When studying the heavens the earliest astronomers and astrologers thought that the sky was a celestial sphere that revolved around the Earth. Taking the Earth as the centre of the sphere the sun seems to travel round it once each year along a regular path called the ecliptic. Following the same route as the sun, the moon and planets seem to trace a similar path in front of certain constellations, but stay within a zone extending 8 degrees on each side of the ecliptic containing the 12 constellations of the zodiac.

The full 360-degree circle of the zodiac is divided into 12 equal regions of 30 degrees. The dates for each sign are based on the position of the sun on 21 March, the spring equinox. In about 900 BC the sun appeared to enter the constellation Aries on that date, and this set the start of the zodiac circle. Astrology takes no account of the fact that the stars from which the signs were named have since taken up different positions relative to the Earth because of the 'wobble' or precession of the Earth's rotation about its axis.

Due to this precessional wobble, the Earth's spinning axis with respect to the stars shifts slightly. Astronomical calculations have arrived at a time-span of about 21,000 years for the Earth to return to exactly the same orientation in the heavens relative to any star, although astrologers take the figure to be 25,868 years. During this movement the Earth is thought by astrologers to pass through the influence of each sign of the zodiac in turn. Thus in the course of recorded history we have passed through the Age of Leo, which corresponds to the Stone Age around 9000 BC, and are now entering the Age of Aquarius which will last 2,160 years. Since Aquarius is associated with the dual concepts of science and humanity, astrologers predict that while man will become increasingly preoccupied with technological advancement, he will also strive towards gaining a victory for peace over conflict.

The apparent movements of the planets relative to the signs of the zodiac are among the cornerstones of astrology. For astrological purposes the sun and moon are included among the 'planets' and both are thought to be powerful

The sun and its retinue of planets are moving through space at a speed of 12.1 miles (19.4 km) per second toward the constellation of Hercules. As a result, the planets orbit the sun in spirals, the diameter of each spiral depending on the planet's distance from the sun Making a cross-section through this long body of the solar system is like stopping it in time and, like a horoscope, it tells the position of the planets at that one moment. The horoscope, *right*, is for the birth of the United States of America, when the Declaration of Independence was signed at Philadelphia at 3 am on 4 July 1776. Astrologers predict the fate of the United States and her people from the fact that Gemini was rising, and from the relationship of the planets to each other and to the signs they were in at that time.

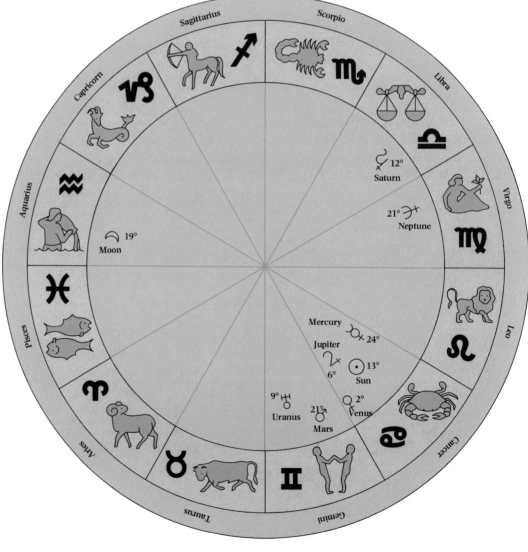

Horoscope of the USA
3 am 4 July 1776 at Philadelphia

influences, especially the sun. The farther a 'planet' is from the Earth, the more slowly it seems to move through the zodiac, so that while the sun takes a year to move in a complete circle relative to the Earth and the 12 zodiac signs, Pluto takes 248 years, and the moon only 28 days.

The most familiar of the cycles used by astrologers is the apparent yearly progression of the sun as it passes 'through' the dozen zodiac signs—in reality, of course, it is millions of miles in front of the constellations. It is from the position of the sun that you arrive at the fact that you were born, say, under the sign of Libra, because the sun is said to pass through Libra between 23 September and 22 October each year. To be born on the cusp means that your birthday is on the dividing line between two adjacent signs.

A person's horoscope, the prediction of his future and an analysis of his personality, is arrived at by more than just the position of the sun and can be thought of as a cross-section through the sun and its attendant planets as they spiral through space. In a true horoscope, therefore, the positions of the planets relative to the signs of the zodiac, and as viewed from Earth, are of great significance. The apparent angles of the planets, as they move in relation to one another and to the Earth, and calculated on the ecliptic, are the aspects. The aspects of the planetary groupings at the time of a person's birth are thought to be essential qualifiers of his 'vital energies', the forces that shape his whole life. When two planets are side by side, in conjunction, in a particular section of the zodiac, they have a combined influence. An angle of 60 degrees (a sextile) between planets, and its double, the 120 degree angle (a trine), are both considered to be beneficial influences, but angles of 90 degrees (a quarter) and

180 degrees (opposition) are disruptive.

As well as its yearly movement, the night sky seems to rotate around the Earth once every 24 hours. Like the yearly 'movement' of the sun, this diurnal rotation is divided into 12, the divisions being known as houses. Within each 24-hour period each planet appears to cross each of the 12 houses in turn. The houses are believed to have particular influences on an individual's life, from his personality and profession to his possessions and pastimes. In an individual horoscope the rising sign is also important. This is the sign of the zodiac on the eastern horizon at the moment a person is born.

There is no doubt that astrology is a fascinating subject, but also one about which it is difficult to be objective, because there always seems to be some part of an astrological assessment that 'fits' with reality. The statistical evidence seems to suggest that there is nothing in it—in one sample of 300 people there was, for example, no clear correlation between actual events and astrological predictions. In an attempt to bring astrology up to date, a new kind of astrology has been advanced. This cosmobiology dismisses traditional astrology but suggests that there is a statistical correlation between career choice or achievement and the positions of the planets. The French researcher Michel Gauquélin contends, for instance, that talented athletes are born when Mars is in the rising sign or in the mid-heaven—that is, has just passed its highest point in the sky. Others dispute their results and so the controversy rages on.

There is, however, no controversy about the fact that the cycles of the heavens are among the most impressive of all the cycles of nature, and it is hardly surprising that for thousands of years man has striven to associate these ▷

The annual revolution of the Earth around the sun is one of the main astrological rhythms, since a person's sun sign is decided by the date on which he was born. Because the Earth is orbiting around it, the sun appears to pass in front of each sign of the zodiac once every year. If a person is born between 20 February and 21 March the sun then is said to be 'in' Pisces as shown by the arrow, *below*, and the influence of the sun's vital force shapes the person's individuality in Piscean fashion. The planets also orbit the sun, but with different orbital periods, their influences changing as they pass from sign to sign. Jupiter, for example, orbits the sun every 11.8 years, spending nearly a year in each sign, so its influence is combined with the characteristics of the sign in which it happens to be.

Jupiter and Saturn were both in the middle of the sign of Taurus when John Lennon was born on 8 October 1940. This conjunction of the two planets occurs approximately every 20 years, since this is the average time of their two orbital periods. When they came together for their next conjunction, in the sign of Capricorn in 1960/61, Lennon was beginning to write, with Paul McCartney, the songs that were to make the Beatles famous. The following conjunction occurred in late 1980 and, just as Lennon was emerging once again into the public eye, he was murdered. The cycle of these successive conjunctions indicates a new start in a relationship and new social adjustments, both of which Lennon was embarking upon, but it is not the cause of the events that occurred.

As planets orbit the sun at different speeds, they regularly make specific angles, or aspects, with the Earth and each other. Planetary aspects present at birth are shown on one's horoscope, and are part of the basis of astrologers' predictions. When planets are within 8° of one another in the sky, a conjunction occurs, *top right*. The planetary influences modify each other and signify the start of a new cycle. When planets form a right angle with Earth, the aspect is square, *middle right*, and the result may be disruptive, although determined people may derive energy from it. When planets form a straight line with Earth they are opposed, *right*, and although their influences complement one another, tension is usually created.

Aries · Taurus · Pisces · Gemini · Aquarius · Cancer · Capricorn · Leo · Sagittarius · Virgo · Scorpio · Libra · Sun · Earth

Conjunction

Square

Opposition

149

▷ cycles with his own destiny. One of the astrological 'planets'—the sun—provides us with yet another natural cycle that seems to influence our lives. The ancient Chinese were probably the first people to notice that the dark spots visible on the sun seem to come and go. The first 'modern' astronomer to see these sunspots was Galileo, who carried out his work in the early seventeenth century, but the most thorough investigation made before the present century was by the German amateur astronomer Heinrich Schwabe in 1843. Since that time scientists and non-scientists alike have searched for a connection between the rhythmic appearance and disappearance of sunspots and cycles and the occurrence of events on earth such as earthquakes, droughts and floods.

Even before Schwabe's discovery that sunspots occur in 11-year cycles, the eminent English astronomer, William Herschel, had noticed that the numbers of spots varied, and in 1801 he suggested that the market price of wheat was closely correlated with sunspot numbers. Over the short-term period studied by Herschel this may have been true, but there was no correlation over a longer period. Since that time the quest for a relation between sunspot cycles and earthly events has been a curious one, and for every suggestion confirmed, several others have fallen by the wayside. The obvious reason for this is that two cyclical patterns may appear to be marching in step for a while, then move out of phase. The other is that human nature finds comfort in connections which explain events we cannot control.

Throughout the second half of the nineteenth century the fact that sunspots do come and go in a more-or-less regular 11-year cycle was common knowledge. Clear associations between sunspot cycles and magnetic storms on

earth had been recorded and have since proved to be totally valid. The establishment of this connection led, however, to an explosion of supposed associations. In this burst of enthusiasm almost any cyclical occurrence could be used to play the sunspot game. Periodic epidemics of Asiatic cholera; mass uprisings around the world; the ups and downs of the world's financial markets; temperatures in tropical regions; wind direction in Argentina; the rabbit population of England; the levels of Lake Victoria; the depths of the Nile and the Thames; monsoons in India; soil temperatures in Scotland—all have been linked to the sunspot cycle at one time or another.

Some of these proposed links had a reasonable basis in fact, while others were totally arbitrary statistical associations with no rational back up. All of them were eventually disproved. Even today there is the occasional suggestion that stock market cycles have something to do with sunspot cycles, but these must be dismissed with the scorn they deserve. In the 1970s, a book was written suggesting that a complex chain of events involving sunspot cycles would lead to a massive series of earthquakes in 1982, when all the planets of the solar system will be within a 60 degree span on the same side of the sun. The combined gravitational effect, named the Jupiter effect, will, it was argued, come at a peak in the 11-year sunspot cycle and result in an especially high degree of solar activity. This will apparently alter the Earth's rotation and cause a devastating series of earthquakes in California. The idea rapidly captured the public imagination, but has since been retracted by its leading author.

Although earthquakes may not be influenced by events on the sun, there are some links between solar activity and earthly happenings that cannot be

dismissed so lightly. Sudden releases of magnetically pent-up energy, or solar flares, send out streams of high-energy particles, intensified X-rays, ultraviolet light and other forms of electromagnetic radiation. These charged particles race through space. Some collide with the earth's atmosphere and, through a series of complex reactions, create the beautiful and awesome spectacle of the aurora borealis or northern lights. The same emanations initiate geomagnetic storms that disrupt radio communications and cause power blackouts.

The frequency of these solar flares tends to ebb and flow roughly in phase with the 11-year sunspot cycle. The most intense and potentially dangerous flares—those that cause 'super' geomagnetic storms—occur mostly in the declining phase of the sunspot cycle, several years after sunspot numbers have reached their maximum. The sun reached a peak in the current sunspot cycle in late 1979 and early 1980 and will be in its declining phase in 1984. There is a real threat that intense solar flares will cause disruptive geomagnetic storms during this period.

As well as the sudden release of energy and particles in the form of solar flares, the sun continually issues forth a stream of charged particles, known as the solar wind. On earth, moderate magnetic storms seem to occur at 27-day intervals and this is roughly the same as the rotation period of the sun as viewed from earth. It has been discovered that openings in the magnetic field of the sun's outer atmosphere or corona, allow the escape of high speed solar wind particles. These coronal holes thus 'point' towards the earth once every 27 days, which explains the cycle. In addition, coronal holes tend to be more prominent during the declining phase of a sunspot cycle.

If events on the sun can cause magnetic storms on earth is there any truth in the rumour that sunspot cycles influence our weather and climate in a more general way? Since 1976, when it was conclusively shown that the sunspot cycle has not always been regular in the past, the search for the answer to this question has become more urgent. During the period 1645 to 1715, for example, practically no sunspots were observed and auroras and other earthly manifestations of strong solar activity were infrequent. This was, however, the time of the 'little Ice Age', so there may have been some connection. It is now known that over the past 7,000 years there have been many interruptions in the solar cycle, but it is not certain whether they have been responsible for the undoubted fluctuations in our climate.

After each 11-year sunspot cycle the magnetic field on the sun reverses, producing a 22-year magnetic cycle, which may prove to be more fundamental than the 11-year one. Certainly a striking and well-documented correlation exists between this 22-year cycle and widespread drought in the western United States. Studies on tree rings have successfully traced this correlation back to the seventeenth century.

Most investigators of the relationship between events on the sun and those on the earth view the connections with cautious optimism mixed with a liberal dose of necessary scepticism. But there is no arguing with the truth that the sun is a rhythmic body. The quest for a solar link-up has recently acquired a new respectability and sophistication, so the future may give us an entirely new perspective on our relationship with the sun's rhythms which will allow our climate to be more accurately predicted.

Sunspot numbers rise and fall every 11.1 years on average, and many attempts have been made to correlate events on earth with them. A Russian scientist, A.L Chizhevsky, studied many such cycles and found a corresponding cycle in the mass movements of people. In the year or two before peak sunspot numbers occur, people are excitable, nations are aroused to their greatest achievements, and revolutions, strikes and wars occur. Twentieth-century examples are the student riots and workers' demonstrations in Paris in 1968, *left*; the Russian Revolutions of 1905 and 1917; the British General Strike of 1926; the Spanish Civil War in 1936, and Iran's revolution in 1979.

Huge amounts of high-energy particles are released from the sun at times of peak sunspot numbers. With X-rays and ultraviolet light, these particles stream through space to interact with the earth's upper atmosphere, causing it to fluoresce. The night sky over the polar regions lights up in a display of bright colours—the aurora borealis of the northern hemisphere, the aurora australis of the southern hemisphere.

A sense of rhythm

Within an hour of birth babies flex their limbs and move their heads in approximate time to the rhythms of the human speech they hear around them. If a different language is used the actual movements of the babies alter slightly but the rhythm of their movements changes to match the language that is being spoken. This fascinating discovery was made in 1974. It is still not known when the sound of a child's native language becomes so firmly locked into his make-up that the switch from one rhythm to another is no longer automatic. What is certain is that human babies begin learning language—and with it culture—from the moment of birth on, and that they first do so through the rhythms,of movement. Babies do not respond in this way to music of any kind, nor to random nor regular tappings.

As well as responding to cultural influences the human body also responds powerfully to the rhythms of the universe, to day and night and the march of the seasons. Most important of these cosmic rhythms are the 24-hour or circadian ones which set the pattern for day-to-day life. By the time a child is three months old circadian rhythms have become part of his physiological constitution, but the exact beat of these rhythms is determined by the cultural environment in which the child is reared. The result of this is that we all respond in roughly the same way to daily and seasonal change, but it is our cultural rhythms that give us an inner sense of feeling 'at home' in certain circumstances and ill at ease in others.

Just as surely as our bodies work according to rhythmic principles, so too do human societies. Cities are social clocks of great accuracy, with morning wakings, raucous noons, declining evenings punctuated by spurts of activity around midnight and, finally, sleep. These rhythms are repeated daily but vary in detail according to the day of the week or the season of the year.

The city is only one sort of cultural environment to which the body can adapt, and throughout our lives we constantly change according to our physical and cultural surroundings. But one of the ironic ambiguities of belonging to a culture is that we can consciously seek out synchronization with the rhythms around us and, at the same time, override some of our body rhythms dictated by the cosmos, such as the rhythm of sleeping and waking. The result is that we can understand when we are in synchrony with the rhythms of the universe and with the rhythms of our own bodies, and seek that kind of harmony, but we can also disregard that synchronization or deliberately put it out of our minds as we obey cultural mores or psychological demands.

The rhythms of the universe are not innately good or bad, it is cultural evaluations that label them—and cultural rhythms too. This means that in the end we all have to make these evaluations and take up a moral position with regard to ourselves, our neighbours and our environment. We can be like the yogi and search for 'oneness' with the universe. Or we can go to the opposite extreme and seek to control or exploit our natural and environmental rhythms (whatever the cost to the environment or the well-being of others). Some cultures, particularly those of the Western world, encourage us to overlook our natural rhythms as we search for some cultural ideal—or merely aim to be like everybody else. Because we have the power to disregard our body rhythms, and those of the world around us, we are masters of our own destiny. We can cooperate with or oppose those rhythms, or we can choose any number of intermediate strategies of life.

Culture penetrates every dimension of our lives. Even the most obvious characteristics which we share with our fellow members of the animal kingdom, such as our need to eat, reproduce or sleep, are in fact closely linked with the cultural requirements of the society in which we live. It is for this reason that actions which fall outside our particular social norms make us uncomfortable. One good reason for the discomfort we experience in foreign situations and environments is that we do not know exactly what is expected of us. Even more significant is the fact that our movements are not cued to the language, nor to the other rhythms around us. Speaking your own tongue after you have been abroad for a long time seems so delicious simply because your movements can again be ordered in the way you first learned them. We can all learn to speak other languages and adopt the mores of other cultures as if they were our own, but most people are most at ease in their 'native' surroundings.

In all cultures there are 'owls' and 'larks'—people who operate best either at night or in the morning. These habits and inclinations are cued to our body metabolism and so cross cultural boundaries, but different cultures make us respond to our basic rhythms in different ways. Most farming people, for example, begin work at dawn, or even before, and in such cultures owls are at a disadvantage. In an office operating flexible hours, on the other hand, the owls can work things to their benefit. Owls come into their own in any culture that adopts the use of artificial light because they can stay up later at night in a way that is not feasible for people reliant on the sun alone as a source of illumination.

The idea of a 'good night's sleep' is essentially a Western one and many Westerners make themselves miserable because they cannot sleep at the particular time, and for the prescribed number of hours, that their culture—and their parents—tell them they should. The Tiv people of Nigeria, however, divide the night into three sleeps and expect to wake first around midnight, again at about 3 am and yet again just before dawn. For the Tiv, as for everyone else, personal biological rhythms are thus made to adjust to cultural rhythms. Many other people of the world sleep only when they are sleepy.

As with sleeping, our culture tells us when to eat, what we should eat and how our food should be prepared. Physiologically the human stomach contracts powerfully every 75 to 115 minutes both day and night, yet almost no human beings eat so often, and, at the other end of the scale, comparatively few adults eat only when they are hungry. The number of meals eaten in a day, and the timing of meals, conforms to many different patterns. In some European cultures four meals are prepared and eaten during the day—an Englishman once invited an American to tea and was taken aback by the response, 'I never eat between meals.' Most Americans eat three meals a day, most Asians eat two—one in the morning, the other in the evening. Some West African peoples produce one huge meal in the late afternoon then finish up the left-overs in the course of the next morning. Other societies allow, or even encourage people to have snacks whenever they feel hungry, yet others forbid it. An enormous amount of psychological motivation is needed to alter the rhythms of eating to which the body has become accustomed once cultural conditioning has done its work.

Both bodily and cultural rhythms work together to affect our moods. Most

people imagine they are in some sort of general command of their moods and do not notice small-scale mood changes, but when they keep meticulous diaries and records of the way they feel, it is sometimes possible to discern roughly rhythmical cycles of mood changes over many days. Scientists have not yet determined the major rhythms of these mood cycles, nor their causes, but certainly many people believe in them and some religions or pseudo-religions have tried to standardize them. The only known cycle which undoubtedly affects moods is the female menstrual cycle.

Many cultures allow their members to do things that have a considerable influence on the body's natural rhythms and many even dictate that such actions are carried out. Drugs, including alcohol, have a powerful effect on natural rhythms, including those of mood, and while some cultures forbid the use of all drugs to men and women of all ages, others encourage them, either overtly or unknowingly. Alcohol is a drug whose effects are both physiological and cultural. Cultural influences dictate how alcohol should make us feel—at a party, for example, we expect it to make us elated—but the alcohol itself affects our body rhythms such as our pattern of sleeping and dreaming. The use of these and other drugs—for example, the nicotine in cigarettes—is totally under cultural control. People of different ages or rank in society may be allowed quite disparate access to drugs, and the drug chosen may depend on one's age group. In Africa, for example, young men were traditionally allowed only a little beer—the bulk of the supply went to the old men and sometimes the old women of the society.

One of the most widespread and potent ways by which the basic body rhythms have been altered is through the use of the birth control pill. Whether or not a woman should take the pill is determined by medical and religious considerations. Some of the medical objections to it arise because doctors are wary about the damage that may be wrought by a long-term manipulation of the basic body rhythm of ovulation.

Natural body rhythms are also disrupted by stress. The stress of an illness or some other psychological pressure, the stress of unaccustomed physical effort or of social disaster may all bring about many changes. Even altering our daily routine is stressful, and travel to east or west puts more strain on the body systems than travel to north or south simply because it involves a disruption of our biological clocks. Furthermore, modern culture allows us to make these changes at great speed. In the days before air travel, when people crossed the Atlantic in liners and allowed their internal biological clocks to adjust day by day, there was no such thing as jet lag. So it may be that the more complex a culture becomes the more strain is imposed on human rhythms.

The internal rhythms of all animals, including man, are influenced or entrained by the rhythms of the environment, and without these external pacemakers a creature's internal rhythms will free-run and may become quite random. The way in which we receive the essential information about our surroundings is through our sense organs. Thus a flock of birds is made a flock by entrainment, each animal using delicate sensory mechanisms to pick up environmental cues, then adjusting its activities accordingly. Herding animals entrain their rhythms and so do people—even newborn babies take the rhythm of the adult speaking voice. ▷

It has become culturally necessary in Western society to sleep only at night and for a prescribed number of hours, instead of falling asleep when we feel drowsy. Cat naps during the working day, *above*, are often frowned upon, but many societies condone 'siestas'.

Alcohol severely disrupts the body's natural rhythms. Hunger, sleep and temperature rhythms are thrown out of gear and the brain's normal perception of passing time is upset. Social and business drinking can become an integral part of the day's rhythm, but when alcohol becomes a problem, *right*, the timing of the next drink dictates the pattern for the whole day.

In all societies the day revolves around mealtimes, *left*, although the number of meals and their timing are usually influenced by class, occupation and religion. High tea in the north of England is often as early as 5 pm, while a restaurant dinner at midnight is normal in Latin cultures. The weekly pattern of Jewish Friday night dinners and the fasting month of Ramadan, however, are firmly rooted in religion.

To observe the entrainment of human rhythms at first hand, stand on a busy street corner and watch the movement of traffic, each driver adjusting to the presence of others and each moving precisely attuned to his fellow road users. Although an accident may occur because of equipment failure, most road accidents arise when the rhythm is broken, either by miscueing, or by errors in perception or action by one or more drivers. Mile for travelled mile, the safest sort of driving is on freeways, motorways or *autobahns*, because these roads are controlled to cut down the number of variables. Speeds on freeways may be faster, but faults of rhythm, and hence accidents, are less likely. When they do happen, however, such accidents are likely to involve more cars.

Ballroom dancing is another good example of the way in which one human being can take the rhythm of another. To the accompaniment of music, which provides the basic beat—the fundamental external stimulus to which the body rhythms respond—the couple make their way around the floor, usually with the male partner in the 'lead' and the female partner 'following'. The skill of the dancers is directly related to the clarity with which the lead dancer can give signals and how skilful the following dancer is in responding to them. The more competent a dancer the more cues he can deal with, the ultimate being the intricate interrelated movements of ballet dancers.

Adopting the rhythm of another person is not limited to physical movements but can also be followed on an intellectual or emotional plane. A lecturer can, for example, sense whether the audience is following his argument, whether he is presenting his ideas too sketchily or too fully, whether those ideas are too difficult or too simple, and whether he is speaking too fast or too slowly. In the best lectures, as in the best theatrical performances, the lecturer or actor gets into conceptual synchrony with his audience. It is this that helps to create an inspired performance, and high points in the theatre always involve a rhythmic interchange which entrains the bodies and minds of the performers with those of their audience.

All human groups impose upon themselves a measure of social restraint but entrainment can work powerfully to break down the barriers of self control. Crowd phenomena, such as riots, arise when people take the rhythms of those around them, and are intricate examples of the way in which we can be entrained by the emotional rhythms of those around us and succumb to an emotional outpouring by means of rhythmic movements.

In our daily lives we can all notice examples of dysrhythmia, times when our fellows either fail to take our rhythms or do not set good rhythmic beats for us to follow. The pace of activity and the rhythms of the very young are, for instance, different from those of adults between the ages of puberty and senility. For a parent in a hurry, being patient while a child ties her shoelaces can almost qualify as sainthood and the natural tendency is, of course, to take over and do the job yourself. The child cannot, however, be expected to do up her laces quickly because she is inexpert in the complex pattern of movements and sequences the task involves. More than this, the basic rhythms she uses are quite different from those employed by an adult.

In a similar way, it is a common experience to stand behind an old lady in a queue who takes too long to find the exact coins she needs in the small purse lost somewhere in the depths of her shopping bag. As we get more and more

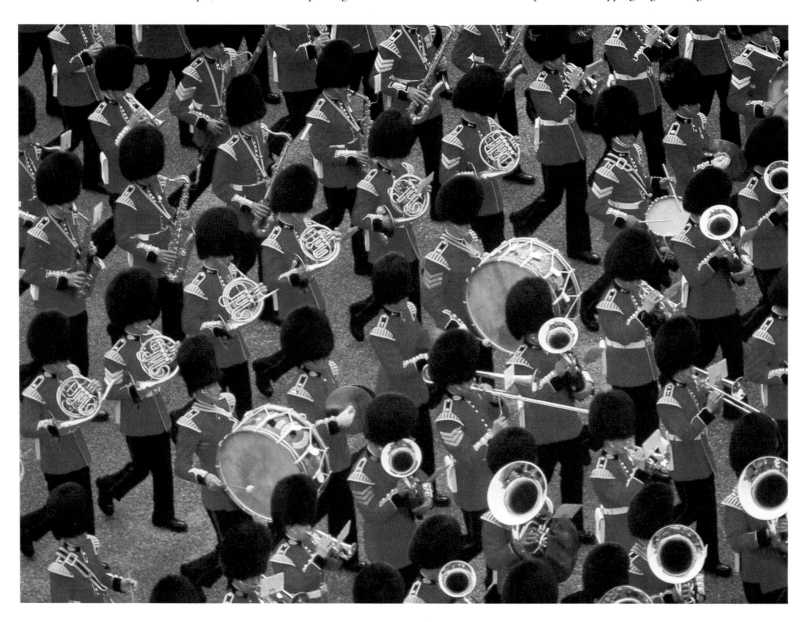

exasperated we fidget while she gets the contents of her bag together again before moving on. Old people boarding buses are often out of synchrony with the movements of other people and can disrupt our own performance, cause themselves embarrassment and even spark off a display of anger by the bus driver, the conductor or other passengers. Unconsciously we all adjust to take account of the different ages of the people around us, but once we reach a threshold of some sort—and this has not yet been seriously studied—we become aware that the other person is out of phase with our rhythms. This realization may be expressed as annoyance, pity or any sort of emotion in between.

Dysrhythmia may also be involved in illness. One of the reasons that hospitals are not better places to be is that, of necessity, the rhythms of the patients and the staff are set by different pacemakers. Illness of any sort upsets the human body rhythms and the patients are, almost by definition, dysrhythmic. The doctors march to their own busy schedules, while the nurses have the thankless—and unthanked—task of trying to adjust the rhythms of the patients to those of the doctors, and to the daily round of hospital activity. Moreover, nurses often have to work different shifts week by week, so their own body rhythms may be disrupted. Such disjointed, unconnected rhythms make it hard indeed for patients to realign their body rhythms. Once a patient leaves hospital he may find it equally hard to readjust his rhythms to those of the healthy people who now surround him.

In the field of mental health, one of the most promising aspects of research involves investigation of the ways in which diseases, such as autism, may display complex forms of dysrhythmia. Autistic children are either unable or unwilling to interact with other people as normal children do. They do not take cues from others and refuse to interact with their neighbours, often seeming as if such interaction is intensely painful. These children seem to be able to recognize no external pacemakers, which puts them out of phase with the rest of society, and the absence of such pacemakers also lets their various body rhythms free-run out of phase with one another. Such complicated dysrhythmia, arising from a lack of entrainment from external pacemakers, is also a symptom of schizophrenia. Psychiatrists and researchers into mental health are only just beginning a programme of research which promises to tell us much more about the way in which mental rhythms can be disrupted, and it is likely that by the end of the 1980s we will have a better understanding of the role of rhythms in individual and social health.

From the moment of birth onwards, external pacemakers help to create the rhythms to which human beings respond internally. At birth, babies have 50 to 60-minute cycles of rest and activity but, unlike adults, the rate at which their hearts beat does not vary between night and day. By the sixth week of life external factors—presumably both human and non-human—have set the rhythms for the child's bodily existence.

Physiologists claim that external factors, such as the patterns of family living, help to put the infant's rhythms into motion and these factors include the attitude of the parents towards feeding during the day and the night, the temperature and noise levels in and around the home and parental patterns of activity and rest. The synchronization of the child's rhythms with both the ▷

Group action is usually organized into a coordinated rhythm by one individual. This may be conscious as in an orchestra or choir where the conductor's role, *above*, is to set the pace and control the rhythm, or unconscious as in riots, *right*, when contagious emotion spreads to the masses from one or two leaders, causing an apparently spontaneous eruption. Similar though less involuntary cooperation is effected by the strict rhythms of a military band. Marching in step imposes discipline, and a sense of unity and purpose, upon a large body of men, who are encouraged by the synchronized movements to suppress their own individuality in favour of group identity. Two people walking along together tend to 'fall into step' and may even synchronize their breathing, to establish a rapport. In competitive sports, one individual sets the pace, which is taken up by the rest of the team or field and reflected in the crowd's movements. Even social and political trends follow the pattern mapped out by a leader.

▷ social and the natural world requires responses to such external pacemakers. The evidence is that without these entrainment forces the child's rhythms become increasingly random—only by entrainment can the infant organize his unordered rhythms to accord with those of his family and of the community of which he is a part.

The human infant is a helpless being who cannot survive unless tended by adults. Given his innate capacity for entrainment, the ministrations of adults provide the child with his first experience of coordination with human movement. So the mother's rhythms become an integral part of the child's experience of growing up and, along with other environmental rhythms, become built into the fabric of his internal physiology, making his 'mother rhythms' as natural as the rhythms established by light, temperature and gravity.

Rhythmic communications are closely associated with a baby's lifestyle in his first few months after leaving the womb. In many African societies, babies are carried on their mothers' backs. Tiv mothers in central Nigeria put their babies in slings. The baby's legs straddle the mother's back and the cloth sling is wrapped tightly round the child so that he can ride along as the mother goes to and fro from the fields or to market.

Among the !Kung Bushmen of southern Africa, a baby is carried in a sling at the mother's side. The baby looks out in the same direction as the mother and is approximately at her height from the ground so that the mother's actions are more or less duplicated in the child's body rhythms. In most Western societies, babies are not carried far by anybody, so they must learn interactional synchrony in other ways; but taking their cues from African mothers, many Western mothers have adopted the practice of carrying their babies in slings. For a considerable part of the day, many eight or nine-month African babies are turned over to children of about ten. The ten-year-old carries the baby, tends to his needs, sees that he does not roam too far and is near his mother when he is hungry. Babies acquire a great deal of cultural information from older children in this way.

As they learn to creep, crawl and walk children take more and more of the rhythms of society into their own bodies. Although each of us has a characteristic gait, for example, there are cultural differences in the exact ways in which people walk: the way they pick up their feet, the precise order in which some of the muscular contractions are performed, the way in which the foot hits the ground, the manner in which the arms swing and how the shoulders are held. During the 1940s and 1950s, a characteristically English masculine walk depended on reducing the amount of movement in the hips. To prevent the hips from swinging as you walk it is necessary to lower your hip bones and shorten your stride. At the same period, a typically masculine walk for an American was associated with broad shoulders. To adopt this sort of gait you have to stand tall, raise your hips and push your breast bone up as high as possible. To add to this, the elbows are turned out slightly, the arms swung vigorously. The difference between these two styles of walking led the English to think that Americans walked like gorillas, while the Americans thought that the English minced. Both were trying to walk like 'real' men, but the resulting body rhythms communicated different messages.

When a baby spends the day strapped closely to his mother's back, side or breast, her natural rhythms become an integral part of the child's early experience. The soothing pulse of her heart, the gentle sway of her walk and the regular swinging of her body as she goes about her daily work, combined with the reassurance of close physical contact, become instilled into the baby's consciousness. The Malinke tribe from West Africa, *left*, who have a rich musical tradition, spend the first few months of life tucked snugly astride their mothers' stomachs. As in most cultures, exposure to music begins very young so that every child develops the taste for rhythm from the start. In some societies, particularly in Southeast Asia, children learn to dance almost as soon as they can walk, so that the rhythms of their own culture become 'second nature', and only when confronted with a different set of rhythms, do they become consciously aware of their own.

Encouraged by group skipping games, roller skating, dancing, basketball and other games, children soon learn that it is more fun and easier to move rhythmically. In later life they will use these rhythms not only in their leisure time but to move in harmony with their workmates. Children are not usually fully coordinated until the age of 12. Those who have difficulty moving rhythmically find visual stimuli more helpful than aural ones, as they can often improve their coordination through imitation.

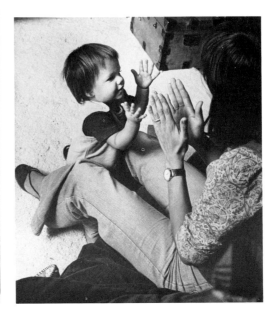

Clapping games are a popular and effective way of teaching children progressively complex rhythms, *left*. 'Pat-a-cake', for example, encourages a very young child to coordinate with other people and to pick up their rhythms. This ability to beat out a steady rhythm develops earlier than the ability to identify the rhythm in a piece of music. Children under eight cannot easily sustain long notes nor distinguish clearly between strong and weak notes. Individual rhythmic ability varies and continues to improve with practice well into adult life.

Culture and the rhythms of body movement are linked in other ways. Young children may learn the various methods of locomotion in a different order, so that crawling does not always precede walking, and some babies shuffle along on their bottoms before learning to walk. The cultural influence on such learning processes is well documented in Balinese babies, who first learn to squat, then to stand up from a squatting position. In contrast, most Western babies first pull themselves up into a standing position from their knees and only later—if at all—learn to sink into a squatting position. For Western people squatting, with the feet fully extended and the heels on the ground, is not usually employed as a way of resting the body and of sitting comfortably. The result is that most Westerners never learn how to get into a squatting position gracefully and rhythmically and, even if they do get there, find it extremely uncomfortable.

In some societies, children learn the rhythms of dance at an early age and babies may even be carried while their mothers dance. Even the smallest toddler can be seen 'bopping' to the beat of pop music, picking up the rhythm in a seemingly innate way. In a court session among the Tiv of Nigeria an energetic 14-month-old was even seen to pound on a drum, an activity normally reserved for adults only. The baby had a perfect right to be in court as long as he did not interrupt anything. On this occasion the judges paused, considered his dance rhythm, helped him to improve it, told him what a great boy he was and only then instructed his nurse to take him away so that the court could resume its deliberations.

As children grow up and are taught the technological skills of their society, they use these skills in concert with the people around them. So they learn the rhythm of hoeing a field by hoeing in time with others, or pound grain in a mortar, lift heavy loads, or move goods along a production line in rhythms that set them in easy relationships with their workmates. By the time they are about 12 years old children are normally fully coordinated and can perform the full concerto of physical rhythms that they will use for the next 50 years or more. Our entire adult lives are rhythmical—whenever the rhythm is lost we become ill or maladjusted. Rhythms lie at the base of all normal social interactions and these cultural rhythms are learned early, usually by the time a child is four or five. Although adult rhythms have not yet emerged these basic social rhythms have, as much as our language, converted us into permanent members of our own particular culture.

The variations in the spectrum of body rhythms reflect the ways in which time can be organized, and we can sense these same differences in the nuances of poetry, dance and music. Poetry can be thought of as a reorganization of the rhythms of speech in such a way that both sound and meaning are heightened. Without changing the words of the language the poet orders the words he uses so that regular patterns emerge. The arrangement of words in poetry can thus become less complex than that of everyday speech, and to this special arrangement of words, which gives poetry its rhythm, is added the quality of reverberation which makes the poetry 'sing'.

Different languages lend themselves to distinct patterns of syllables and beats. The classic English iambic pentameter has five 'feet', each one made up of an unaccented syllable followed by an accented one. The French alexandrine ▷

▷ has six iambic feet arranged in such a way that a natural pause follows the third foot—it is almost impossible to write sensible alexandrines in English. Japanese, which lacks the stressed accent and depends on tone to do the same job, measures the number of syllables rather than the stresses. 'Free' verse may lack surface regularity, and also rhyme, but good free verse depends even more on meter and rhythm than rhyming verse.

The writer of good prose attends as closely to rhythm as the poet, but because prose does not have such an obvious pattern as poetry it does not have the same capacity to create a heightening of meaning. Both poetry and prose depend, however, on the rhythms of the spoken language, although the prose rhythm is more subtle and less restricted. One of the marks of poor prose is that it makes the tongue do unexpected things. A good lecture or political speech stands out because the rhythms of speech are controlled so that they flow nimbly and allow climaxes to be reached without noticeable effort on the part of the speaker.

The written word is frozen speech absorbed by the eye rather than the ear. People who do not read well 'translate' the words back into the 'proper' or spoken mode and may move their lips as they read. Even fast readers 'hear' a page of words at some level, despite the fact that they can read far faster than anyone can speak, and the eye can trip over dysrhythmias in the language as readily as the ear. The skill involved in writing prose that is easy to read comes less from vocabulary and sentence structure than from the mastery of prose rhythm, so that good writing is attuned to the rhythms of the human voice— just as the rhythmic movements of babies are.

Rhythm is only one element of music, but it is the element that ties music to the human body most directly: music 'soothes the savage breast' precisely because it provides us with a pacemaker to which we can entrain our rhythms. It can also be a moving emotional or intellectual experience. Whether music is peaceful or agitating depends in part on its relationship to body rhythms. Many people find that when music is playing they entrain any other activity they are involved in to its beat. Other musical components, such as scale, melody or timbre, do not travel as well across cultural boundaries as rhythm. Some Europeans cannot hear the intricate polyrhythms of African music so that it becomes just a blur, but the rhythm of foreign music is far more likely to be appreciated than its melody. The very way in which music is perceived differs from one culture to another and some languages do not have a word that translates the word 'music' found in all Indo-European tongues. Instead they have a plethora of words to indicate types of song and instrument.

The emotional effect of music is strongly linked with its relationship to body rhythms. The most natural rhythm is 80 to 90 beats a minute and any figure higher or lower than this sets up a state of inner tension. Many people react violently against both 'unnatural' and experimental rhythms when they first hear them. It was thus not merely the untraditional harmonies and unconventional dancing that created the riotous reaction to the first night of Stravinsky's ballet, *The Rite of Spring*. The rhythms of the music, which most of us have learned to appreciate today, were strange and disturbing. The spontaneous reaction of the audience was not distaste but rage. A similar sort of reaction is felt by some older people to hard rock music. What distresses them—

apart from the volume—is the demanding pulsation of the beat. Subtle it is not, but coordinating it certainly can be, if one is not deafened in the process. However, some young people say that the purpose of the high decibels is to get the body membranes vibrating with the beat.

The choreography of the dance makes use of natural movements much as poetry employs natural language. These movements are, however, exaggerated and, most important of all, the pace is changed. Alteration of natural pace—usually a slowing down—and often the simplification of natural rhythms to a totally overt pattern, are the basic characteristic of all dancing everywhere. When we watch a prima ballerina we may admire the speed of her movements, but to see her gradually raise and unfold one leg in a *développé* can be a breath-taking experience. The wonder is not so much that she can do it, but that she can do it so slowly, and with such consummate grace and ease of movement.

African and Indonesian dances show us most clearly the nature of dance as a total experience. African dance, and music, concentrates on polyrhythms and the African orchestra is built around drums. A large drum with a fairly deep note is often the one that carries the basic beat while perhaps as many as a dozen smaller drums each have a different beat. The whole forms a pattern that is repeated after a given number of 'bars', but each drum comes in at a different place and often with a different rhythm. In most parts of Africa hand gongs or iron bells are added to the drums, while shawms (relatives of the oboe) and flutes may carry a tune.

African dancers make use of all the orchestra's rhythms. One part of the body—often the feet but equally often the trunk—maintains the basic beat, then the shoulders pick up part of the 'counterpoint', the arms and perhaps the knees another part. The head may move to the rhythm of the gongs or bells. Sometimes a number of people, often dancing in a ring around the orchestra, all perform the same steps and move in unison, creating an overwhelming impression of unity, but on other occasions the point of the dance may be the individual, creative use of rhythms intended to blend all the rhythms of the drums into a dramatic and pleasing whole. The best dancers are those whose bodies are filled with and express all the rhythms of the orchestra and who are graceful and agile. The result is both fast and subtle, but the audience must understand the process to appreciate the skill.

The *gamelan*, the orchestra of Bali and Java, is also built around percussion instruments, although flutes and stringed instruments may be added. The instruments of a Balinese *gamelan* are arranged in pairs with one member of the pair echoing the other. The basic pair of drums is called the male and the female (a similar division is also true in much of Africa), but they are slightly different in pitch and are beaten to create interlocking rhythms so that it sounds as if only one drum is playing. The two drummers control the tempo. The lead melodic instrument is the metallophone, which is like a xylophone but has metal bars rather than wooden ones. The first metallophone player is the melody maker and, especially if the male and female drums are silent, may influence the rhythm by becoming the pacemaker as well. The second metallophone is tuned a quarter tone from the first, but struck a fraction of a second after it. Thus every note has an answering call and this sets up a ▷

The tension caused by dancing to an accelerating rhythm can lead to a cathartic state of 'ecstasy'. The whirling dervishes in Turkey form a clockwise-moving ring, and as the music speeds up the dancers whirl wildly around in a trance. In this spiritually receptive state, these *sufi* seek total union with God.

The magic of the Fred Astaire and Ginger Rogers partnership was their perfect synchrony with each other and the music. They visualized the music's exaggerated rhythms in an exhilarating display of stylish coordination.

▷ reverberation—itself rhythmic in its vibrations—as well as complicating the rhythms set by the drums.

The cymbal players now come in to underline and shade the drums and metallophones, resulting in music that makes use of counterpoint of tone, melody and rhythm. The *gamelan* sits in a rough square and the rhythms echo across it from one side to the other and diagonally from corner to corner, so that the entire area reverberates. The whole is finally embroidered with smaller instruments playing fast, cascading 'flower parts' which provide the colour and make Balinese music so unmistakable to the ear.

Balinese music is arguably the most vibrant dance music in the world and young Balinese girls are among the most graceful dancers. The gestures of the dancers of Bali and Java, in which the shoulders shiver to the beat of the drums and the head and neck are moved in sinuous undulations, are carefully controlled, and each has a traditional meaning usually connected with religious myths and legends. The dancer is, therefore, not only unifying rhythm and motion but also the religion, history and literature of her people. Balinese dancers are taught by skilled instructors, usually themselves retired dancers. Student and teacher have a relationship of intense and close physical contact so that the student can sense and take the rhythm of the teacher in a type of entrainment process.

While the essence of a culture can thus be found in its music and dance, both are also tightly bound to the rituals of religion. Ted Shawn, the American dancer of the 1920s and '30s was an assiduous student of forms and styles of dancing throughout the world. His feeling on the matter was, 'I could not conceive of God being without rhythms, without grace, without intelligent expression, nor without possession of infinite forms of beauty through which to express His infinite Rhythmic Being.'

It is no accident, and certainly not a profane sentiment, that dance and religion have been intimately linked from man's earliest relationships with supernatural gods. Dance lies at the basis of religion because of its capacity to unleash our awareness of interlocking rhythms and to slow them down so that they are brought under conscious control. Music, always a dimension of dance, also lies at the root of religious experience. The cosmic rhythm is represented by the drums, the breath of the human spirit, by the woodwind, man's ability to be creative with his hands and mind, by the strings. The whole dance represents man's capacity to control his movements and express his own feelings.

Through its inextricable link with music, dance intermeshes the internal rhythms of the body with the immediate rhythms of the environment and ultimately with cosmic and celestial rhythms. The physical sensations that arise from controlling the coordination of one's own body are in themselves exhilarating, so that it is not surprising that in many eras and cultures dancing has been, and still remains, the primary means of communicating with the spirit world. Nor should it astound us that in some religions the gods who control the universe are thought of as dancers, or that one's own dancing body can be a receptacle for God himself to enter.

Yoga, a sort of static dancing, begins with breathing. As you prepare to assume the yoga positions you first become aware of the rhythm of breathing. Then you centre your mind and associate it with your body by consciously

At African curing ceremonies, the deep rhythm of a hand-beaten drum and the haunting sound of rhythmic chanting is believed to draw out evil spirits and cure illnesses.

In South Africa, Zulu mineworkers re-enact their ancestor's ritual war dance, *left*. Frenetic leaping and stamping dances used to build up the warriors' aggression to fever pitch, until they became thirsty for battle.

Muslims express their submission to God by bowing down in prayer five times a day. At dawn, noon, mid-afternoon, dusk and after dark, they remove their shoes and, facing towards Mecca, humbly pay homage to Allah.

slowing down, deepening your breathing and becoming quietly aware of your heartbeat. As concentration continues your heartbeat slows, your shoulders relax and so does the rest of your body, except for those muscles needed to hold the body in a particular yoga position. With an easy mind you can then use your brain to control your internal rhythms so that yoga becomes a spiritual and intellectual way of experiencing your intimate association with the rhythms of life.

Established Eastern religions, especially Buddhism, are highly sensitive to the impact of simplified, ordered rhythms on the state of body and mind. Many such religions use the mantra, or 'instrument of thought', as a sacred prayer or incantation to achieve a rhythmic 'self-centredness'. The mantra is personal— it is usually presented as a gift by a master—and is repeated in a single rhythm, over and over again. One of the aims of this monotonous, monorhythmic exercise is to disentangle the self from the disrupted rhythms of daily life. Using a mantra it is possible to capture an awareness of your own body rhythms, which are slightly different from those of anyone else.

In meditation the aim is for the body to be rested and the mind both rested and enlightened. In Eastern religion the mental state of meditation is thought to be essential if one is to re-establish contact with the rhythms of the universe, that is, with God. Western versions of meditation have lost their religious element, but make use of the same techniques to counter stress and to achieve peace of mind. The mental state of meditation has some similarities with the state of hypnosis. During meditation the electrical rhythms of the brain become more even, the heartbeat slows and the skin temperature drops, yet the meditator is aware of the world around him. In hypnosis the same physical changes may occur, but the state of the mind is quite different from that of meditation because consciousness and memory are drastically altered.

A trance is quite opposite in its effects from meditation and can be accompanied by increased rapidity of body rhythms. The resulting feeling of ecstasy may take the form of a religious experience, and ecstasy is also associated, in religious history, with prophecy. The practice of prophecy during an ecstatic or hypnotic trance is still carried on in North and West Africa. Spirits, or *bori*, are said to enter or 'ride' human priests and allow them to prophesy or divine the cause of illness. At the insistence of Islamic officials, *bori* were banned in West Africa during the colonial era, but *bori* dancers kept appearing because their divinations were thought to be able to determine the cause of illness in a way that is beyond the scope of both modern and traditional medical techniques. These causes stem most often from ruptured social relationships, so that as well as being concerned with curing individual physical ills, *bori* seances, during which the spirits are summoned up, are said to heal social wounds.

Prophecy, music and dance all have their place in the Bible, the book that charts the history of the Jewish people and is the cornerstone of Christian faith. King David and the Israelites, for example, danced abandonedly before the Ark of God, and the Hebrew words originally used to describe the event can be translated as dancing, skipping, leaping and turning. Of all religious dancing, that of the 'devil dancers' of Sri Lanka is among the best documented and observed. In the area of Kandy, religious officials dance until they become ▷

▷possessed by spirits, which the early Christians (missionaries) named devils. Once inside the dancers these spirits can be made to remove the illnesses from the beneficiaries of the performances, so that when the spirit leaves the priest, the disease also leaves the patient.

The religious practice of becoming entrained to the rhythms of the universe can be turned on its head so that, via religion, mankind can take upon his shoulders the task of being a pacemaker. As a result, many peoples of the world believe that if they do not perform certain specified rituals the seasons will not turn, crops will not ripen and the rains will fail to arrive. If our culture allows us we can, therefore, use our body rhythms to take on the responsibility for environmental rhythms. In some religions it is ritual and dance that are the pacemakers. The trouble with such dogmas is that there must always be an 'escape clause' to accommodate failure. Most usually the 'excuses' are that the practitioners did not keep taboos, were inaccurate in their ritual or were the unknowing targets of witches or evil spirits.

While rhythmic harmony is the essence of internal peace of mind, almost all human work is easier to do and produces more efficient results when performed in tune with a rhythmic pacemaker. Hunting and gathering people take their basic economic rhythms from the animals they hunt and the food they gather. The more hunters adapt to the rhythms of their prey, the better their results, because to be successful a hunter must observe his quarry and plan his actions accordingly. It is not surprising that game animals were worshipped by early hunters, because these men believed themselves to become the animal as they entrained their actions in order to kill it. Equally, survival of both the hunter and the hunted are interdependent, because if an animal is overhunted it becomes extinct, and this adds to the religious significance of prey species.

The stock farmer or herdsman also entrains his rhythms to those of his animals, but over a longer period of time. African herdsmen are in constant contact with their beasts, and in their dancing imitate the movements and rhythms of their cattle. Like the herdsman, the peasant farmer synchronizes his daily, seasonal and annual rhythms to those of the living world, but adjusting to the rhythms of a plant is far more difficult than becoming entrained by an animal, because the psychological feedback of success may not arrive for a year or more. Nevertheless, such entrainment is vital to the subsistence farmer and even today is what gives a gardener his 'green fingers'.

With the coming of the industrial revolution, which began in Europe in the nineteenth century, man's rhythmic unity with his environment changed. Instead of entraining his rhythms to other living organisms, the worker had to adjust his body rhythms to machines, and his daily routine to the production line. The synchronization of factory workers with their machines meant that human rhythms had to be suppressed in the service of the machine. The human body was designed, over centuries of evolution, to detect and deal with environmental change. In the early days of life with the machine demands were totally different, and the constant repetition of identical movements at a fairly rapid rate became the imperative of work.

The machine that is the human body finds it difficult to maintain a single, repetitive motion in a monotonous rhythm. It is far better adapted for a variety of tasks in more complex rhythms. When a person entrains on a machine it is

the person who does all the adapting, so that work on an assembly line is both mechanical and monotonous. The lack of synchrony between human and mechanical rhythms was brilliantly caricatured by Charlie Chaplin in the film *Modern Times*: his muscles kept twitching long after he had finished work, while his body rhythms went completely crazy when the production line was speeded up. Chaplin thoroughly understood the rhythms of the body, and those of society, and used his knowledge to create a comic masterpiece.

The machine is the pacemaker to which the worker adapts but each machine has its own rhythm, making the array of industrial rhythms even more exhausting. In the office, the typewriter is a rhythm machine. The typist whose rhythm is even and secure not only types much faster but makes fewer mistakes than one whose rhythm is choppy. And the typing rhythm changes with content, which explains why copy typing, demanding little thought, is more rhythmic than composing copy straight on to the keys. Just as with reading, if you stumble over the rhythm, the words may well be written in the wrong order.

Hundreds of the machines we commonly use demand our rhythmic attention and entrainment. Sewing machines, food processors, agricultural machines—all demand careful integration of human rhythms with those of the machine and the result may be disastrous if we miss a beat. Of all man-made machines, the jack hammer or pneumatic drill is among the most rhythmic. You cannot control its rhythms, and must join in with them if the machine is to be of any use. The jack hammer shakes the human body into its own rhythm which can be felt for hours afterwards in the muscles of the arms and hands.

As well as the rhythms of machines, industrial society also imposes a diurnal nine-to-five rhythm on its workers, which is at odds with the diurnal rhythm of seasonal change. Most industrial rhythms do not alter with the seasons and the uncompromising clock has replaced the tempo dictated by the year. The weather outside the factory or office window makes little difference as we struggle to keep to a regular schedule, even if we are impeded by snowstorm or tropical hurricane. Whatever the weather, the production deadline will not wait. In many industries, work is organized in shifts which overlook the body's 24-hour circadian rhythms; although for 'owls', working an evening shift may be ideal and all but a few of us can eventually adjust to working a night shift, even if it takes several weeks to do so. For the purposes of production, however, one hour on the clock is no different from the next.

As more and more industrial processes become fully automated, and as more working men and women become involved in service industries rather than manufacturing, so the rhythm of work is changing. Compared with our fathers or grandfathers, we have more freedom to choose our times and hours of work and do not have to perform such demanding repetitions. Staggered working hours allow us to fit our working lives into the rest of our existence so that the pace of life is more comfortable. And today's innovators have begun to take human capacity into consideration when they design machines or machine systems. The work of the air traffic controller at a modern airport is an excellent example. Because one mistake can kill hundreds of people, the controller's rhythms must be correctly attuned to his equipment and his job must be planned with infinite care so that he works to maximum efficiency. ▷

Since the industrial revolution, man has had to submit his own rhythms to the overriding rhythms of machinery. On a production line each worker must adopt the pace set by the machine to maintain a smooth and effective flow. While it is less tiring to wield a scythe or an axe rhythmically, the unbroken monotony of factory rhythms can be exhausting. Music can delay fatigue and increase enthusiasm and thus the working speed of those performing mechanical tasks, but the inaccuracy that results from this distraction can prove dangerous and inefficient.

▷ Music is sometimes used to provide workers with a pacemaker. Sea shanties evolved to ease the task of sailors rigging a vessel, and for tribal and peasant peoples, singing together and working in rhythm to a song also seem to lighten the load by creating a sort of community rhythm. Even in modern offices and factories 'musak'—part tune, part gentle rhythmic beat—provides a background and increases output. Even without music, workers are usually rhythmically entrained. The rhythms of a good cook or the rhythms of cane cutters are a joy to the eye, and one of the most beautiful of all work rhythms is a crew of some 16 African fishermen handling a large canoe in the surf as they take each others' rhythms and constantly adapt to the ever-changing patterns of the sea.

Away from the rigours of work, men and women retreat to their homes and families. To provide themselves with living accommodation, people in different cultures build houses to all kinds of design. These houses, which may be large or small, rectangular or round, have a few large rooms or many small ones, are also home to a variety of family members—sometimes only a couple and their children, sometimes sets of siblings and all their children, sometimes all the members of several generations. All these kinds of households impose different rhythms upon the people who live in them. Every household has a daily round, which changes at weekends or on feast days, holy days or holidays, thus creating a weekly rhythm, while the cycle of the seasons may set the pace of yearly rhythms.

Division of labour among the many members of a household occurs worldwide. Careful coordination is essential to get all these jobs interlinked so that household life carries on smoothly. The rhythms of these tasks, and the tasks themselves, change from one culture to another, but the result is always a sort of domestic choreography which, when it works well and everybody fits comfortably together, can be both beautiful and rewarding.

Almost all the members of a household have schedules outside its four walls and leave it for work, to go to school, to participate in community activities, for sport or entertainment. Each member of the household must coordinate his or her personal schedule with the activities and schedules of others. The coordination of all this coming and going may take considerable skill, if tempers are not to become frayed and rhythms disrupted.

In traditional Western middle class families the husband or father is the only member of the family who 'works', which means that for five or more days a week he leaves the home. At that time, for some culturally decided period, he engages in another sort of activity, possibly alone but more probably with a small group of people who have no contact at all with the members of his family. His activities are directed toward goals quite different from those he aims for during his 'family time', but at the back of his mind he is aware that one of the purposes of his work is the economic support of his family. Increasingly in the West, wives, mothers and adult, unmarried children also leave the family home to work. The children go out to school or to a child minder, and so the basic rhythm of the household is created.

There are two points in this cycle at which the personal rhythms of the individuals involved must adapt, namely departure and re-entry. Like sliding your car into the pace and rhythm of the traffic on the freeway, human

rhythms must intermesh so that the activities of the household can continue on beat until all its members have returned. When a family reassembles a period of readjustment is both necessary and inevitable. However welcoming a home, or however much we have longed to be there, we may still experience a re-entry problem.

The best way of investigating the delicate intermeshing among family members is to examine families who have unusual rhythms. Oil workers on the pipeline and pumping stations in the snowy wastes of the northern shelf of Alaska are a perfect example. In this Arctic desert there are two people for every job—one is 'boss' and has a 51 per cent say in how the job is performed. The boss goes to the shelf and works for a week. With a 2½-hour overlap the boss and his alternate—and indeed the whole team—hand over their jobs before flying off home for a week.

During their week on the job, every person in the team has a single bunk room with a bath shared by one other person in the next room. Each room has two sets of lockable cupboards and drawers and is occupied for alternate weeks by members of the two different shifts. Work is arranged on the basis of a 12-hour day, but members of the management team are on call round the clock. At the end of the week everyone must re-enter their family world, and this is much more difficult after a week at work than just a day away from home. Ten minutes after his father walked into the house for his week 'at home' the eight-year-old son of one pipeline worker did something for which he needed a reprimand. The parents looked at each other but neither made a move. Then the mother said, 'Well, you're his father—are you going to let him get away

with that?' To which the father retorted, 'You want me to beat the child ten minutes after I get into the house?' The parents then talked over the problem, not just on that day but for the best part of the week. Since there were no cultural rules on which to base their behaviour they began to make their own rules and thus solved their own re-entry problem.

Fishermen, sailors and soldiers have to make a similar sort of adjustment at far longer intervals, as do men who choose to work in the all-male communities which have sprung up in the countries of the Middle East. A businessman who travels a lot may experience the same sort of problems as the Alaskan oil worker and even feel estranged from his wife and family until he readjusts his rhythms to those of his kin, rather than taking his cues from his workmates.

Every household also has a long-term rhythm. In Western society, newly married couples usually move into new households—if they do not they can severely disturb the normal family rhythm. When a child is born the rhythms of family life change, and this alteration will be largest if the mother gives up a job outside the home when the baby arrives. With the birth of subsequent children the rhythms change again, as they do when children start school and when they become adolescents. On each occasion the internal rhythms of the family have to adapt to the changing rhythms in the comings and goings of its members. When the children leave home the rhythms 'settle down' until the parents retire from full-time work. The final desolation of being a widow or widower is not only because of the loss of a loved one, but because the remaining member of the partnership is deprived of being in synchrony with another human rhythm.   ▷

All working people have to adapt to a different set of rhythms when they return home after a day's work. For coalminers, *left*, and others working anti-social hours and in unnatural conditions, the problems of adjustment are even greater. Food and drink may be the miner's first priority when he enters his front door, but the rest of his family may have conflicting physical and emotional demands. An acclimatizing period is essential for each individual before the household can regain its normal pattern. Family relationships may even begin to disintegrate if there is insufficient time for a harmonious rhythm to establish itself. But for many people, the best part of the day begins when the 're-entry' problems have been solved, and the familiar home routine takes over. This daily cycle of going to work and coming home again, with its inbuilt pressures, is itself a rhythm and when it is broken by retirement, unemployment, illness, or even weekends and holidays, there is a transitional period of re-orientation before a new set of rhythms can take over.

The intense elation and relief of returning home from war and reuniting with loved ones will inevitably be followed by a long period of readjustment. The soldier's long absence, coupled with his shattering experiences, which alienate him from those who have not shared them, make his integration into the family a long and difficult process. The household rhythms may no longer seem familiar or, like anyone returning home from an eye-opening trip, he may find them too familiar, too slow, too unchanging. Whether initially he rebels instinctively against the rhythms established by the rest of the family or tries consciously to fall in with them, it will take a while for both the 'returnee' and his family to regain some form of synchrony. During his time away, the children may have left home, creating a different pattern in the household to which he must adjust, or the family may have held firmly to the same reassuring daily cycle of activities, which they, in turn, may have to adapt to suit the different needs of the homecomer.

▷ In societies, such as those of many African and Asian peoples, in which households are large and are either polygamous or contain members of several generations, the family rhythm is less disrupted by arrivals and departures than it is in smaller units. Even in the West, comings and goings tend to disrupt rhythms more in small families. Paradoxically, in the big, easy-going family it seems much easier for everyone to intermesh their activities than it is in the small, tightly knit family unit.

The family is the basic unit of the community, but human relationships also involve the forging of bonds with other members of our species. And like the individuals of which they are composed, all human groups have their own rhythms. For a group to be welded into a unit, the rhythms of the single human components must blend easily with one another. The anthropologist Edward T. Hall, in his book *Beyond Culture*, notes that once he was sitting in an outdoor cafe on the Greek island of Mykonos when he saw a group of young people apparently listening to rock music on their portable transistor radio. As he observed them more closely, Hall concluded that the youngsters were not 'listening' to the music, rather they were using its rhythm as a 'sine wave' that would synchronize their movements with one another and, at the same time, increase group feeling and make the group-bond stronger.

A student of the same anthropologist once made a film of children in a playground. As he worked through the film he ran it at different speeds and examined it on a time-motion analyzer, a machine that makes it possible to run a film at any speed and stop at any frame. Gradually, he began to realize that a complex rhythm was taking place in that playground which had a 'tune' and a 'leader'. This leader was an extremely active child who moved long distances over the playground, interacting with each small group of children. Hall named that child the 'director or orchestrator of the playground rhythm'.

With the help of some musician friends the student then found a rock tune that fitted the complex rhythm of the children's play. The moment the music was put to the $4\frac{1}{2}$ minutes of the film sequence the rhythmic pattern was clear for all to see. What everyone asked was, 'How did you get the children to do all that in time to the music?' Some of the audience completely refused to believe that the music was added long after the film was made, and so was entrained to the 'music' that the playing children were making.

In a similar way, every traveller who goes from one city to another recognizes the changes he must make in the rhythms of his movements, if he is to fit in with the rhythm of the people in his new surroundings. The pace of some cities, such as New York, is fast and furious. London is a little slower, but still falls in the fast category. For a slow pace you must travel to a city such as Abidjan in West Africa. Some cities have mixed rhythms—Los Angeles, for example, is fast on the freeways but slow in the downtown areas. The rhythms within each city are many and complex. These rhythms change by the hour as people go about their daily routines, starting work, breaking for lunch and going home. Cities have high-energy areas that buzz with activity during daylight hours but are sleepy at night, while streets that seem dead during the day come alive when darkness falls. Even the rhythms of a neighbourhood change from street to street and block to block. Many of these fascinating city rhythms are as predictable as the daily and yearly activities of animals and

plants, but they remain largely unstudied.

At a much slower pace, long-term rhythms of activity and change make an indelible mark on the development, growth and decline of neighbourhoods. The linking of these developmental rhythms to the history of a particular city are purely incidental. Once an area has been built it seems to be typical of the very nature of its materials that it must become 'run down' before it can be rebuilt. Any idea of 'city maintenance' may even be incompatible with our present ideas of expansion and renewal because the rhythms do not blend with one another. On an even larger scale, countries and empires experience growth and decline through history and many democracies build a governmental rhythm into their constitution—thus the United States elects a new president every four years, France every seven years, while British governments can run for a five-year term before they must seek re-election.

All the rhythms of a community have their basis in the activities of the people that compose it, their goal, their jobs and the culturally determined way in which they go about the day-to-day business of living. The rhythms of the individuals that make up a city are an integral part of that city and the rhythms change with age. The alterations in the rhythms of any individual depend not only on his age but also on his health, the kind of work he does, the phase of family life he happens to be in, and on the many other environmental pacemakers to which his rhythms become entrained.

In concert with the changes in its inhabitants, the rhythms of the city also shift. If taken in tandem the rhythms of development, decay and renewal of a city, and the rhythms of ageing, health and work of the individuals in that city,

form a kind of infrastructure. It is the atmosphere of this infrastructure, determined by the rhythms of bricks and mortar and of living beings, that attracts you to one neighbourhood rather than another and so influences where you live, where you raise your family, where you play and, to a lesser extent, where you work.

The combination of cultural rhythms within one territory creates a number of habitats or niches within a city. The niche is the place in which any group, or an individual member of that group, finds security, enough to eat and sufficient services to satisfy at least a minimum of social needs. Survival of a population depends, at least in part, on the capacity—or luck—of its members to find a niche in which to prosper. The city is the sum of all the niches it provides, from Brooklyn to Greenwich Village in New York and from Soho to Hampstead in London.

It is the rhythms of our bodies, our culture and our cities that combine to make us what we are, just as we, in turn, make them. We may all think that our own methods of interaction and communication are universal, but that is far from the truth. One reason why we cannot understand why we are out of phase, and thus out of sympathy, with people from other cultures is that we overlook the differences in our basic rhythms, learned from our first experiences of language as babies and subconsciously adopted as normal. The rhythmic imperative is the essence of life, health and harmony, and the key to human well-being is the synchrony that can be achieved by keeping in step with the rhythms of our neighbours and also the rhythms inherent in the environment.

The market, which recurs every four or five days in many African societies, *far left*, divides time into a weekly rhythm. Not only a meeting place for social and business affairs, the market provides a regular focal point in the lives of every member of the community. In Western societies, weekends and holidays form similar focal points, drawing people out into the community, *left*, in contrast to the working week. Saints' days, religious festivals, school and summer holidays, pleasantly divide up the annual cycle.

Freeways and motorways not only reflect the increasing pace of modern society, but speed up the development of the world's big cities. 'Modernization' slowly regenerates declining neighbourhoods, new rhythms replace old, and growing areas rub shoulders with decaying ones. By its very nature, a city encompasses many diverse rhythms. Commercial, artistic, industrial, residential and municipal sectors each have their own pace, and come alive at different times of the day and the week. Diverse though less complex rhythms also exist side by side in rural communities.

Rhythms of time

From long before the beginnings of recorded history man has perceived that from day to day and year to year time moves with a cyclical rhythm. As civilization has developed, the desire to name and measure the natural cycles of time has grown, and as civilizations have become more complex so these accountings and measurements have become more exact—and ever more necessary. The natural days, months and years have been rigidly defined and reinforced by other measurements such as seconds, minutes, weeks, decades and centuries. Despite the fact that artificialities have been imposed upon it, the natural cyclical rhythm of time remains. As they continuously tick away, the cycles of time act as the lubricant that keeps modern society in motion. All our journeys, meetings, entertainments, plans and ceremonies are arranged and performed with reference to time.

Of all the natural cycles of time the most obvious one is that of day and night. Early man needed no special powers of intelligence to realize that he must gather food during the day and be sure of a safe, warm place to sleep at night. If he lived outside the tropics he could see that the relative lengths of day and night varied according to a regular pattern, but this mattered little until after civilization had become established.

To most primitive tribes the year was a relatively obvious cycle—not at first as any exact number of days, but in terms of the changing pattern of the weather and the relative abundance of both animal and vegetable foods. In temperate zones the regular procession of the seasons was unmistakable. The most accurate method of determining the length of the year, for the primitive and technologically aided observer alike, is to study the cycle of star patterns in the night sky. If, for example, you note the rising of a particular star above the horizon immediately before sunrise, exactly a year will pass before that star again rises immediately before the sun. It was this heliacal rising of Sirius or the Pleiades that enabled some early civilizations to calculate the length of a year to the nearest whole day.

Lunar months can be observed with no trouble, and for many tribes and civilizations have been the standard unit of timekeeping, each phase of the moon being used to name the parts of the month. Although the lunar cycle equates reasonably well to the menstrual cycle, and is a handy means of fixing the dates of feasts or religious services, it fits awkwardly with the solar year and is not allied to other natural cycles to the same extent as the day and the year.

From the natural cycles they perceived, early civilizations derived their various concepts of time. They learned to divide up the day into hours and to aggregate days into months and years. From the early Middle Ages men built increasingly sophisticated mechanical clocks so that today the accuracy of these timepieces exceeds that of the Earth's orbit around the sun. With greater precision in the measurement of the passing seconds time has become increasingly important, so that twentieth-century man is enormously dependent upon cycles of time.

In an attempt to rationalize the world we live in we have gradually become divorced from many of the natural seasonal and cosmic rhythms. Our festivals of Christmas and New Year, for example, are no longer linked to the shortest day but have artificially convenient dates. At the same time, split-second accuracy has gained an exaggerated importance. There seems to be so much to fit into our lives that they have become a race against time, with each passing second mourned as an irredeemable loss.

Assisted by a plethora of instruments and devices, modern man has developed a sophisticated way of telling the time. But the evolution of the concept of time, and the measurement of the passing minutes, have a long history that goes right back to man's beginnings. The least common concept of time used today is that of the Hopi Indians of Arizona. Unlike our compartmentalized or 'diary' time, the Hopi see time as an unquantifiable 'getting later'. For them, time is not a loss but a gain, so that each day is like a person who disappears and returns at sunrise in the same shape as yesterday but in a slightly older, more experienced form. In the timeless, Hopi view of the world, distinctions are not made between past, present and future but between momentary, continuing and repeated occurrences.

The ability of the Hopi to describe the universe without reference to either time or space is unusual, but it may well be far from primitive. It is much more likely that the time sense of the Australian Aborigines, who recognize the distant past as 'dreamtime' and treat everything else as a continuous present, is much nearer to man's earliest concept of time. Certain African and South American tribes believe in a concept of oscillating time. That time moves back and forth like a pendulum is an obvious, but simplistic, conclusion drawn from the observation of night and day, cold and heat, drought and flood, youth and age and so on.

From man's emergence up to the fourteenth century or later, the common people of most societies tended to view time as a cyclical progression of birth, maturity and death. Until this period men and women had no concept of the past or of the future stretching away from them, particularly as there was little change in their lives from year to year or from generation to generation. People would remember the recent past—the time of their own childhood or some important event—and would have hopes of a better future, but no more definite time sense than that. Thus there was little reason for such people to attempt to measure time for themselves.

Whatever the theory, it was chiefly for religious reasons that man began to count the passing days. For the men who directed the religious institutions of the common people there was good cause to divide up the days and to keep track of the years. Official religious views of the nature of time varied with the faith involved. According to most oriental faiths, and to the ancient Greeks, time was cyclical or, in the words of Aristotle, 'time itself is thought to be a circle'. This view was probably derived from observation of the cyclical rhythms of the patterns made by the stars in the sky, and of the seasons. In Greek mythology the idea of cyclical time is manifested as Oceanus which was, according to Homer, an 'immense stream' encircling the Earth which ebbed and flowed twice each day. Another chronological character of Greek mythology was the tail-eating serpent which encircled the universe and bore the cyclical zodiac on its back.

The idea of cyclical time is closely bound, in the Buddhist faith, with the idea of a continuous ever-rotating reincarnation. This concept of time is psychologically comfortable because it implies that there is no death of the body or of the universe. In ancient India time itself was supposed to progress in huge cycles, each lasting 1,080,000 years. Four of these cycles represented an even larger cycle of destruction and re-creation beginning with a Golden Age and

declining into dissolution before the next Golden Age 4,320,000 years later. The ancient Chinese also had a more positive circular model of time which included the cycle of years named after animals that persists today—the year of the rat, the year of the tiger and so on. In contrast to this concept of cyclical time, the Aztecs of South America believed in time as a constant linear progression. Linear time is also a tenet of the Christian faith, which charts a steady progression of mankind from the Fall in the Garden of Eden through to the final Day of Judgement. Linear time is traditionally symbolized as an 'ever-flowing stream'.

Most religions contain elements of both linear and cyclical views of time and the practicalities of measuring time began largely as a result of the inextricable link between time and religion. In many early cultures religious festivals were celebrated at times of equinox and solstice—or at some other specific time of year—and it was the responsibility of members of the priesthood to calculate exactly when these occurred. Priests were also responsible for marking off the years to record the length of a monarch's reign and the timing of certain important events, such as natural catastrophes, battles, eclipses and the like.

Religion was not always the impetus for measuring time. Sometimes the passing years were marked off just for the sake of keeping account. An outstanding example of this form of timekeeping is the Long Count of time kept by the Maya and the Aztecs. This was a running total of the number of days that had passed since a particular starting point. For the Maya this was the year 3113 BC, although the counting was begun some 2,000 years or so after that date. By counting in units of 20 the Maya were able to deal easily with numbers

up in the millions, although it was probably only relatively few priests who really understood the system. The Long Count came to an end when the Spaniards destroyed the Mayan empire in the seventeenth century AD.

Although the day, month and year were natural divisions of time, the peoples of the earliest civilizations found it desirable to divide the day into hours and the month into weeks. The reason for having a week—a cyclical period of a few days—was both religious and commercial. It was widely agreed by the priests of many religions that one regular day in every so many should be kept free from the normal working routine and devoted to religious services and prayers. It also came to be a widespread practice, not tied to any one religion or culture, that on one particular day all people within walking distance would gather at an agreed central place to sell, buy and barter goods, an arrangement that was advantageous to both buyers and sellers. So the Sabbath and the market day came into being, and their periodicity gave rise to the week. This was not necessarily seven days in duration—although seven was well known as a magical number of great significance, probably related to the seven known planets. Weeks of as few as four days and as many as 10 have been known, but seven is a reasonable compromise used in Europe since Roman times and now universally accepted.

For thousands of years hours were a much more fluid concept than they are today. They have been used—at least in a primitive form—since before 2500 BC, but until the fourteenth century AD they were merely a convenient means of dividing up the varying period of daylight. Some peoples measured the span of daylight in six equal parts while others divided it into 12. At the latitude of ▷

The Taoists in ancient China believed that time consisted of recurrent cycles. The priest's satin robe, *far left*, symbolizes Taoist belief. The Pearl of Beginning created the cosmic rhythms of yin, the receptive, feminine force bringing matter to completion, and yang, the creative, masculine force acting with time and yin to initiate life and its cycles. Earthly yin, represented by the phoenix, and heavenly yang, symbolized as dragons and clouds, refer to the regular oscillations of night and day, negative and positive, feminine and masculine.

The Buddhists' search for a state of oneness with the universe and also for reincarnation through a cycle of lives is consonant with their view that time progresses in massive cycles or yugas of 1,080,000 years. These are grouped into larger recurring cycles of four yugas, which decline from Golden Age into dissolution. Buddha appears bottom centre in the Nepalese picture, *left*.

The death of Christ, depicted in Giotto's *Deposition, right*, initiated the concept of both eternal, linear time and our dating system.

▷ Britain the daylight 'day' can be as long as 16 hours in summer and as short as eight hours in winter, yet the day was still divided into 12 equal parts, irrespective of the season. These time divisions of varying length were traditionally known as hours but today are called 'temporary' hours.

Without reservation, we accept certain conventions concerning the way that time is divided up, for example, that the new day begins at midnight and the new year about 10 days after the winter solstice. In earlier societies, however, the conventions were different from those of today. The new day most frequently began at sunrise but, according to the practice still used by astronomers, it sometimes began at noon. In ancient Greece and Israel, and in Italy up to the Renaissance, sunset was taken as the start of a new day and practising Jews still adhere to this principle. Months have traditionally been counted from new moon to new moon or full moon to full moon, rather than in our more artificial system. Year end has been fixed at many different times in different societies and has been allied to the rutting of certain animals or to one of the equinoxes or solstices. Or the length of the year may bear no relation to the seasons and end simply after 12 lunar months have passed.

Early civilizations developed several ways of keeping track of the passing hours. Most of these were inaccurate by modern standards although some of them were quite ingenious. The earliest and most obvious method was to make use of the sun by measuring the changing length of the shadow it casts. The shadow clock was developed independently by the Chinese earlier than 2500 BC and by the Egyptians and Babylonians before 1000 BC. As with the counting of the days, the people who were responsible for this task seem to have been priests, who were probably anxious to fix the times and duration of religious services. By the end of the fourth century BC the Babylonians had developed a much more complex and efficient version of the shadow clock known as the hemicycle. In the hemicycle a hemispherical opening was cut into a clock of stone or wood and in the centre of the hemisphere was a pointer whose shadow travelled in an arc. The hours, indicated as a series of arcs, each with 12 parts, were engraved around the curved surface. These hours varied in length with the season and so were temporary hours. Sundials, which strictly tell the time by the angle, not the length of the shadow, were a common feature of the Roman world. Affluent families had one in the courtyard of their home and rich travellers carried portable models with them.

The Arabs made more advances in accuracy and used their sundials for astronomical purposes. Following an ordinance by Pope Gregory I in AD 600, most Christian churches had a simple scratch dial fixed to their walls. Up to the fourteenth century all sundials were inscribed in temporary hours, which meant that the hours spent in working, prayer and so on fluctuated with an annual rhythm dictated by the sun. It was only after mechanical clocks came into use during the fourteenth century that sundials were inscribed with equal hours. After 1300 their use persisted and their importance actually increased because, for at least three centuries, they were more accurate than their mechanical successors and were employed as a regular check on the time. Many different types of sundial were constructed, most of them extremely accurate—at least in a particular latitude—and some of them beautifully ornamented.

Even the earliest of civilizations recognized the need for an alternative to the shadow clock or sundial, namely, a method of timekeeping that would still work efficiently on a dull day and during the night. First conceived before 1400 BC, and employed throughout the Mediterranean lands and China, was the water clock or clepsydra. In principle the clock was designed to produce a gradually rising or falling water level which would show the time by revealing or concealing lines marked on its container. The idea was neatly symbolic of linear time as a flowing stream, but in practice it was virtually impossible to maintain a constant flow or drip of water. The reason for this was that the water evaporated or became frozen, and that changes in the viscosity of the water (which varies according to temperature) altered the rate of flow. Water clocks were thus notoriously inaccurate. They would have worked better if the markings had been in standard hours, but temporary hours were used to make them compatible with sundials. This meant that a series of different scales had to be inscribed, and that one needed to know the date in order to read off the correct time—or at least a rough approximation of it.

Some improvement in the clepsydra was made by the Greeks and Romans who managed to control the water outflow and thus increase its accuracy, but in order to avoid the problems of freezing, evaporation and changing flow rates the hourglass came into use, with sand representing the flow of time in place of water. The Greeks and Romans had clocks of this sort and, although they disappeared for a while during Europe's Dark Ages, they were reinvented during the eighth century AD. While temporary hours were the rule, different hourglasses were needed for summer and winter because of the different

daylengths. Hourglasses were in regular and common use for many centuries and did not disappear as soon as mechanical clocks were introduced because they were cheaper, initially more accurate and, for several more centuries, superior under stormy conditions at sea.

Several other systems of non-mechanical timekeeping have been used over the centuries, including those that involve a constant rate of burning. Of these, candles with hour scales painted on the wax are the most familiar. Because the rate at which the candle burns depends upon the ambient draught, it was found to be much more satisfactory if the marked candle was enclosed in a lantern. In the Dark Ages, little glass was made, so such lanterns usually had a 'glass' face made of horn. In China more use was made of slow-burning ropes, incorporating tiny bells which would fall off and so strike each hour, and of powdered incense which was burned in shallow, mazelike containers. In sixteenth-century Japan, incense clocks were sometimes designed for 'smelling the time'. Small pieces of incense were burned one by one, each giving off a different smell and so indicating the time.

A later device, specifically designed for telling the time at night, was the aptly named nocturnal. This was an inscribed disc, normally made of brass, with scales and pointers attached to its hub. If one knew the date and the latitude the pointers could be set to the pole star and the Plough (the Great Bear) and the hour read off from the instrument's calibrations.

While it was relatively easy for man to devise the means of charting the rhythm of the days and hours reasonably accurately, designing a workable calendar was altogether a much more arduous task. Nature seems to abhor

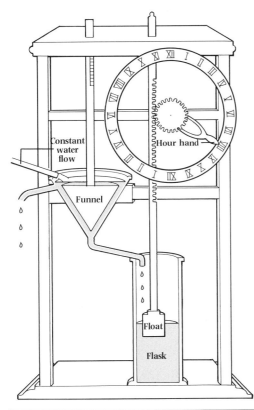

Early man found the time of year written in the stars. The distinct rising and setting of the Pleiades, an open cluster with seven of its stars normally visible to the human eye, was often used to calculate planting and harvest times. In Southeast Asia, the Pleiades appear on the horizon around 1 June and at their zenith from the end of August and mid-September.

Water clocks, used by the ancient Chinese and Egyptians, were improved by the Romans who added a 24-hour dial. As water dripped constantly from the funnel into the flask, the float would rise, moving the clock hand slowly round the dial. Less accurate than shadow clocks, water clocks needed frequent adjustment.

The nocturnal, *right,* was designed for telling the time at night from the positions of the pole star and the Plough. The inner plate is set for the date, the pole star is sighted through the hole and the long pointer aligned with the Plough. The angle between the pole star and the two stars in the Plough which always point towards it, gives the time to the nearest hour.

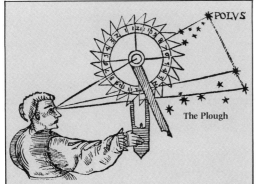

In AD 600 Pope Gregory I ordered a sundial to be put on every church. Since the Earth's orbit around the sun is not exactly circular, the speed of its rotation changes, and the $23\frac{1}{2}°$ tilt means that the height of the sun in the sky alters with the season. A solar day, therefore, varies during the year. A sundial can thus be up to 15 minutes behind or ahead of clock time at Greenwich.

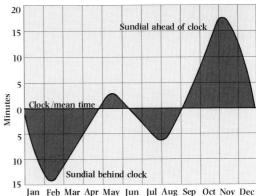

▷ round numbers. So while the rhythm of the Earth's orbit is almost constant, its period is an awkward one in terms of days (365.242) or of lunar months (12.368). For this reason considerable sophistication—particularly of astronomical observation—is required to design a calendar that will keep in step with the apparent motion of the sun over hundreds of thousands of years. All early civilizations tried and, to a greater or lesser extent, failed. Not until the sixteenth century was a system put into practice that was both workable and accurate.

It was probably the Egyptians who set up the earliest calendar, which was based on the heliacal rising of Sirius, the brightest star in the sky, and accepted the length of the year as 365 days. This calendar was used for the civil purposes of government and administration—the lunar calendar was still retained for the business of everyday life and agriculture. According to the civil calendar the year was divided into 12 unnamed months, each lasting 30 days, with the extra five days being tacked on all together at the end of the year,

The priests, whose job it was to mark off the days of the year, must soon have realized that their year was a little short compared with the solar year as the heliacal rising of Sirius would have been a day late at the end of every fourth year and another day later still with every successive four-year period. Yet no attempt was made to correct this error and it was allowed to build up over the centuries until, after 1,460 years (365 × 4), the heliacal rising of Sirius coincided once more with the beginning of the year. Even then the system was not changed. Because the Egyptians did not number their centuries or eras it is difficult to ascertain the starting date of the system. What is known is that their

new year coincided with the heliacal rising of Sirius in AD 139, so it must also have done so 1,460 years earlier in 1321 BC. The origin of the Egyptian calendar is thought to have been two cycles earlier still, in 4241 BC.

Among other ancient civilizations of the Mediterranean there was a great diversity of calendars. From about 2000 BC the Assyrians had a 360-day year of a dozen 30-day months, to which 15 days were added every third year. The Assyrians must have realized that they were getting out of step with the heliacal risings upon which their system was based, and it was altered in a number of ways before being merged with the Babylonian calendar soon after 1100 BC. The Babylonian calendar had evolved from being a lunar-based six-month year to a system of 12 lunar months. This made a total of only 354 days a year, so an extra month was inserted whenever dates seemed to be getting too much out of step. Later, in the fifth century BC, the Metonic cycle was imported from Greece. It was reasonably accurate but complex, employing both lunar and solar data and involving the addition of seven extra months into the calendar over a 19-year period.

Since the fourth century AD the Jewish calendar has also been based on the Metonic cycle of 19 years. In each cycle years 3, 6, 8, 11, 14, 17 and 19 all have 13 months. The extra month, inserted between months five and six, has 30 days. Because of religious festivals or fasts the year cannot begin on a Sunday, Wednesday or Friday so the length of the year is altered by one day if necessary, the addition or subtraction being made to the third month. The ancient Greek calendars, which existed from about 1200 BC, were peculiar to different city states and had various lengths and bases, but the 19-year Metonic cycle became

The ancient Egyptians originally used a lunar calendar with named days, which they synchronized with the rising of the star Sirius and the annual flooding of the Nile. They later devised a civil calendar with numbered days, based on the 365-day solar year. As the civil calendar did not take into account that the solar year was in fact $365\frac{1}{4}$ days long, the months were never truly synchronized with the seasons. Eventually an extra 30-day month was added to the civil calendar every 25 years to correct the error. The Egyptian calendar, *left* from 1230 BC marks the lucky named days in black and the unlucky ones in red.

The French Revolutionary calendar was introduced in 1793 to break away from the ecclesiastical Gregorian calendar. It consisted of twelve 30-day months, each with a name suitable for its season, and each divided into 10-day weeks.

generally accepted from the third century BC and many improvements were made to it.

Beyond the lands of the Mediterranean, other cultures attempted to devise calendars. In Central America the Mayan and Aztec civilizations had similar systems dating from around 3000 BC. Each recognized a 365-day solar year divided into 18 months, each lasting 20 days, plus five final 'evil' days; but despite this they regarded a 260-day cycle, which ran parallel to the solar year, to be of greater importance. These two cycles were meshed with great exactitude to produce a cycle of 52 years—the Calendar Round—having 18,980 days each with a different name. It is remarkable that the Mayan calendar is, in fact, more accurate than the one we use today as it loses only two days every 20,000 years compared with the loss on our calendar of three days every 10,000 years.

In Asia several different early calendar systems evolved independently. From the eleventh century BC, and possibly much earlier, the Chinese adopted a 60-day cycle of named days combined with a form of lunar calendar. At some time during the seventh century BC this was allied with a 12-year cycle. In order to keep the lunar cycles in step with the solar years, extra months were added on when necessary, but not in any planned way. The Hindus, Buddhists and Brahmans used lunar calendars—although with different month names—from as early as 1500 BC, adding days and months during each five-year period to keep in step with the solar year.

Circles and rows of standing stones, set up in northern Europe during the period 1500 to 500 BC, which had religious significance, were also used to identify the summer solstice. A number of other solar, lunar and stellar alignments have been claimed for these megaliths as the positions of the stone blocks can be used to measure the movements of the sun and moon and to chart the occurrence of eclipses. The best-known megaliths are at Stonehenge in southern England and Carnac in Brittany in northwest France.

Many of the early calendar systems have been replaced by the Gregorian calendar we use today. The Gregorian calendar is a direct descendant of the republican calendar of ancient Rome, which had 12 months, with names that mostly persist today, and a total of 355 days. Until 153 BC the year began on 1 March, but the change was then made to 1 January. Any extra days continued to be inserted into the 'last' month, February, although originally this was done after 23 February. In practice the republican calendar was interfered with for political ends so that by 46 BC it had fallen behind the seasons of the year by three months. The essential reform was carried out by Julius Caesar who, with the good advice of the astronomer Sosigenes, added 90 days to the year 46 BC and accepted a $365\frac{1}{4}$-day year. The accumulated quarters resulted in an extra day being added to February every fourth or leap year.

The resulting Julian calendar was so nearly correct that no discrepancies were noted for several centuries. The accumulated error was only one day in 128 years but nothing was done to correct it. In the 1470s, when the discrepancy totalled nine days, moves were made toward reform but the Pope's chosen astronomer was murdered and nothing was achieved. A century later Pope Gregory XIII appointed the German Jesuit mathematician Christopher Clavius to calculate the necessary changes. In October 1582 the calendar, now ▷

The Aztec calendar stone, from the Pyramid of the Sun in Mexico City, *left*, features symbols of all the previous epochs of the world, as well as a 260-day calendar. The Sun God is surrounded by symbols of the 20 day-names (reed, monkey etc), which were used with 13 day-numbers. The 260-day year ran parallel to the 365-day solar year, which was divided into 18 months of 20 days with 5 extra 'unlucky' days. The two serpents with touching fangs represent time.

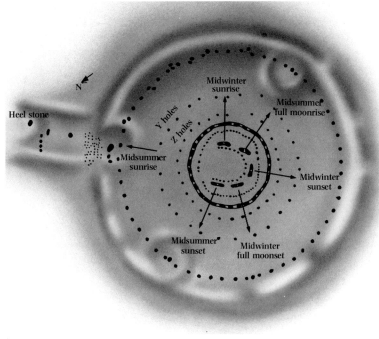

Stonehenge, *above right, and right*, is a remarkably sophisticated 3,500-year-old megalithic calendar clock. From the remaining central stones and the pits of others, it has been possible to detect alignments pointing to sunrise and sunset at both winter and summer solstices, to the extreme positions of moonrise and moonset, and even to its eclipses. These positions have moved a few degrees since 1500 BC, but the readings were accurate at the time the stones were erected. The 30 Y and 29 Z holes which encircle the ring of stones may have been used by the pagan priesthood to count alternate months.

▷ known as the Gregorian calendar, was righted by advancing the date 10 days. At the same time a new rule was introduced to drop three leap days every 400 years, the change being made in century years unless they are divisible by four.

In 1582 it was only Roman Catholic countries which adopted this change. Protestant Europe persisted with the Julian calendar for almost another century. Britain only changed to the Gregorian calendar in 1752, and Russia only in 1917 after a Soviet government had replaced the rule of the Tsars, which means that the Soviet Union now celebrates the October revolution on November 7th. During the twentieth century the Gregorian calendar has been slightly amended so that the error will be reduced to only one day in 20,000 years—an excellent example of the way man has harmonized his timekeeping with the natural rhythm of planet Earth.

In parallel with the evolution of the calendar came a technological advance in clock design which allowed the passing hours and days to be recorded more accurately. The most significant advance came with the invention of the mechanical clocks, machines with integral power systems driven—at least for the first six centuries after their introduction—by falling weights or an uncoiling spring. The various timekeeping devices that operated by burning, or by the flow of water or sand, although they were often so complex that they struck a bell on the hour or triggered an alarm system telling the clock keeper to do so, were not truly mechanical.

No trial models or prototypes of the first mechanical clocks exist, nor are there any surviving descriptions of them. The earliest known public striking mechanical clock was built in 1335 and installed at the palace of the Visconti in Milan. Like most fourteenth-century clocks it had no face or hands but was placed in a tower and equipped with a bell to strike the hours. Several other clocks are known from before 1350—most are Italian and show considerable sophistication in their design. Commissioned and set in towers by noblemen, they struck the passing hours of day and night from one to 24, beginning at sunset. This was the period during which standard hours, unvarying throughout the year, came to replace temporary hours. Slowly the use of mechanical clocks spread across Europe. The earliest surviving example in England is the turret clock of Salisbury cathedral, installed in 1382, but it is not as complete as the one from Wells cathedral which dates from 1392 and is now displayed in the Science Museum in London.

Until the beginning of the sixteenth century all mechanical clocks were powered by falling weights. The principle of the mechanism is that as the weights fall they set in motion a system of gears to which a hand indicating the hour is attached. For the clock to run for any length of time the power in the weights must be used up slowly, not all at once. To achieve this one of the gears is first 'held' then released by means of an escapement device. In the first clocks this consists of a crossbar, carrying a weight on each side, which swings to and fro. Below the crossbar is an escape wheel with toothed notches attached to a vertical bar. Each time the crossbar swings, the vertical bar twists from side to side causing one tooth to be released and the next to slot in. The time of the swing of the escapement depends on the positioning and heaviness of the weights. Due to the inefficiencies of the escapement, these clocks frequently lost or gained as much as 15 minutes a day—a process exacerbated, no doubt, by

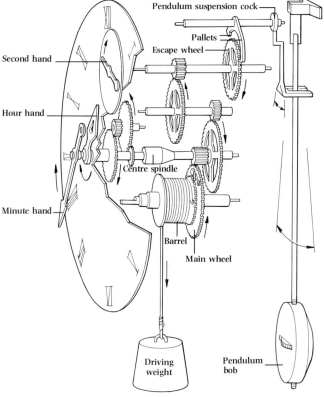

The ornate dial of the clock at Hampton Court Palace, which dates from 1540, indicates the hour, the day, the month, the number of days since the beginning of the year, the phases of the moon, the time of high water at London Bridge and the zodiacal signs. The original mechanism of the clock was replaced in 1835.

Falling weights were used as a driving force in clocks from the early 14th century, and towards the end of the 17th century the one-second pendulum was added as an accurate regulator. The pendulum's movement is transmitted through a pair of pallets which allow the escape wheel to rotate at the rate of one tooth per second. The power of the falling weights is thus only slowly consumed and the force of the escape wheel against the pallets keeps the pendulum swinging. The escape wheel is directly connected to the clock's second hand and, through a series of linking wheels, to the minute and hour hands. A slight loss or gain can be corrected by moving the pendulum bob up or down on the rod.

the fact that the mechanism was not cased, but open to dirt and the public. To set them right these clocks were checked against sundials.

Domestic clocks, with a face and a single hand to indicate the hours, were few and far between in the fourteenth century and were large, ugly and expensive. All these early clocks were made from wrought iron by a single workman who would complete one clock before starting the next. During the late fifteenth century there was a trend towards clocks with faces, but the first real improvement in clock design came with the use of a coiled metal spring to provide the motive power in place of falling weights. The development of these clocks is also shrouded in uncertainty, but the first working, spring-driven clock was probably made by Peter Henlein of Nuremburg in or around 1510. Spring-driven mechanisms were not only more accurate but allowed smaller timepieces to be built. And because spring-driven clocks were more likely to keep going if they were moved around, they led to the introduction, during the sixteenth century, of the first travelling clocks and watches.

The watches of the sixteenth century were certainly not pocket-sized, being typically 9 in (22.8 cm) in diameter by 5 in (12.7 cm) high. Known from their shape as drum clocks, they had a dial on the upper face and most were equipped with a hinged metal cover to protect the hour hand. Some were furnished with alarms, and almost all were beautifully finished, with an engraved case often embellished with gold or silver. Although little more accurate than their fourteenth-century predecessors, and only within the means of the rich, these watches represented a change in the order of things—man had begun to make the rhythm of his life fit in with an artificial division of night and day.

While previously only kings and princes had been able to afford them, during the sixteenth and seventeenth centuries there was a great demand for public and domestic clocks from the upper and middle classes of society. Gradually brass replaced wrought iron as the material used for clock making, making clocks lighter in weight and, because the metal was easier to work, more reliable. In Britain it took centuries for clock making to become an established craft. Most clocks were made by blacksmiths or locksmiths—often with mediocre results—but despite this a few handsome and reasonably accurate English clocks remain from the reign of Elizabeth I (1558–1603).

The most important single advance in the history of mechanical clocks was the introduction of the pendulum as a precise regulator of the escapement. The idea of the pendulum must be credited to Galileo, who was inspired in the late sixteenth century after watching a lamp swinging on a chain in Pisa cathedral, but the first clock to be regulated by a pendulum was designed by the Dutchman Christiaan Huygens in 1656. The first man to produce a pendulum with a beat of exactly one second was Dr Robert Hooke the English physicist, in about 1660.

The time of the pendulum's swing depends on its length and the one-second or Royal pendulum must be about 3 feet (90 cm) long and can be finely adjusted by means of a screw set in the bob. The significance of the one-second pendulum was that it enabled an enormous advance to take place in the accuracy of clocks. Minute hands were added to clock faces and the idea of dividing the minute into 60 seconds was readily accepted because it was possible, at last, to count off the seconds with great precision. Another of ▷

The oldest surviving clock in England *below* was installed in the tower of Salisbury cathedral in 1382. Although like other early European public clocks it had no dial, hands or case, it had a bell mechanism for striking the hours. It has been modernized over the centuries but much of the original wrought iron structure and verge escapement remain. This verge or crown wheel escapement mechanism, *right*, was the earliest means of regulating the speed of mechanical clocks. The crown wheel, which could be mounted vertically or horizontally, was prevented from using up all the force of the driving weights (not shown) by the foliot balance which swivelled from side to side, its pallets alternately releasing and blocking the teeth.

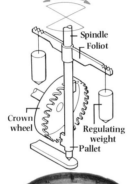

When spring-driven clocks were invented in the 1550s, a device was needed to produce a constant degree of power from the uncoiling spring. The most inspired method was the fusee, a tapered drum which housed the spring and allowed the force of the spring to be transmitted via a length of gut or cord. When fully wound, the cord winds first off the thin end of the drum, producing only a little power. As the day progresses, however, it begins to wind off the thick end, producing greater force to compensate for the reduced power of the uncoiling mainspring. Although the Italians were believed to have originated the fusee, the first man to insert this system into a clock was Jacob Zech of Prague in 1525.

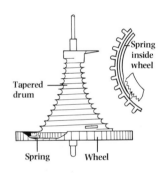

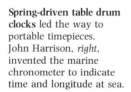

Spring-driven table drum clocks led the way to portable timepieces. John Harrison, *right*, invented the marine chronometer to indicate time and longitude at sea.

Our day begins at midnight, but until 1925 astronomers counted the days from noon. Until the last century, the nautical day ended at noon, *left*. So 6 am Monday civil time was also 6 am Monday in nautical time (but only 4 hours before Tuesday) yet it was 6 pm Sunday to astronomers.

Civil

SUNDAY			MONDAY			TUESDAY		
6 am	noon	6 pm	6 am	noon	6 pm	6 am	noon	6 pm

Astronomical

SAT	SUNDAY			MONDAY			TUES	
6 pm	6 am	noon	6 pm	6 am	noon	6 pm	6 am	

Nautical

SUN		MONDAY			TUESDAY			WED
6 am	noon	6 pm	6 am	noon	6 pm	6 am	noon	6 pm

▷ Hooke's inventions, made at about the same time, was the recoil or anchor escapement which allowed a heavy pendulum to swing in a small arc.

These innovations changed the shape of clocks: the longer pendulum meant a taller clock and the English longcase or grandfather clock came into being. This period, the last quarter of the seventeenth century, was a time during which English clock and cabinet making were the best in Europe, and there are many grandfather clocks from the seventeenth and eighteenth centuries still keeping good time all over the world. Another effect of more accurate timekeeping, but one more difficult to quantify, was an increased awareness of small time intervals and the punctuality that they made possible.

Many refinements to clock mechanisms were made during the eighteenth century. Hooke's anchor escapement was replaced with the dead-beat escapement which made the swing of the pendulum more accurate by cutting out sway or vibration, and several solutions were found to the problem of a pendulum expanding—and thus moving more slowly, so causing the clock to lose time—in hot weather. If a steel pendulum is heated by 2 degrees centigrade (4 degrees F) it will lose one second a day. To compensate for this, metals were used whose expansion and contraction would cancel each other out.

The construction of ornamental clocks to act as decorations for the palaces and mansions of the rich was a lucrative business that required great imagination and reached its peak in France during the second half of the eighteenth century. Several French kings kept court clockmakers who produced many elaborate metal-cased clocks. Less tall than the English longcase clocks, these were designed to be placed on tables or mantlepieces.

Clocks have undoubtedly played a large part in influencing man's attitude to the rhythm of time but it is arguable that watches have been even more influential. The drum clocks of the sixteenth century were gradually improved in accuracy and reduced in size but watches did not become truly pocket-sized until the invention, around 1660, of the hairspring or balance spring, which was the essence of the seventeenth-century watch. The hairspring is a steel band or spring stressed by being bent or coiled which, when wound, acts as a store of energy. In early watches the spring was attached to the crossbar of a balance which, under the tension of the spring, swung first one way and then the other. The accuracy of the watch was controlled by a regulator which shortened or lengthened the spring and thus varied its tension.

As pocket watches became more popular, new types of escapement were developed specifically for them—the cylinder escapement in 1726 and the lever escapement in 1765. By this time most watches were almost as accurate as domestic clocks and by the end of the nineteenth century had become accessible to all. The wrist watch of today, if it is of the traditional type, has a balance in the form of a wheel with a heavy rim and a spring providing a restoring force. Electronic watches use another type of mechanism. A miniature high-density battery is used to power a minute tuning fork or a magnet and coil, and the vibrations produced are translated into digital signals that glow in gas-filled valves or activate liquid crystals.

The mass production of domestic clocks began in the nineteenth century and made such timepieces cheap enough to be standard items in every household. The factory manufacture of clocks started in America, but before

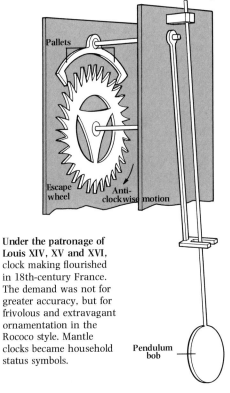

Under the patronage of Louis XIV, XV and XVI, clock making flourished in 18th-century France. The demand was not for greater accuracy, but for frivolous and extravagant ornamentation in the Rococo style. Mantle clocks became household status symbols.

The anchor or recoil escapement, *above,* invented in 1660 by Robert Hooke, was a great improvement on the verge escapement. The pallets are 'anchored' by a joining yoke, and alternately engage and release the excape wheel once per second to keep the pendulum swinging. The precision of this mechanism allowed seconds to be accurately counted. A second hand was therefore frequently incorporated in a smaller dial within the face of the clock, as in the early 18th-century English longcase clock, *right.* This escapement, however, tended to recoil against the pallets, so in 1715 a modified version—the dead-beat escapement— was developed by George Graham which corrected the defect by changing the shape of the pallet heads.

the end of the century cheap clocks were being made in quantity throughout Europe. Clocks were installed in every public building, in town squares and railway stations. Most men carried pocket watches and in the industrialized nations of the world people were becoming subservient to the rule of the clock.

The first electric clocks were designed and built in the 1840s, but it was another 50 years before electrical components were reliable enough for commercial production. In such an electric clock, a pendulum is kept in motion by regular electric impulses and, as a result, the hands move once a minute. This is the master clock and it can be used to drive many slave clocks, an invaluable system for factories and similar building complexes. As soon as a regular frequency of current was established—in 1918 in the USA and nine years later in Britain—it became possible to make synchronous clocks, powered by an alternating current of mains electricity. Such clocks are now in common use all over the world and should be accurate to within a few seconds over extended periods. Yet since this accuracy depends largely on the exactness of the voltage generated by power stations, these electric clocks do not really qualify as timepieces at all. They contain no timekeeping mechanism, only gears for reducing the frequency of current until it matches the required frequency of rotation for their hour, minute and second hands.

The twentieth century has witnessed a vast increase in the accuracy of timekeeping. The scientific clocks of today are far removed from even the most accurate wrist watches or domestic clocks, although there has been a significant spin-off from the laboratory to everyday life. In 1921 William H. Shortt invented the ultimate in mechanical clocks, the Shortt free pendulum

clock which varies by less than a few thousandths of a second a day. The clock consists of two separate clocks, one synchronizing the other. Every 30 seconds the swinging pendulum receives an impulse from a lever, which is released by an electric current transmitted from the secondary or slave clock. A synchronizing signal is then returned to the slave clock, so ensuring that the next impulse will follow exactly half a minute later.

More accurate clocks than the Shortt free pendulum have been based on an entirely different concept, that of the rhythm of natural vibrations. The quartz clock, for example, makes use of the oscillations given off by the mineral quartz crystal when alternating current is applied to it. These oscillations act in place of a pendulum to govern the motion of the clock. The principle is essentially the same as the mains electric clock, except that a much higher frequency of vibration or oscillation allows much greater accuracy—so much so, in fact, that a quartz crystal is more accurate than the rotation of the Earth around the sun. First developed in 1929, the quartz clock consists of a ring of quartz $2\frac{1}{2}$ inches (6.35 cm) across suspended on threads in an insulated chamber. Electrodes are attached to the ring and connected to a current to produce oscillations. The frequency of the oscillations is then reduced by a ratio of six million to one to make the ring rotate once every 60 seconds. This ring is then connected to a clock dial by means of mechanical gearing.

The atomic clock has evolved from the quartz clock and makes use of the natural oscillation frequency of atoms—usually those of the element caesium —as its pacemaker. The atomic clock is accurate to within one second over 30,000 years, a precision that the human mind finds hard to grasp. Thus in its ▷

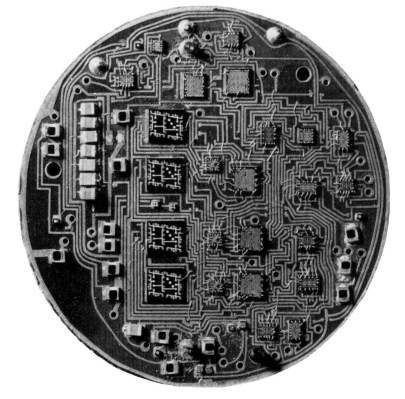

Astronomical dials, showing the date, phase of the moon and time in different parts of the world, were popular features of grandfather clocks, *top.* Early watches were also ornate, often beautifully enamelled and gilded, *right,* with a metal lid to protect the hand. They became increasingly accurate and developed into 'fob' or pocket watches.

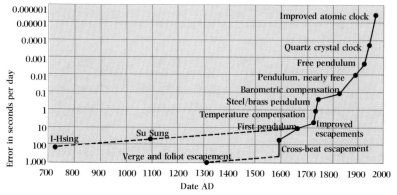

Accuracy of timekeeping improved only very gradually during the Middle Ages, and the earliest mechanical clocks were even less precise than sundials. Since the late 17th century accuracy has increased dramatically, particularly in the last 50 years, when modern atomic clocks losing less than a second every 30,000 years were developed.

Quartz crystal was first used in clocks in 1929 and was applied to electronic watches in the 1950s. In the watch, *above,* the crystal, supported inside a metal holder and sealed in a vacuum, has properties which cause it to vibrate at a steady frequency, when an alternating current is applied. The current is supplied by a long-lasting battery through an integrated circuit (a silicon chip). The frequency of the oscillation of the quartz crystal is converted into electronic impulses with a constant frequency determined by the size, shape and cut of the crystal—in this case 34,000 cycles per second. The impulses pass through another circuit which reduces the impulses to a frequency (eg 1 per second), which can be transmitted to the digital display, composed of tiny red dots.

▷ most rarified and artificial form, twentieth-century timekeeping takes its cue from the rhythm of the Earth's own building blocks.

With each year that passes, human society is becoming more and more time orientated. For most of us, there is a greater need than ever before to know the exact time of day, and to perform particular tasks at particular moments. Yet we possess no inborn sense of the passage of time. Our bodies maintain certain rhythms, which may be extremely accurate, but these rhythms cannot tell us precisely when a minute, an hour or a day has passed. As a result of this paradox, we all have our own subjective view of the rate at which time is passing and this, in turn, depends on our occupation, age and state of health. Also, we need continual prompting to tell us how quickly time really is ticking by. This means that watches and clocks are everywhere, that time-based systems have come to govern our lives and that references to time have become an important ingredient of our language.

Occasionally an individual can cultivate an accurate time sense. A soccer referee, for example, knows when 45 minutes have passed, while many professional broadcasters are skilled at judging the passage of time, from one minute to 30. But both referee and broadcaster have a clock or watch to back up, or correct, their estimate. Any of us can, if we wish, count off the passing seconds by muttering, 'one mississippi, two mississippi, three mississippi' and so on, but only when we are not otherwise occupied. It is much more usual for time to surprise us—sometimes going much faster than we think and sometimes much more slowly.

When we try to estimate the amount of time that has passed while we have been engaged in some activity or another, we are usually wrong, having been misled by our internal, subjective clock, of which the brain is a vital part. If the brain is dealing with new material, such as a movie or foreign surroundings, the amount of time spent is always overestimated—the more so if the movie is complex or if the trip abroad involves many varied activities. Conversely, the duration of those periods during which the brain receives little stimulus are underestimated. The best evidence of this is provided by volunteers who have spent periods of 100 days or more in deep caves. Cut off from all external stimuli, including the natural rhythm of day and night, such volunteers always estimate their spells of isolation to be shorter than they have been, sometimes by as much as 50 per cent.

The subjective perception of time varies with age so that the days seem to pass slowly for children, but rush past faster and faster with increasing age. The degree of mental stimulation in one's life is also involved in this phenomenon because for children everything is new, while old age brings few surprises. Another cause of this age-related difference in time sense is metabolic rate, that is, the rate at which the processes of body chemistry take place. Children have a slightly higher metabolic rate, and consequently a higher body temperature than adults, and for this reason can actually process more information than an adult in any given time, although they have a greater liability to boredom. The lower than average metabolic rate of the elderly, which gives them a lower body temperature, means that all their organic processes slow down and time seems to move correspondingly fast.

This relationship between body temperature and the subjective measure-

Ruled by schedules and high precision clocks, timetables and stop-watches, modern life has become a race against time. 'Time is money', 'time to get up' 'time for lunch', 'time's up' are constant reminders that we belong to a frenetic, time-ordered society which largely ignores natural human rhythms. It is one of the frustrating paradoxes of contemporary life that, having rushed to 'beat the clock' and arrive in time for a flight, train or bus, we must then queue and wait. Wrestling with time has increased efficiency but often at the cost of high blood pressure.

ment of time is equally valid when you are ill. When you have a fever your body temperature rises and time seems to pass slowly, but if you are subjected to extreme cold, time seems to go more quickly. There is even a mild but regular variation in our subjective clocks over the course of the day, which explains why time appears to go most slowly in the mornings.

The perception of time can be altered by the use of drugs. Stimulants, such as heroin, cannabis, LSD and other psychedelic drugs, and even the caffeine in coffee and tea, all speed up our subjective view of time and make us overestimate time intervals. Barbiturates, alcohol and other drugs that cause drowsiness are likely to have the opposite effect, so that considerable lengths of time can slip by unnoticed if you are under their influence. A state of trance, as produced by transcendental meditation and some forms of rhythmic dancing, also leads to underestimation of passing time and the same effect is produced by schizophrenia and some other mental disorders.

All around us are reminders of the extent to which we are obsessed with and governed by time. This is inevitable since a complex technological society with a high population density can only organize itself by conforming to timetables, schedules and timed appointments. Despite our criticism of public transport systems, most of them do run to time and, whether they are trains or aircraft, it is essential that they should do so, if fatal collisions are to be avoided. Our television and radio services work to exact schedules, while all entertainments, such as sporting events, films and plays, have pre-announced starting times and known durations.

The whole of industrial production is based on scheduled flows of manufacture and delivery. We are expected to be at our places of work between fixed hours and arrange to meet people at particular times. 'Time is money' is a catch phrase often used in business and for almost every moment of our waking lives there is some reference to time—it is coffee time, lunch time, tea time, going home time, bed time and so on. Alarm clocks wake us in the morning and breakfast is accompanied by radio or television shows which include frequent time-checks. Many of us jostle in a rush hour on the way to and from work and this may mean catching a bus or train at a particular time. When we do arrive at our work place, hopefully at the proper time, we may have to 'clock in' especially if we work 'flexitime' hours.

It seems to be an accepted part of the twentieth century ethos that 'faster is better'. World records are not only universally acclaimed but, for sports such as swimming and athletics, measured by split-second timing. Many companies have a department whose job it is to speed up all the others, while computers, and their software systems, are designed to process data more and more swiftly. All this adds up to an invitation to rush through life at breakneck pace, and results in a steady build-up of pressure on members of industrialized societies. To cope with this problem of 'future shock'—the disorientating effect of excess change in all areas of society—it is helpful to have the steady rhythm imposed by a time schedule, because it can act as a cushion by counterbalancing the effects of change in one or two areas of life. It may make it easier to move house or change jobs, for example, if you can maintain the rhythm of a particular journey to and from work, keep to the same mealtimes and watch the same television programmes in the evening.

The Hopi Indians have no word for past, present or future, but express time in terms of momentary, continuing or repeated events, and cannot conceive of things happing simultaneously. Their language suggests rather than defines the timing of an event. For example, 'a light flashed' to us states that the flash is already in the past, but the Hopi would simply say 'flash', to describe the duration of the flash. They also see time as varying with each observer, a view of time which does not lend itself to objective timekeeping.

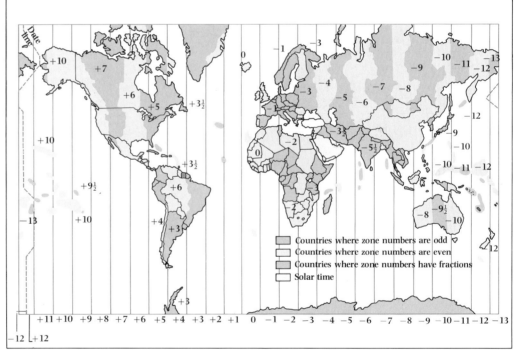

Noon (when the sun reaches its zenith) occurs at different times according to position on the earth's surface, hence the need for International Time Zones. High speed travel produces some bemusing effects—if a traveller crosses the International Date Line eastwards on his birthday, he will relive it, but if he crosses westward, he loses his birthday altogether.

Our perception of time changes considerably with age. Young children tend to overestimate short time-spans, but underestimate long ones, so that a week can seem as long as a year. A mature working man sees today as longer than the previous week because so much current information crowds his mind. Old people, on the other hand, look more to the past and feel time passes very fast. One explanation for this is that a slower metabolism and lower body temperature influence the consciousness of time.

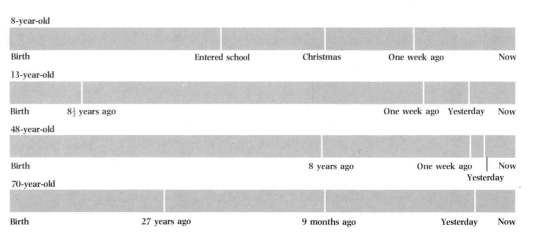

183

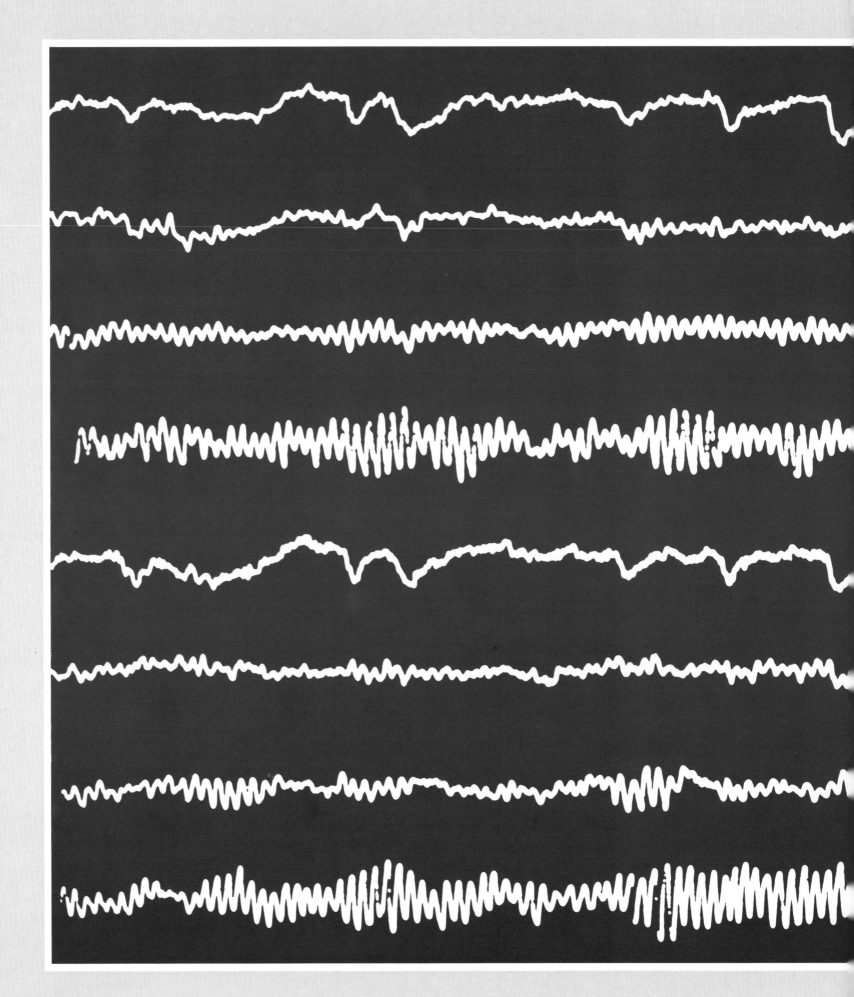

Rhythm data

Confronted with the awesome diversity of the universe and all it contains, the human intellect strives to recognize some sort of order. From the apparently ever-changing and complex physical and living worlds around him, man grasps at patterns. Of all these patterns the one that emerges most clearly and most often is the pattern of rhythm. As a result, rhythmic and cyclical phenomena are seen to be the central feature of a huge number of systems, processes and activities.

Over many thousands of years man has gathered the data that lends support to the idea of rhythms. This data concerns subjects as diverse as the revolution of galaxies and their stars and the rhythmical process of breathing. Although they might seem unconnected, these phenomena are inextricably linked because the celestial dance performed by the sun, moon and Earth has profound consequences for all earthly life. All life has evolved to respond to the complex, polyrhythmic beat of the physical world and the echoing counterpoint rhythms of biological clocks are to be found in almost all living things. In addition, all organisms demonstrate intrinsic activities and processes that are themselves rhythmic. From the growth of a tree to the flight of a bird, the processes of life are centred upon repetitive, rhythmical acts. And human life itself, constructed from a collection of biological and social components, reveals layer upon layer of rhythmical and cyclical patterns.

This final stage in the journey through the rhythms of life attempts to compress part of the enormous bank of data from which the preceding pages have drawn their inspiration. It is a catalogue of the types of events that most clearly demonstrate physical or living rhythms. Most of the lists extend or expand areas considered earlier, while a few embark on new areas of study. Most of the lists are entirely factual—they describe what certainly exists. A few others, such as the cycle of historical events, represent our desire to find patterns in everything, despite the fact that the evidence still remains insubstantial.

Planetary rhythms

The period of each planet's orbit round the sun is the length of its 'year'; the seasons on each planet are determined by the inclination of its Equator to the plane of its orbit; the length of the day on any planet is the time it takes to rotate about its axis. The distance of each planet from the sun determines its temperature and the composition of its atmosphere.
(*see pages 14–33*)

Planets	Period of orbit around the sun	Inclination of Equator to orbit	Period of rotation about axis	Distance from sun (miles)	Distance from sun (km)
Mercury	87.9 days	7°	58.5 days	36,000,000	57,920,000
Venus	224.7 days	3° 24'	243 days	67,200,000	108,100,000
Earth	365.2 days	23° 27'	24 hours 4 mins	92,957,200	149,600,000
Mars	687 days	23° 59'	24 hours 37 mins	141,600,000	228,000,000
Jupiter	11.8 years	3° 05'	9 hours 51 mins	484,300,000	779,100,000
Saturn	29.5 years	26° 44'	10 hours 14 mins	886,100,000	1,426,000,000
Uranus	83.8 years	97° 55'	10 hours 48 mins	1,783,000,000	2,870,000,000
Neptune	164.8 years	28° 48'	14 hours	2,793,000,000	4,493,000,000
Pluto	247.7 years	c.50°	6 days 9 hours	3,666,000,000	5,898,000,000

Comet cycles

Comets, like planets, orbit around the sun but, unlike them, have orbits that are often very large and eccentric. Comets are only visible when they come close to the Earth, and Halley's comet is the only bright comet to have an orbital period of less than a few centuries. Several other comets are listed here, although these are so faint that they cannot be seen with the unaided eye.

Comet	Orbital period (years)
Encke	3.3
Finlay	6.9
Tuttle	13.6
Crommelin	27.9
Halley	76.0
Grigg-Mellish	164.3

Meteor showers

Meteors are associated with comets and travel around the sun in swarms. When the Earth passes through a swarm, the meteors burn up in the Earth's atmosphere. They occur frequently throughout the year but, except for the Perseids, are not always consistently rich. Every 33 years the Leonids reach a peak of activity when as many as 100,000 shooting stars an hour can be seen.

Meteor showers	Dates
Quadrantids	1—4 January
Lyrids	19—22 April
Aquarids	1—13 May
Perseids	27 July—15 August
Orionids	15—25 October
Leonids	17 November
Andromedids	26 November—4 December
Geminids	9—13 December

Miscellaneous cosmic rhythms

Apart from the planets, comets and meteors, there are various other cosmic rhythms, a few of which are given here.
(*see pages 14–33*)

The period of the sun's revolution around the centre of the Milky Way	200 million years
Interval between one new moon and the next	29.53 days
Monthly revolution of moon around Earth	27.32 days
Daily revolution of moon around Earth	24 hours 50 mins
Interval between two perfectly straight alignments between Earth, moon and sun responsible for the solar eclipse cycle and for atmospheric tides produced by the moon	18.6 years
The meteorological drought cycle of the American Midwest: east of the Rocky Mountains, west of the Mississippi	18.6 years
Rotation of Earth's spin axis	21,000 years
Change in shape of Earth's orbit	c.100,000 years

Rainy seasons

Many tropical areas have a distinct annual season of heavy rainfall while the rest of the year is dry. Other areas, such as Lagos and parts of Ghana on the coast of the Gulf of Guinea, have two rainy seasons each year. Much of the life in these areas is shaped by the rhythm of the rainy season. The seasons given here are taken from the average monthly rainfall figures for the area concerned.
(*see pages 14–33 and 34–51*)

Africa

Conakry (Guinea)	June—October
Half Assini (Ghana)	March—July October—December
Kano (Nigeria)	May—September
Khartoum (Sudan)	July—September
Lagos (Nigeria)	April—July September—October
Mangoche (Malawi)	December—April
Wenchi (Ghana)	April—June September—October

Asia and Australasia

Bombay (India)	June—September
Broome (Australia)	December—April
Calcutta (India)	May—October
Colombo (Sri Lanka)	October—December March—May
Darwin (Australia)	November—April
Mandalay (Burma)	May—October

South and Central America, Caribbean and Hawaii

Havana (Cuba)	May—October
Honolulu (Hawaii)	November—March
Manaus (Brazil)	November—May
Mexico City (Mexico)	June—September
Port of Spain (Trinidad)	June—December

Breeding seasons

An animal's breeding season is timed so that the young are born at a time of year when they have the best chance of survival, that is, when there is plenty of food around and the climate allows them to complete the early stages of growth. The chart shows those times of year for various vertebrates, although the exact times vary with latitude. (*see pages 64–81*)

Animal	Scientific name	Breeding season
MAMMALS		
Armadillo	*Dasypus novemcinctus*	June—August
Bactrian camel	*Camelus bactrianus*	All year
Chimpanzee	*Pan troglodytes*	All year
Chipmunk	*Tamias striatus*	March—July
Common seal	*Phoca vitulina*	June—August (Atlantic) September (Pacific)
Common shrew	*Sorex araneaus*	March—September
Fin whale	*Balaenoptera physalus*	November—March
Goat	*Capra hircus*	September—January
Hedgehog	*Erinaceus europaeus*	March—September
Lemur	*Lemur catta*	Late March—early July
Man	*Homo sapiens*	All year
Mandrill	*Papio sp*	All year
Mink	*Mustela vison*	March—April
Musk rat	*Ondatra zibethicus*	April—October
Platypus	*Ornithorhyncus anatinus*	July—October
Porpoise	*Phocoena phocoena*	July—August
Raccoon	*Procyon lotor*	January—June
Virginia opossum	*Didelphus marsupialis virginiana*	January—October
BIRDS (in NW Europe)		
Canada goose	*Branta canadensis*	March—April
Eider	*Somateria mollissima*	Late April—early September
Greylag goose	*Anser anser*	April—May
Mallard	*Anas platyrhyncos*	March—October
Mandarin	*Aix galericulata*	April—May
Mute swan	*Cygnus olor*	March
Shelduck	*Tadorna tadorna*	May—August
Teal	*Anas crecca*	Mid-April—July
Tufted duck	*Aythya fuligula*	Mid-May—early September
Wigeon	*Anas penelope*	June—August
REPTILES AND AMPHIBIANS		
American toad	*Bufo americanus*	February—July
Clawed toad	*Xenopus laevis*	All year
Eastern box turtle	*Terrapene carolina*	April—May
Mississippi alligator	*Alligator mississippiensis*	January—September
Moccasin snake	*Agkistrodon contortrix*	May
Prairie rattlesnake	*Crotalus viridis*	April—May
FISH		
Atlantic salmon	*Salmo salar*	September—February
European carp	*Cyprinus sp*	Spring—early summer
King salmon	*Oncorhynchus tschawytscha*	April—June
North Sea cod	*Gadus morhua*	February—April
Plaice	*Pleuronectes sp*	December—March
Sea lamprey	*Petromyzon marinus*	Spring
Yellow-fin tuna	*Thunnus albacares*	Spring—Summer

Estrous cycles and gestation periods

When a female mammal comes into 'heat' she is in estrus and capable of producing eggs for fertilization by a male. The estrous cycle is the sequence of biological events leading to egg production and fertilization. Many mammals have one or two estrous cycles a year, while the cycles of others such as our own species are continuous. (*see pages 64–81*)

Animal	Scientific name	Average estrous cycle (days)	Average gestation period (days)
Bactrian camel	*Camelus bactrianus*	15	400
Bushbaby	*Galago senegalensis*	32	115
Chimpanzee	*Pan troglodytes*	36	235
Cow	*Bos taurus*	21	284
Goat	*Capra hircus*	21	151
Golden hamster	*Mesocricetus auratus*	4	16
Gorilla	*Gorilla gorilla*	39	270
Guinea pig	*Cavia porcellus*	16	68
Horse	*Equus caballus*	22	336
Man	*Homo sapiens*	28.3	294
Mink	*Mustela vison*	8	53
Mouse	*Mus musculus*	4	25
Musk rat	*Ondatra zibethicus*	4	30
Pig	*Sus scrofa*	21	114
Platypus	*Ornithorhynchus anatinus*	60	12
Rabbit	*Oryctolagus cuniculus*	0	31
Rat	*Rattus norvegicus*	4	21
Red kangaroo	*Megaleia rufa*	35	33
Rhesus monkey	*Macaca mulatta*	28	170
Sheep	*Ovis sp*	16	148
Wallaroo	*Macropus robustus*	33	32

Bird migration times

Many birds that spend the winter in the tropics migrate north to breed in temperate regions during the summer. Birds that breed in the Arctic during the northern hemisphere's summer migrate south to temperate regions for the winter. The times of year are given when these birds are present in full force in the British Isles, California and Ottawa. (*see pages 64–81*)

Common name	Scientific name	Arrival	Departure
BRITISH ISLES			
Swallow	*Hirundo rustica*	Early April	Late October
Cuckoo	*Cuculus canorus*	Mid April	Early September
Nightingale	*Luscinia megarhyncha*	Mid April	Late August
Osprey (northward)	*Pandian haliaetus*	Late April	Early June
Quail	*Coturnix coturnix*	Mid May	Late September
Osprey (southward)	*Pandian haliaetus*	Late August	Mid October
Snow bunting	*Plectrophenax nivalis*	Early October	Late March
Little auk	*Plantus alle*	Early November	Mid February
CALIFORNIA			
Northern cliff swallow	*Petrochelidon albifrons*	Late February	Late October
Barn swallow	*Hirundo erythrogaster*	Early March	Early October
Northern black swift	*Nephoecetes niger*	Mid April	Mid September
Western tanager	*Piranga ludoviciana*	Early May	Late September
OTTAWA			
American woodcock	*Rubicolor minor*	Late March	Late October
Eastern kingbird	*Tyrannus tyrannus*	Early May	Late September
Ruby-throated hummingbird	*Archilochus colubris*	Early May	Mid October
Bobolink	*Dolichonyx oryzivorus*	Mid May	Mid September
Black-polled warbler	*Dendroica striata*	Mid May	Mid October

Flowering seasons

Many wild plants flower at the same time every year and their flowering often distinguishes the gradual changes of the temperate seasons. (*see pages 34–51*)

Garden flowers are the cultivated relatives of wild plants. Although selective breeding can produce plants that will flower at any time of the year, garden plants have specific times when they will flower. This chart is divided into annuals, which must be sown each year and perennials, which die down each winter and then grow again the following spring. (*see pages 34–51*)

Plant	Scientific name	Season
WILD FLOWERS:		
Bramble	*Rubus fruticosus*	May—September
Buckwheat	*Fagopyrum esculentum*	July—August
Butterbur	*Petastites hybridus*	March—May
Common sundew	*Drosera rotundifolia*	June—August
Deadly nightshade	*Atropa belladonna*	June—August
Dog's mercury	*Mercurialis perennis*	February—April
Fennel	*Foeniculum vulgare*	July—October
Gorse	*Ulex europaeus*	February—June
Heather	*Calluna vulgaris*	July—September
Hop	*Humulus lupulus*	July—August
Ivy	*Hedera helix*	September—November
Love-in-a-mist	*Nigella damascena*	June—July
Marsh mallow	*Althaea officinalis*	August—September
Marsh marigold	*Caltha palustris*	March—July
Mistletoe	*Viscum album*	February—April
Primrose	*Primula vulgaris*	December—May
Shepherd's purse	*Capsella bursa-pastoris*	January—December
Snowdrop	*Galanthus nivalis*	January—March
Sweet violet	*Viola odorata*	February—April
Touch-me-not balsam	*Impatiens noli-tangere*	July—September
TREES:		
Aspen	*Populus tremula*	March
Beech	*Fagus sylvatica*	March—April
Common maple	*Acer campestre*	May—June
Hazel	*Corylus avellana*	March—April
Ma-dake bamboo	*Phyllostachys bambusoides*	Every 120 years
Muli bamboo	*Melocanna* sp	Every 30 years
Rowan	*Sorbus aucuparia*	May—June
Sessile oak	*Quercus petraea*	April—May
Silver birch	*Betula pendula*	April—May
Small-leafed lime	*Tilia cordata*	July
Umbrella bamboo	*Thamnocalumnus spathaceus*	Every 90 years

Plant	Scientific name	Season
GARDEN FLOWERS:		
ANNUALS		
Bell flower	*Campanula medium*	July
Californian poppy	*Eschscholzia californica*	June—July
Hollyhock	*Althaea rosea*	June—July
Honesty	*Lunaria annua*	May—July
Lobelia	*Lobelia erinus*	June—October
Snapdragon	*Antirrhinum majus*	July—October
PERENNIALS		
Alyssum	*Alyssum saxatile*	April—June
Aster	*Aster novi-belgii*	September—October
Autumn crocus	*Crocus sativus*	October
Bougainvillea	*Bougainvillea spectabilis*	August—September
Columbine	*Aquilegia* sp	May—June
Cyclamen	*Cyclamen coum*	December—March
Bleeding heart	*Dicentra formosa*	April—June
Freesia	*Freesia* sp	March—April
Fritillary	*Fritillaria imperialis*	April
Gentian	*Gentiana asclepiadea*	August—September
Ginger lily	*Hedychium densiflorum*	May—June
Grape hyacinth	*Muscari armeniacum*	April—May
Hellebore	*Helleborus orientalis*	February—April
Lion's tail	*Leonitis leonurus*	October—December
Lotus	*Nelumbo nucifera*	July—September
Lupin	*Lupinus* sp	May—June
Mallow	*Malva moschata*	May—October
Mint	*Mentha spicata*	August
Oregano	*Origanum marjorana*	June
Primula	*Primula denticulata*	March—May
Sage	*Salvia officinalis*	May—June
Ice plant	*Sedum spectabile*	September—October
Lemon verbena	*Lippia citriodora*	August
Wisteria	*Wisteria sinensis*	May—June

Effects of daylength on flowering

All flowering plants can be divided into three categories: those that flower in autumn when the days are becoming shorter; those that flower in spring and early summer when the days are becoming longer; and those that do not depend on the length of day to flower. In the first two categories, the short day and long day plants, the plants will flower only when the length of the day has reached a certain critical value. (*see pages 34–51*)

SHORT DAY PLANTS	
Aster	*Aster novi-belgii*
Balsam	*Impatiens balsamina*
Chrysanthemum	*Chrysanthemum* sp
Coffee	*Coffea arabica*
Corn	*Zea mays*
Cosmos	*Cosmos sulphureus*
Lima bean	*Phaseolus lunatus*
Love-lies-bleeding	*Amaranthus caudatus*
Soy bean	*Glycine max* 'Biloxi'
Sweet potato	*Ipomoea batatus*

LONG DAY PLANTS	
Black-eyed Susan	*Rudbeckia speciosa*
Carnation	*Dianthus superbus*
Cloud grass	*Agrostis nebulosa*
Dill	*Anethum graveolens*
Gardener's garters	*Phalaris arundinacea*
Italian rye grass	*Lolium italicum*
Peppermint	*Mentha piperita*
Radish	*Raphanus sativus*
Scarlet pimpernel	*Anagallis arvensis*
Spinach	*Spinacia oleracea*

DAY NEUTRAL PLANTS	
Carrot	*Daucus carota*
Cayenne pepper	*Capsicum frutescens*
Cucumber	*Cucumis sativus*
Holly	*Ilex aquifolium*
Honesty	*Lunaria annua*
Marigold	*Calendula officinalis*
Onion	*Allium cepa*
Potato	*Solanum tuberosum*
Spanish clover	*Gomphrina globosa*
Tomato	*Lycopersicon esculentum*

Seasons to propagate cultivated plants

To multiply already existing stocks of cultivated plants certain methods and times of propagation are most successful. The most widely used method is given here, together with the optimum season to apply it. Many plants are not propagated by seed, since the seed may produce a new plant that differs from its parent. (*see pages 34–51*)

Plant	Scientific name	Propagation time and method
Alder	*Alnus* sp	Sow seeds in autumn
Beech	*Fagus* sp	Sow seeds in autumn
Cedar	*Cedrus* sp	Sow seeds in autumn
Chrysanthemum	*Chrysanthemum* sp	Take cuttings all year round
Dahlia	*Dahlia* sp	Make root division in spring
Elm	*Ulmus* sp	Sow seeds in spring
Eucalyptus	*Eucalyptus* sp	Sow seeds in spring
Fig	*Ficus carica*	Take cuttings in autumn
Fir	*Abies* sp	Sow seeds in autumn or spring
Juniper	*Juniperus* sp	Take cuttings in late summer
Magnolia	*Magnolia* sp	Sow seeds in autumn or spring
Maidenhair	*Ginkgo biloba*	Sow seeds in autumn or spring
Maple	*Acer* sp	Sow seeds in summer or autumn
Oak	*Quercus* sp	Sow seeds in autumn
Olive	*Olea europaea*	Take cuttings in winter or summer
Shagbark hickory	*Carya ovata*	Make bark graft in winter to spring
Strawberry	*Fragaria chiloensis*	Lay runners in summer
Vine	*Vitis* sp	Take cuttings in winter
Walnut	*Juglans* sp	Patch bud in spring to summer
Yew	*Taxus* sp	Take cuttings in autumn

Human circadian rhythms

The activity of many of our biological functions rises and falls every 24 hours. The timing of these ups and downs has important implications for our lives. Alcohol, for example, is more easily metabolized in the evening than in the morning. Colds are more easily caught at night when our immunity to infection is lowest. (*see pages 130–41*)

	Time of maximum	Time of minimum
Sleep	Night	Day
Activity	Day	Night
Heart rate	Day	Night
Immunity to disease and infection	Day	Night
Utilization of proteins	Morning	Evening
Blood cortisol level	Mid morning	Evening
Blood sugar level	Middle of the day	Night
Body temperature	Middle of the day	Middle of the night
Production of urine	Middle of the day	Night
Mental performance	Middle of the day	Middle of the night
Blood pressure	Late afternoon	Early morning
Blood amino acid level	Evening	Morning
Rate of alcohol metabolism	Evening	Morning
Blood clotting speed	Midnight	Midday
Cell division rate	Midnight	Midday
Growth hormone secretion in children	Midnight	Midday
Keenness of the senses	Middle of the night	Middle of the day

Linnaeus's flower clock

Various flowers open or close at particular times of the day, and Carl Linnaeus, an eighteenth-century botanist, was the first to arrange some of these into a flower clock. By simply glancing at the clock one could tell the time to within half an hour. (*see pages 52–63*)

Time	Flower	Scientific name	Activity
6 am	Spotted cat's ear	*Hypochoeris maculata*	Opens
7 am	African marigold	*Tagetes erecta*	Opens
8 am	Mouse-ear hawkweed	*Hieracium pilosella*	Opens
9 am	Prickly sowthistle	*Sonchus asper*	Closes
10 am	Common nipplewort	*Lapsana communis*	Closes
11 am	Star of Bethlehem	*Ornithogalum umbellatum*	Opens
12 noon	Passion flower	*Passiflora caerulea*	Opens
1 pm	Childing pink	*Dianthus* sp	Closes
2 pm	Scarlet pimpernel	*Anagallis arvensis*	Closes
3 pm	Hawkbit	*Leontodon hispidus*	Closes
4 pm	Bindweed	*Convolvulus arvensis*	Closes
5 pm	White water-lily	*Nymphaea alba*	Closes
6 pm	Evening primrose	*Oenothera erythrosepala*	Opens

Blood pressure rhythm in man

Blood pressure is measured in an artery in the arm and is a reflection of the heart's action. As the ventricles contract and pump blood around the body, the blood pressure rises. This is the systolic pressure. The pressure then falls to a resting level, when the atria pump blood into the ventricles. This is the diastolic pressure. (*see pages 102–15 and 130–41*)

Average blood pressure (mm mercury)		Systolic pressure	Diastolic pressure
Newborn		80	46
6 months/1 year		89	60
3 years		100	67
8 years		105	57
13 years		115	60
17 years	male	121	74
	female	116	72
20–24 years	male	123	76
	female	116	72
30–34 years	male	126	79
	female	120	75
45–49 years	male	130	82
	female	131	82
55–59 years	male	138	84
	female	139	84
70–74 years	male	145	82
	female	159	85

Hibernation

The metabolism of hibernating animals falls to a low level during the cold winter months. This slow metabolic rate is shown by the low rectal temperatures and low heart rates of some hibernating mammals and one bird, the American poor-will. (*see pages 34–51*)

Animal	External temp		Rectal temp		Heart rate (beats/min)
	°C	°F	°C	°F	
Golden hamster	5	41	6	42.8	14
Hedgehog	−20	−4	3.7	38.7	6
Poor-will	4.8	40.6	4.8	40.6	18
13-lined ground squirrel	5	41	5.5	41.9	8
Woodchuck	−1	30.2	5	41	18

Human heart rates

The heart beats more slowly as people get older. As a general rule, the heart rates of women are slightly higher than those of men. The figures given here are for normal resting people. (*see pages 102–15 and 130–41*)

Age (years)		Heart rate (beats/min)			
			12	male	70
				female	71
			15	male	65
At birth		140		female	67
1	male	116	20	male	64
	female	122		female	69
2	male	104	30	male	59
	female	103		female	65
3	male	92	40–60	male	57
	female	86		female	70
6	male	87	60–70	male	66
	female	80		female	71
9	male	81	70	male	65
	female	85		female	73

Animal heart rates

As a general rule, the smaller an animal the faster is its resting heart rate. Voles, shrews and bats are all tiny mammals with dramatically fast heart rates. Every animal's heart rate increases rapidly when it is active or afraid. Reptiles and amphibians, which are cold-blooded, have heart rates that rise and fall with the external temperature. (see pages 102–15)

Animal	Scientific name	Heart rate (beats/min)
MAMMALS		
Bat	Pipistrellus pipistrellus	660
Bat (hibernating)	Pipistrellus pipistrellus	30
Adult camel	Camelus bactrianus	28
Newborn cat	Felis domesticus	168
Young cat	Felis domesticus	300
Adult cat	Felis domesticus	200
Newborn cow	Bos taurus	150
Adult cow	Bos taurus	50
Adult elephant	Elephas indicus	30
Adult giraffe	Giraffa camelopardalis	66
Adult goat	Capra hircus	81
Adult hare	Lepus europaeus	64
Hedgehog	Erinaceus europaeus	246
Adult hyena	Hyaena sp	55
Adult lion	Felis leo	40
Mouse	Mus musculus	376
Newborn pig	Sus scrofa	227
Shrew	Sorex cinereus	782
Vole	Microtus arvalis	522
White rat	Rattus norvegicus	305
BIRDS		
American robin	Turdus migratorius	570
Cassowary	Casuarius casuarius	70
Hooded crow	Corvus cornis	379
Mallard	Anas platyrhynchos	320
Mourning dove	Zenaidura macroura	135
Ostrich	Struthio camelus	65
Peregrine falcon	Falco peregrinus	347
Starling	Sturnus vulgaris	388
Turkey	Meleagris gallopovo	93
REPTILES AND AMPHIBIANS		
Crocodile	Crocodylus sp	30 (at 15°C)
		70 (at 23.5°C)
Leopard frog	Rana pipiens	7 (at 2°C)
		37 (at 22°C)

Heart rates of diving mammals

The slowing of the heart rate is an adaptation to diving and ensures a reduced rate of blood flow to all organs except the brain. Along with other adaptations, it allows these mammals to remain below the surface of the water for long periods. (see pages 102–15)

Animal	Scientific name	Normal Heart Rate (beats/min)	Heart Rate while diving (beats/min)
Beaver	Castor canadensis	140	10
Bottle-nosed dolphin	Tursiops truncatus	110	50
Dolphin	Delphinapterus leucas	150	15
Manatee	Trichechus sp	60	30
Seal	Phoca vitulina	100	10
Whale	Beluga sp	145	16

Human respiration rates

The respiration rate slows down as people grow older, but for adults averages about 12 breaths per minute. Heavy work, such as lifting large weights, nearly doubles the rate. For someone who is not in training, heavy exercise, such as running or climbing a steep hill, increases it to nearly one breath per second. (see pages 102–15)

	Breaths/min
Premature baby	34
Newborn baby	29
Adult male, 150 lbs (68 kg)	12
doing light work	17
doing heavy work	21
doing heavy exercise	up to 53
Adult female, 119 lbs (54 kg)	12
doing light work	19
doing heavy work	30

Animal respiration rates

The smaller an animal, the faster the rate at which it breathes air in and out. The respiration rate also increases with the amount of activity. It decreases during hibernation and when marine mammals dive. (see pages 102–15)

Animal	Scientific name	Respiration rate (breaths/min)
Beaver	Castor canadensis	16
Blue whale	Balaenoptera musculus	4
Chipmunk	Tamias striatus	65
Deer mouse	Peromyscus leucopus	135
Dog	Canis familiaris	18
Dolphin	Tursiops truncatus	2
Flying squirrel	Glaucomys volans	91
Giraffe	Giraffa sp	32
Goat	Capra hircus	19
Golden hamster	Mesocricetus auratus	74
Guinea pig	Cavia porcellus	90
Horse	Equus caballus	10
Jersey cow	Bos taurus	27
Marmot	Marmota marmota	8
Marmot (hibernating)	Marmota marmota	0.7
Mouse	Mus musculus	163
Rabbit	Oryctolagus cuniculus	37
Rat	Rattus norvegicus	86
Rhesus monkey	Macaca rhesus	33
Sheep	Ovis aries	20
Short-tailed shrew	Blarina brevicauda	186
Spiny ant-eater	Tachyglossus aculeatus	14
Australian lungfish (in water)	Neoceratodus forsteri	28
Canary	Serinus canarius	57
Cardinal	Richmondena cardinalis	45
House wren	Troglodytes aëdon	83
Mallard	Anas platyrhynchos	42
Ostrich (12°C, 53.6°F)	Struthio camelus	5
Ostrich (25°C, 77°F)	Struthio camelus	45
Sparrow	Passer domesticus	50

Wing-beat frequencies

As a general rule, the larger the surface area of a bird's wings, the slower its wing-beat frequency. This physical relationship holds good for flying animals as varied as birds, bats and insects. (*see pages 116–29*)

Animal	Scientific name	Beats/second
BIRDS		
Hummingbird, 0.07 oz (2 gms)	Family Trochilidae	50
Hummingbird, 0.12 oz (3.5 gms)	Family Trochilidae	32
Hummingbird, 0.21 oz (6 gms)	Family Trochilidae	24
Sparrow	*Passer domesticus*	14
Swift	*Apus apus*	10
Owl, 14 oz (400 gms)	Family Strigidae	4
Pigeon	*Columba livia*	4
Crow	*Corvus corone*	3.6
Buzzard	*Buteo buteo*	3
Gull, 35.3 oz (1,000 gms)	*Larus canus*	3
Heron	*Ardea cinerea*	2.5
Stork	*Ciconia ciconia*	2
Pelican	*Pelicanus sp*	1
INSECTS		
Biting midge	Family Ceratopogonidae	1,000
Honey-bee (drone)	*Apis mellifica*	235
Cuckoo bee (queen)	*Psithyrus rupestris*	123
Common rose beetle	*Cetonia aurata*	101
Red-winged hadena	*Apamea laterita*	45
Cockroach	*Periplaneta americana*	35
Privet hawkmoth	*Sphinx ligustri*	30
Dragonfly	Order Odonata	25
Swallowtail butterfly	*Papilio machaon*	5.5
BATS		
Lesser horseshoe bat, 0.88 oz (25 gms)	*Rhinolophus hipposideros*	17
Mouse-eared bat, 0.74 oz (21 gms)	*Myotis myotis*	11.5
Spear-nosed bat, 3.17 oz (90 gms)	*Phyllostomus hastatus*	10

Cell division cycles

Many of the microscopic single-celled organisms that cause diseases and infections multiply very rapidly. Most of their reproduction is a simple division into two of each microbial cell. The time they take to complete this cell division cycle may be only a matter of minutes or hours at body temperature (37°C, 98.6°F). (*see pages 94–101*)

Organism	Action/environment	Cycle length	°C	°F
Amoeba proteus	Found in freshwater ponds	5 hours	27	80.6
Clostridium botulinum	Causes botulism	35 mins	37	98.6
Diplococcus pneumoniae	Causes pneumonia	24.5 mins	37	98.6
Entamoeba histolytica	Causes amoebic dysentery	5 hours	37	98.6
Escherichia coli	Found in human intestines	16.5 mins	37	98.6
Influenza A, PR-8 virus	Causes influenza	7 hours	37	98.6
Paramecium aurelia	Feeds on bacteria and yeast	5 hours	27	80.6
Rhizobium leguminosarum	Fixes nitrogen in root nodules of plants	130 mins	25	77
Salmonella typhimurium	Causes dysentery	29 mins	37	98.6
Staphylococcus aureus	Causes boils	27 mins	37	98.6
Trichomonas vaginalis	Found in woman's vagina	5.5 hours	37	98.6
Trypanosoma mega	Causes protozoan infection	19 hours	23	73.4

Human cell and tissue cycles

Many human body cells divide continually and different cell types divide at different rates. The cells of the stomach lining divide every two days. As a result of these cell cycles (C), whole tissues are renewed (T): the gums are renewed every 94 days on average. (*see pages 94–101*)

	Cycle length
Lining of the mouth	5 days (T)
Gums	94 days (T)
Oesophagus	6 days (T)
Stomach	2 days (C)
Duodenum	4 days (T)
Ileum (part of small intestine)	2 days (C)
Colon (part of large intestine)	30 hours (C)
Colon (part of large intestine)	4 days (T)
Rectum	13 hours (C)
Cells that make red blood cells	20 hours (C)
Red blood cells	120 days (T)
Skin	308 hours (C)
Skin with psoriasis (disease causing red scaly patches)	37.5 hours (C)
Skin	26 days (T)
Hair cycle (on arms, chest, ears, eyebrows, hands and legs)	$5\frac{1}{2}$ months (T)

Development of human alpha rhythms

The brain's electrical activity is measured with an electroencephalograph, which records the type of rhythm and the voltage that generates it. Recordings from children of different ages reveal that the alpha rhythm begins to emerge at 20 months and is developed by the age of four. (*see pages 102–15 and 142–51*)

Age	Voltage	Frequency of dominant rhythm
3 days	Low	Irregular
3 months	Low	$1\frac{1}{2}$–3 cycles per sec
5 months	Increasing	$1\frac{1}{2}$–4 cycles per sec
6 months	High	4 cycles per sec
11 months	Moderate	4–5 cycles per sec
20 months	High/moderate	High at 4 cycles per sec with moderate at 8 cycles per sec (alpha rhythm)
4 years	High	9 cycles per sec alpha rhythm
8 years	Moderate	9–10 cycles per sec alpha rhythm

Length of sleep cycles

Sleep is a cyclical phenomenon. Every 90 minutes we go through one complete four-stage sleep cycle, each stage having a typical brain rhythm. Before beginning the next cycle we go through a light dream-filled sleep characterized by rapid eye movements. (*see pages 102–15*)

	Total sleep (per 24 hours)	Cycle length (mins)	Percentage of REM sleep
Cat	14	26	28
Hamster	14	12	23
Man	8	90	23
Mole	8	10	25
Mouse	13	12	10
Opossum	19	23	29
Rabbit	7	42	11
Rat	13	9	20
Squirrel	14	13	25

RHYTHM DATA

Animal life spans

The life span of an animal is the period of its life cycle from birth to death. The figures quoted here are the maximum periods that have been authentically recorded, although the average life cycle periods are obviously much shorter. (*see page 82–93*)

Animal	Scientific name	Years
MAMMALS		
Man	*Homo sapiens*	115
Fin whale	*Balaenoptera physalus*	80
Asiatic elephant	*Elephas maximus*	70
Hippopotamus	*Hippopotamus amphibius*	51
Ant-eater	*Tachyglossus aculeatus*	$49\frac{1}{2}$
Horse	*Equus caballus*	46
Chimpanzee	*Pan troglodytes*	$44\frac{1}{2}$
Rhinoceros	*Rhinoceros unicornis*	40
Gorilla	*Gorilla gorilla*	$39\frac{1}{2}$
Bear	*Ursus arctos*	37
Seal	*Phoca vitulina*	34
Giraffe	*Giraffa camelopardalis*	$33\frac{1}{2}$
Cow	*Bos taurus*	30
Camel	*Camelus bactrianus*	$29\frac{1}{2}$
Wild cat	*Felis catus*	28
Pig	*Sus scrofa*	27
Deer	*Cervus elaphus*	$26\frac{1}{2}$
Tiger	*Panthera tigris*	26
Beaver	*Castor canadensis*	20
Dog	*Canis familiaris*	20
Sheep	*Ovis aries*	20
Goat	*Capra hircus*	18
Platypus	*Ornithorhynchus anatinus*	17
Bat	*Pipistrellus subflarus*	15
Squirrel	*Sciurus carolinensis*	15
Fox	*Vulpes vulpes*	14
Rabbit	*Oryctolagus cuniculus*	13
Mink	*Mustela vison*	10
Chipmunk	*Tamias striatus*	8
Guinea pig	*Cavia porcellus*	$7\frac{1}{2}$
Golden hamster	*Mesocricetus auratus*	4
Hedgehog	*Erinaceus europaeus*	4
Mouse	*Mus musculus*	$3\frac{1}{2}$
Rat	*Rattus norvegicus*	$3\frac{1}{2}$
Shrew	*Sorex palustris*	$1\frac{1}{2}$
BIRDS		
Raven	*Corvus corax*	69
Condor	*Gymnogyps californicus*	65
Ostrich	*Struthio camelus*	50
Chicken	*Gallus gallus*	30
Swift	*Apus apus*	21
Mallard	*Anas platyrhynchos*	$20\frac{1}{2}$
Starling	*Sturnus vulgaris*	16
REPTILES AND AMPHIBIANS		
Eastern box turtle	*Terrapene carolina*	85
Alligator	*Alligator mississippiensis*	56
Leopard frog	*Rana pipiens*	6
FISH		
Sturgeon	*Acipenser fulvescens*	152
Carp	*Cyprinus carpio*	47
Pike	*Esox lucius*	24
Salmon trout	*Salmo trutta*	18
Mackerel	*Scomber scombrus*	15
Salmon	*Salmo salar*	13
Electric eel	*Electrophorus electricus*	$11\frac{1}{2}$

Signs of the Zodiac

From the point of view of people on Earth, the sun appears to make an annual passage in front of the 12 constellations called the signs of the zodiac. The table lists the dates when the sun enters and leaves these signs. People's characters are believed to be shaped by the sign in which the sun happens to be at the time of their birth. (*see pages 142–51*)

Annual apparent passage of the sun through the 12 constellations (signs of the zodiac)	
Aries	22 March—20 April
Taurus	21 April—21 May
Gemini	22 May—22 June
Cancer	23 June—23 July
Leo	24 July—23 August
Virgo	24 August—23 September
Libra	24 September—23 October
Scorpio	24 October—22 November
Sagittarius	23 November—22 December
Capricorn	23 December—19 January
Aquarius	20 January—19 February
Pisces	20 February—21 March

Astrological ages

Astrologers believe that every 25,868 years the Earth completes a Great Year which is divided equally into 12 astrological ages. We are now nearing the end of the Piscean age and entering the age of Aquarius. The age of Pisces is symbolized by Jesus, whose disciples were fishers of men and who fed the multitude with loaves and fishes. The age of Aquarius is symbolized by humanity, communication, and the beginning of space travel. (*see pages 142–51*)

Astrological Ages: the 12 Great Months of the Great Year (Each Age is approximately 2,160 years)	
Age	**Approximate span**
Age of Leo	10000— 8000 BC
Age of Cancer	8000— 6000 BC
Age of Gemini	6000— 4000 BC
Age of Taurus	4000— 2000 BC
Age of Aries	2000— 0 BC
Age of Pisces	AD 0— 2000
Age of Aquarius	AD 2000— 4000
Age of Capricorn	AD 4000— 6000
Age of Sagittarius	AD 6000— 8000
Age of Scorpio	AD 8000—10000
Age of Libra	AD 10000—12000
Age of Virgo	AD 12000—14000
Age of Leo	AD 14000—16000

Astrological cycles of planetary conjunctions

Planets are in conjunction when they are within 8° of each other in the sky. The influences of the planets then combine and are astrologically significant. When the same planets return to the same positions they complete a cycle of conjunction. Astrologically, there is a cycle of relationship between the planets. The meaning of the relationship depends not only on the planets concerned, but also on the sign in which the conjunction occurs, and the position of the planets in a person's horoscope. (*see pages 142–51*)

Planets	Period of conjunction (years)
Pluto and Neptune	492
Uranus and Neptune	171
Saturn and Uranus	91
Saturn and Neptune	35
Saturn and Jupiter	20
Jupiter and Uranus	14
Jupiter and Neptune	13
Jupiter and Pluto	12

1 Pluto cycle =	$1\frac{1}{2}$ cycles of Neptune
=	3 cycles of Uranus
=	$8\frac{1}{2}$ cycles of Saturn
=	21 cycles of Jupiter
=	132 cycles of Mars
=	248 cycles of Earth
=	400 cycles of Venus
=	$1,033\frac{1}{3}$ cycles of Mercury

The astrological ages and life periods of man

Astrologers divide a person's life into seven ages and assign a planet to rule over each age. The character of each planet governs and shapes the age over which it rules. The three life periods of man approximately coincide with the Saturn cycle at the end of which—at the ages of 28, 56, 84—great changes may occur in a person's life. (*see pages 142–51*)

Age (years)	Planetary ruler	Characteristics
1— 4	Moon	Total dependence of infant on its mother
5—14	Mercury	Curiosity and education
14—22	Venus	Adolescence; sexual maturity; entrance to society
23—41	Sun	Peak time of virility, health and activity
42—56	Mars	Increase of sphere of influence; realization of ambitions
57—68	Jupiter	Reflection on earthly events; contemplation of spiritual reality
68—death	Saturn	Life-assessment and preparation for death
After death —before birth	Pluto	The after life

LIFE PERIODS Life period (years)	Characteristics
1—28	Establishment of the self; growth from birth to adulthood
28—56	Expansion of the self; cultivation of one's self in a chosen field; role as parent and spouse
56—84	Contraction of the self; goals are achieved or not achieved, rewards are reaped, debts paid; preparation of the soul for physical death

Yin and Yang

According to the Chinese principle of opposites, everything is composed of yin and yang which are both antagonistic and complementary. (*see pages 142–51*)

	Yang	Yin
In the natural world	Day	Night
	Clear day	Cloudy day
	Spring/summer	Autumn/winter
	East/south	West/north
	Upper	Lower
	Exterior	Interior
	Hot	Cold
	Fire	Water
	Light	Dark
	Sun	Moon
In the body	Surfaces of the body	Interior of the body
	Spine/back	Chest/abdomen
	Male	Female
	Clear or clean body fluid	Cloudy or dirty body fluid
	Energy	Blood
In disease	Acute/virulent	Chronic/non-active
	Powerful/flourishing	Weak/decaying
	Patients feels hot or hot to touch or has a high temperature	Patient feels cold or cold to touch or has below normal temperature
	Dry	Moist
	Advancing	Retiring
	Hasty	Lingering

Sunspot cycles and historical events

It has been controversially suggested that a cycle of historical events is synchronized with the sunspot cycle, and that nearly all major historical events have occurred in the three or four years leading up to and including the time of peak sunspot numbers. The dates of recent sunspot maxima are given with some of the major events of the time. (*see pages 142–51*)

1778	The American War of Independence (1775–83) led to the Declaration of Independence (1776) and to the defeat of the British.
1788–89	The American constitution was drafted (1787); the French Revolution succeeded (1789) and led to the rise of the middle classes.
1802–3	An Irish rebellion was suppressed (1799); parliamentary union of Great Britain and Ireland—the United Kingdom formed (1800).
1817	Duke of Wellington ended the Napoleonic era, defeating the French at Waterloo (1815) but Britain entered an economic depression.
1829–30	Irish Catholics were emancipated (1829); the first police force was established in London (1829); Greece became independent (1830).
1837	Trade unionism was born (1834); the People's Charter was drawn up by the Chartists (1836); Victoria became Queen of England (1837).
1849	Communist Manifesto produced by Marx and Engels (1848); French Republic proclaimed (1848); gold found in California (1848).
1860–61	Uprisings occurred in India (1857–8); Darwin published *Origin of Species* (1859); American Civil War broke out (1861).
1871	Westernization of Japan began (1868); education was made available to all British children (1870); Paris Commune was declared (1870).
1883–84	Germany, Austria and Italy signed Triple Alliance (1882); terrorism struck Britain and Russia (1880–2); Krakatoa erupted (1883).
1893	Uprisings by socialists and anarchists occurred in France, Spain and USA (1890–2); a great famine spread across Russia (1892).
1905–7	South Africa erupted with the Boer War (1899–1902); church and state separated in France (1905); Russian Revolution failed (1905).
1917	The Easter uprising occurred in Dublin (1916); Russian Revolution succeeded (1917); Palestine became national home for Jews (1917).
1926	The General Strike occurred in Britain (1926); Stalin assumed power in Russia (1926); women in Britain were enfranchised (1928).
1937	Hitler became dictator (1934); Spanish Civil War broke out (1936); Keynesian economics began (1936); Japan attacked China (1936).
1948	The United Nations was formed (1945); the state of Israel was proclaimed (1948); revolution succeeded in China (1946–9).
1957	Israel invaded Egypt which led to the Suez crisis (1956); Russia invaded Hungary (1956) and launched the first space satellite (1957).
1968–69	The Cultural Revolution occurred in China (1965–8); worldwide protests take place (1968); Irish troubles began (1968–9).
1979–80	Democracy began in Spain (1976); drought struck Britain (1976); revolution in Iran (1979); Russia invaded Afghanistan (1979).

Page numbers in bold type refer to subjects mentioned in captions to illustrations. The majority of these subjects also appears in the main text (eg Adrenalin, **133**). Page numbers in ordinary type refer to entries in the main text only (eg Aborigines, 75, 172). Page numbers in italic refer to subjects contained only in charts (eg *Aix galericulata, 187*).

A

Aborigines, 75, 172
Acer campestre, 188
Acipenser fulvescens, 192
Activity patterns, **12–13**, 52–63, 103–15, **134–5, 154–5**
Acupuncture, 143, 146, 147–8
Adaptation, 35, 54
Addison, Thomas, 138
Addison's disease, **136–7**, 138, 140
Adenosine triphosphate (ATP), **100**, 105, 109, 110
Adrenalin, **133**
Adrenocorticotrophic hormone (ACTH), **132, 133, 137**
Aestivation, 51
Agkistrodon contortrix, 187
Aix galericulata, 187
Alcohol, 63
effects of, **155**
metabolism, **134**
tolerance levels, **135**
Algae, 30, 32
blue-green, 105
fucoid, **30**
green, **30**
laminarian, **30**
red, **30**
Allard, H.A., 45
Allergies, 141
Alligator mississippiensis, 187, 192
Alpha rhythms, 146, *191*
Alternation of generations 86, 87
Althaea officinalis, 188
rosea, 188
Alyssum saxatile, 188
Anas crecca, 187
penelope, 187
platyrhyncos, 187, 190, 192

Anchovies, **90**
Anemone, 43
Anguilla anguilla, 79
rostrata, 79
Anser anser, 187
Ant, **55**
honeypot, **110**
Ant-eater, *190, 192*
Antelope, 68, *124*
Antherea pernyi, 62
Anthrenus verbasci, 61
Antihistamines, 141
Antirrhinum majus, 188
Aphid, 83, 87, 90
oat, *87*
sycamore, *87*
Apollo 11 mission, **20**
Apple, 43
Apricot, 43
Aptenodytes patagonica, 72
Aquilegia sp, 188
Arapaho Indians, 76
Archilochus colubris, 187
Ardea cinerea, 191
Aristotle, 79, 172
Armadillo, *187*
Asia, 26
Aspen, *188*
Astaire, Fred, **161**
Aster, 45, *188*
Aster novi-belgii, 188
Asthma, 138–40
treatment, **139**, 140
Astrology, 143, **148–9**, *192, 193*
ages of, *192*
cycles of planetary conjunctions, *192*
Atropa belladonna, 188
Auk, little, *187*
Aurelia, 128
Aurora australis, **151**
borealis, **151**
Autism, 157
Autumn, 22, **23**, 35–51, **36, 49, 60, 61, 70, 87**
Aythya fuligula, 187
Aztecs, 173, 177
calendar stone, **177**
Sun god, **177**

B

Baboon, gelada, 68
Babyhood, biological rhythms in, **12, 13**
growth rates, 95
Babylonians, 174
calendar, 176
Backswimmer, **128**
Bacteria, 84, **97**, 101
Badger, **67**
Balaenoptera musculus, 190
physalus, 187, 192
Bamboo, Chinese umbrella, **45**, *188*
Barley, 47

Bat, 51, **71**, *190, 191, 192*
hammerhead, 74
horseshoe, 112, *191*
Bear, *192*
Beaver, *190, 192*
Bee, 86, **105**
flight, 120
Beech, European, **40**, *188*
Beet, 45
Beetle, **122**
carpet, 61
Colorado, **40**
Khapra, 51
water, 85, 128
whirligig, 128
Bell flower, *188*
Beluga sp, 190
Bering Strait, 26
Beta rhythm, 146
Bettmeralp glacier, **27**
Betula pendula, 188
Beyond Culture, E.T. Hall, 168
Bilharzia, 85
Biofeedback, 143, 145, **146**
Biological clock, 13, 53–63, 132–3, **140**
Biomate, **144**
Biorhythms, 143, 144–5
calculations, **144**
Bird, activity, **112**
bipedal running, 124
breeding seasons, 24, 65–81, **72**
feeding, **111**
migration, 70, **71**, 76, 77, *187*
moulting, 49
respiration, **106**
song cycle, 73, **74**
torpor, 50–1
wing-beats, 118, 119, *191*
Bison bison, 76
Blackbird, **74**, 112
Blackcock, 74
Black Cuillin Mountains, **27**
Blackfly, 92
Blaring brevicauda, 190
Bleeding heart, *188*
Blood, 106–9
circulation, 134
hormone content, **10**
Blood pressure, 10, 11, **133**, *189*
high, **133**
of astronauts, 131
Blowfly, 84, **89**
Bobolink, *187*
Body rhythms, human, 9–13, 103–15, 131–41, 143–7, 153–9, *189*
Body temperature, 13, **50, 51, 66, 113, 134**, 135, *189*
Bonasa umbellus, 111

Bos taurus, 187, 190, 192
Bougainvillea spectabilis, 188
Boxfish, 126
Brain, 10, 11, **62–3, 113, 134, 135**
Brain waves, **114–15**, 146
Bramble, *188*
Branta canadensis, 187
Breathing, 9–13, **105–107, 138–39**, *190*
Breeding seasons, 40–41, 65–81, **68, 70, 75, 76, 80, 81**, *187*
Bronchi, 138
Buckwheat, *188*
Buddhism, 163, **172**
Budmoth, larch, **91**
Bufo americanus, 84, 187
Bufo bufo, 76
Bulb, 43
Bushbaby, *187*
Buteo buteo, 191
Butterbur, *188*
Butterfly, 85, *191*
Buzzard, *191*

C

Caesar, Julius, 177
Calendar, 176–8
Assyrian, 176
Aztec, 177
Babylonian, 176
Chinese, 177
civil, 176
early systems, 176–7
Egyptian, **176**
French Revolutionary, 176
Greek, 176
Gregorian, **176**, 177, 178
Jewish, 176
Julian, 177–9
lunar, **176**, 177
Mayan, 177
Calendar Round, 177
Californian poppy, *188*
Calluna vulgaris, 188
Caltha palustris, 188
Camel, 123, *190, 192*
Bactrian, 48, *187*
Camelus bactrianus, 187, 190, 192
Camouflage, animal, 48, 49
Campanula medium, 188
Canary, *190*
Cancer, drug treatment of, 141
Canis familiaris, 190, 192
Capra hircus, 187, 190, 192
Capsella bursa-pastoris, 188
Carbon cycle, 100–1, **101**

Carcinus maenas, 58
Cardinal, *190*
Caretta caretta, 129
Caribou, 76, 77, 110
Carnac, France, **177**
Carnivores, 36, **101**
Carp, European, *187, 192*
Cassowary, 124, *190*
Castor canadensis, 190, 192
Casuarius casuarius, 190
Cat, big, **69**
domestic, 69, **70**, 115, *124, 190, 191*
wild, *192*
Caterpillar, western tent, 91
Cavia porcellus, 187, 190, 192
Cell cycle, 11, 96–7, *191*
division, 11, 95, **96, 97**, 98, *191*
plant, 100–1
Cellulose, **100**
Centipede, 122
Cerastoderma edule, 98
Cerebellum, **114**, 134
Cerebral cortex, **134**
Chameleon, 71
Chaplin, Charlie, 165
Cheetah, *124*
Chelonia mydas, 79
Chicken, prairie, 74
Chimpanzee, 68, **113**, **115**, *187, 192*
China, ancient, 174, **175**
Chipmunk, *187, 190, 192*
Chizhevsky, A.L., **151**
Chlorophyll, 49, 100, **104**
Cholera, 93
Chronometer, marine, **179**
Chrysanthemum, 45
Cicada, 83
Ciconia ciconia, 191
Circadian rhythms, 11–13, 15, 21, 56–7, **62, 63**, 111, 113, 131–41, **137, 139, 140, 141**, 153, *189*
significance in disease, 136
Circannual rhythms, 15, 60–1
Citellus lateralis, 50
Citellus tridecemlineatus, 71
Clam, 127
Clavius, Christopher, 177
Clepsydra *see* Water clock
Climate, changes in, 23–30, **28**, 38–9
effects on breeding, 70
Climatic zones, 17, 22, 24, 36–40, **39**

Clock, 174–5, 178–81
  atomic, 181
  drum, **179**
  early, **174–5**
  electric, 181
  escapement, 178, **179**, **180**
  falling weights, **178**
  grandfather, **181**
  pendulum, **178**, 179
  quartz, **181**
  spring-driven, **179**
  water, 174, **175**
Clover, 45
Clunio marinus, **59**, 60
Cockle, 98
Cockroach, 56, 122, **122**, *124*
Coe, Sebastian, **144**
Coffea arabica, *47*
Coffee tree, **47**
Columba livia, *41*, *191*
Columbine, *188*
Comet cycles, *186*
Common maple, *188*
Community rhythms, **168–9**
Continental drift, 20, 28
Convoluta roscoffensis, **59**
Copulation, animal, *67*, 69
Corm, 43
Corpus luteum, **66**, 67
Corticosteroids, 134, 136–7
  as drugs, 138, 141
Cortisol, **132**, **133**, 134, **136**, **137**, 189
Cortisone, 134, 136
Corvus cornis, *190*
Corylus avellana, *188*
  corone corone, **29**, *191*
  corone corvix, **29**
Cosmic rhythms, 15–33, *186*
Coturnix coturnix, *187*
Courtship rituals, 65, 73, **74–5**
Cow, *67*, *187*, *190*, *192*
Crab, European shore, **58**
  fiddler, **58**
Crepidula fornicata, **86**
Crepuscular animals, 21
Cricetus cricetus, 72
Crocodile, *190*
Crocodylus sp, *190*
Crocus, 43
  autumn, *188*
Crocus purpureus, **43**
  sativus, *188*
Crommelin comet, *186*
Crop plants, **46**, 47
Crossbill, 76
Crotalus viridis, *187*
Crow, **118**, *191*
  carrion, **29**
  hooded, **29**, **118**, *190*
Crustaceans, 33, 80

Cuba, 47
Cuckoo, *187*
Cuculus canorus, *187*
Culex pipiens fatigans, **111**
Cultural rhythms, 153–69
Cushing, Harvey, 137
Cushing's syndrome, **136**, **137**
Cyclamen coum, *188*
Cygnus olor, *187*
Cyprinus sp, *187*, *192*

D

Daffodil, 43
Dahlia, 45
Dance rhythms, 156, 158, 159, **161**
  African, 161
  Indonesian, 161
Dandelion, 45
Dasypus novemcinctus, *187*
Daylength, changes in, 22–3, **23**, 38, 49, 51
  effects on plants, **44–5**, 48, *188*
Daylight, changes in, 40–1
  cycles, 56
  effect on breeding, 71
  in polar regions, *40*
Day-night cycle, 20, 21, 22–3, 55, 56
  and temperature changes, 20, 21, 22
Deadly nightshade, *188*
Deer, 60, 61, **70**, 71
  courtship display, 68
  sika, **60**
Delphinapterus leucas, *190*
Delta waves, 146
Dendroica striata, *187*
Deoxyribonucleic acid (DNA), 62, **96**, **97**, 98
Deposition, Giotto, **173**
Dervishes, **161**
Desert, 24, **39**, 45
'Devil dancers', Sri Lanka, 163
Devonian, 20
Dexamethasone, **137**
Diadema setosum, 80, **81**
Dicentra formosa, *188*
Dicrocoelium dendriticum, 86, 87
Dictyota dichotoma, **59**, 60
Didelphus marsupialis virginiana, *187*
Digestion, 111
Disease, 83, 89, 91, **92–3**, 131–41
  treatment of, 140–1
Display, sexual, 68–9

Diurnal animals, 21, 53, **113**
DNA *see* Deoxyribonucleic acid
Dog, 66, **67**, 68, **69**, *190*, *192*
  hunting, 76
Dog's mercury, *188*
Dolichonyx oryzivorus, *187*
Dolphin, 68, **106**, **107**
  bottle-nosed, *190*
  swimming pattern, 129
Dormouse, 50
Dorylus, 55
Dove, mourning, *190*
  rock, 41
  stock, **118**
Dragonfly, 84, **85**, **119**
Drepanosiphum platanoides, 87
Drosera rotundifolia, *188*
Drug treatment, 140–1
  of cancer, **141**
Duck, 70, 71
  tufted, *187*
Dysrhythmia, 156, 157

E

Earth, 15, 17–25, *186*
  age of, 16
  atmosphere, **17**, **24**, 53
  crust, **17**
  curvature, **23**
  distance from sun, 22, 23, 24
  evolution, 16
  heat radiation, 25
  humidity on, 20, **21**
  orbital plane, 21, **22**, **28**, **29**
  precessional wobble, 30
  revolution around sun, **149**
  rotation, **20**, **21**, 22
  spin axis, **20**, **21**, **25**, **28**, **29**, 37, 39
  temperature of, 17, 24
Earthquake, 20, 150
East Africa, rainfall belt, 39
Eclipse, moon, 22, 32
  sun, 22, 32
Ecosystems, 24, 36, 38, 100
Eel, 78, 79, *124*, **126**
  electric, *192*
Efficiency levels, human, **113**
Egyptians, ancient, 174, **175**, 176
Eider, *187*
Electroencephalogram (EEG), **115**
Elephant, 84, **85**, *190*
  Asiatic, *192*

Elephantiasis, 92, 93, **111**
Elephas indicus, *190*
  maximus, *192*
Emotional rhythms, 156
Emu, 124
Enchelyopus cimbrus, 81
Encke comet, *186*
Energy, **100–1**, 103–15
  production of, 104–5, 110
Englemann, T.G., 13
Enterobius sp, 85
Entobdella soleae, **57**
Equator, 22, **22**, **23**, 24, 25, 30, 37, 38, **39**, **40**
Equus caballus, *187*, *190*, *192*
Erinaceus europaeus, **51**, *190*, **192**
Erithacus rubecola, 73
Eschscholzia californica, *188*
Eskimos, **110**
Esox lucius, *192*
Estrogen, **11**, 66
Estrous cycle, 66–9, *67*, *69*, *187*
Eunice viridis, 59, 80
Euphorbia pulcherrima, **44**

F

Fagopyrium esculentum, *188*
Fagus sylvatica, **40**, *188*
Falcon, peregrine, *190*
Falco peregrinus, *190*
Family rhythms, 166–7
Feeding, 110–2
Felis catus, *192*
  domesticus, *190*
  leo, *190*
Felix canadensis, 88
Fennel, *188*
Fern, 86, 87
Ferret, 70, 71
Fertilization, 65
Finlay comet, *186*
Firefly, 75
Fish, bony, 98, **106**
  growth rings, **98**
  respiration, **106**, 107
  sexual cycle, 72, **73**
  spawning, 80–1
  swimming, **126**
Flea, 120
Fliess, Wilhelm, 144
Flight, 118–20
  bird, **118**, **119**
  insect, **119**, 120
Flowering seasons, **44–5**, *188*
Fluke, lung, 84, 86
  sheep liver, 86
Fly, tsetse, 86
Flycatcher, pied, 112
Foeniculum vulgare, *188*

Follicle stimulating hormone (FSH), **66**, 71
Food chains, 95, 112
Forest, rain, **39**
  temperate, **39**
  tropical, **39**
Fox, arctic, 88
  rabies cycle, **93**
  red, 88, 89, **93**
Freesia, *188*
Fritillaria imperialis, *188*
Frog, 21, 73, **85**, **113**, **120**, *190*, *192*
  leopard, *190*, *192*
Fundulus heteroclitus, 72

G

Galago senegalensis, *187*
Galanthus nivalis, **37**, *188*
Galileo, 150, **179**
Gamelan, Balinese, 161, 162
Garner, W.W., 45
Garrya, 44
Gasterosteus aculeatus, **68**
Gauquélin, M., 149
Gentian, *188*
Gentiana asclepiadea, *188*
Gestation period, 70, *187*
Ginger lily, *188*
Giraffa camelopardalis, *190*, *192*
Giraffe, 72, 73, 123, 124, *190*, *192*
Glaciers, effects of, 26–8, **27**, **28**
Glaucomys volans, **56**, *190*
Glossina, 86
Goat, 70, 71, *187*, *190*, *192*
Golden rod, 45
Goose, Canada, *187*
  greylag, *187*
Gorilla gorilla, *187*, *192*
Gorse, *188*
Goshawk, 89
Graham, George, **180**
Grain crops, **46**, 47
Gramineae, 47
Grape, 47
Grebe, crested, 74–5
Greece, ancient, 174
  calendar, 176
Gregory I, Pope, 174, 175
Gregory XIII, Pope, 177
Greyhound, *124*
Grigg-Mellish comet, *186*
Groundnut, 47
Group patterns, human, 168–9
Grouse, 74
  black, **89**
  ruffed, **111**
  willow, 89

Growth, 95–101
Grunion, 81
Guinea pig, 6, *187, 190, 192*
Gull, *191*
  great black-backed, **118**

H

Hadley circulation, **24**
Haematoloechus medioplexus, 84, 86, 87
Halley's comet, *186*
Hampton Court Clock, **178**
Hamster, 50
  European, 72
  golden, **68**, 69, *187, 190, 192*
Hare, **70**, *190*
  snowshoe, **88**, 89
Harrison, John, **179**
Harvesting, 47, 48
Hawk, **113**
Hay fever, 141
Hazel, *188*
Heart, **108–9**
Heartbeat, 9, 53, 134, **158**
  in infancy, 9, *189*
  rates of, 9, **11**, 55, 108, **109**, 131, **133**, *189, 190*
Heart disease, **133**
Heath hen, 74
Heather, *188*
Hedera helix, *188*
Hedgehog, 50, **51**, 70, *187, 190, 192*
Hedychium densiflorum, *188*
Hellebore, *188*
Helleborus orientalis, *188*
Henbane, **44**
Henlein, Peter, 179
Hepatitis, 93
Herbivore, 36, **101**
Heron, *191*
Herring, 99, **126**
Herschel, William, 150
Hibernation, **50–1**, *189*
Himalaya mountains, 25
Hippopotamus amphibius, *192*
Hirundo rustica, 77, *187*
Hives *see* Nettlerash
Hollyhock, *188*
Holly tree, 44
Holocene, 25, **26**
Honesty, *188*
Honey-bee, **57**, **105**
  flight, 120
Hooke, Dr Robert, 179, **180**
Hop, *188*
Hopi Indians, 172, **183**

Hormones, 133, 134
  levels of, 11
  output, **132–3**
  production cycles, 11, 66–7, 131
  sex, **66–7**, 69, **132**
Horse, **67**, *187, 190, 192*
  locomotion, 123, **123**, **124**
Horsechestnut, **48–9**
Hourglass, 175
Housefly, **63**
Humboldt current, **90**
Hummingbird, **118**, 119, *191*
  ruby-throated, *187*
Humulus lupulus, *188*
Huygens, Christiaan, 179
Hyaena sp, *190*
Hyalophora cecropia, **62**
Hydrogen, 18, 104
Hyena, 76, 123, *190*
Hypothalamus, **63**, 134, 135, **136**, 137
Hypsignathus monstrosus, 74

I

Ice Ages, 25–9, **26, 27, 28, 29**
Ice caps, **28**
Ice plant, *188*
Ilex, 44
Impala, 76
Impatiens noli-tangere, *188*
India, **25**
Indian Ocean, 25
Industrial rhythms, **165**
Insects, breeding seasons, 24, 75
  diapause, 51
  social, 86
  ventilating mechanisms, **10**
  wing vibration, 9, **119**, 120
International Date Line, **183**
International Time Zones, **183**
Ipomoea, **57**
Iris, 43
Israel, ancient 174
Italy, 174
Ivy, *188*

J

Jackal, 76
Jellyfish, 80, 127, **128**
  European, 128
Jerboa, **112**, **120**, 121
Jet lag, **135**, 136, 155

Jet propulsion, in aquatic creatures, 127
Jumping, **120–1**
Jupiter, 16, *186*

K

Kangaroo, 72, 120, **121**
  red, **121**, *187*
Kelp, laminarian, 33
Killifish, 72
Kingbird, eastern, *187*
Kleitman, N., 13

L

Lambing season, **41**
Lamprey, *187*
Lampyridae, 75
Larus canus, *191*
Legionella pneumophila, **97**
Legionnaire's disease, **97**
Lek, **74**
Lemming, 88
Lemon verbena, *188*
Lemur, 68, *187*
Lemur catta, *187*
Lennon, John, 149
Leodice fucata, 80
Leonitis leonurus, *188*
Leptinotarsa decemlineata, **40**
Lepus americanus, **88**
Lepus europaeus, *190*
Leukemia, studies of, 141
Leuresthes tenuis, **81**
Lichen, **30**, 32, 33
Life span, 83–7, *192*
Lime, small-leafed, *188*
Limpet, slipper, 86
Linnaeus, Carl, 57, *189*
  flower clock, **57**, *189*
Lion, *190*
  marsupial, 121
Lion's tail, *188*
Lippia citriodora, *188*
Lizard, 124, **125**
  basilicus, 125
  collared, 123, **124**
Loa loa, 92, **93**
Loaiasis, 92
Lobelia, *188*
Lobelia erinus, *188*
Locomotion, **117–29**
  amphibian, **120**
  bird, **118–9**, **124**
  fish, **126**, 127
  frog, 120
  insect, **119–20**, **122**, **128**
  invertebrate, **122**, **128**
  lizard, **125**
  mammalian, **120–1**, 123
  snake, **127**
  turtle, **129**

Locust, **105**
Lotus, *188*
Love-in-a-mist, *188*
Lucilia cuprina, **89**
Lugworm, marine, 122
Lunaria annua, *188*
Lunar rhythms, 60
  effects on spawning, 80–1
Lung, **138**, 139
Lungfish, **51**, *190*
Lupin, *188*
Luscinia megarhyncha, *187*
Luteinizing hormone (LH), **66**
Lynx, 88, **89**

M

Macaca mulatta, *187*
  rhesus, *190*
Macaque monkey, 68
Mackerel, *192*
Macropus robustus, *187*
Magicicada spp, 83
Maize, **42**
Malacosoma sp, 91
Mala moschata, *188*
Malaria, **92**
Malinke tribe, West Africa, **158**
Mallard, *187, 190, 192*
Mallow, *188*
Mammals, activity, **112–13**
  breeding seasons, 24, 66–72
  camouflage, **48**
  hibernation, **50–1**
  locomotion, **120–25**
  respiration, **107**
  sleep, **114–5**
Man, beliefs, **143–51**, **162–3**, **172–3**
  biological cycles, **9–13**, **131–41**
  discovery of agriculture, 46
  eating, **110**
  evolution, 35, 46
  health, **131–41**, **146–7**, **154–5**
  life cycle, 83
  life-span, 35, *192*
  locomotion 117, 123, **124–5**
  mental performance, **113**, **134–5**
  reproduction, 66, 67, 69, 75, *187*
  respiration, **107**, *190*
  sexual behaviour, **69**, 75
  sleep, **115–6**
Manatee, *190*
Mandarin, *187*

Mandrill, *187*
Mangebey, 68
Mansfield, Jayne, **144**
Mantis, 122
Marmot, *190*
Marmota marmota, *190*
Mars, 16, **17**, 18, *186*
Marsh mallow, *188*
Marsh marigold, *188*
Marten, 88
Mayan timekeeping, 173, 177
Mayfly, 60, **81**
Meal patterns, 154, **155**
Measles, 83, **93**
Meditation, **146**, 163
Medulla, **136**
Megaleia rufa, **121**, *187*
Megaptera novaeangliae, **78**
Meleagris gallopovo, *190*
Melocanna sp, *188*
Menstrual cycle, **11–12**, 59, 66, 67, 69, 155
Mentha spicata, *188*
Mercurialis perennis, *188*
Mercury, 16, **17**, *186*
Mesocricetus auratus, **68**, 69, *187, 190, 192*
Meteor showers, *186*
Metonic cycle, 176
Microtus agrestis, **111**
  arvalis, *190*
Midge, 59, 119, *191*
Migration, **75–9**, 85, **112**
  bird, 9, 24, 76, **77**, *187*
  fish, **78**, 79
  human, 77
  mammalian, 76, 77, **78**
  turtle, **79**
  see also Birds
Milankovitch, Milutin, 29
  theory of variable planetary climate, **29–30**
Milky Way, **16**
Millipede, 122, **124**
Mimosa pudica, **54**, **55**
Miniopterus australis, **70**
Mink, 67, *187*
Minnow, 72
Mint, *188*
Mirounga leonina, **75**
Mistletoe, *188*
Moa, 124
Modern Times (film), 165
Mole, **115**
Mollusc, 33, 80
Mongoose, 70
Monkey, rhesus, *187, 190*
Monsoon season, **25**, *186*
Moon, craters, 32
  eclipses, **32**
  gravitational pull, 30–31, 32, **33**
  influence on sexual cycles, 65

Moon (*cont.*)
phases of, **32–3**
see also Lunar cycles,
Tides
Morning glory, **57**
Mosquito, **92**, 111
Moth, 85
Mouse, 67, 113, *187,
190, 191, 192*
estrous cycle, 67
Muscari armeniacum, 188
Music, 160–2
Balinese, 161, 162
emotional effects of,
160
Musk rat, *187*
*Mus musculus, 187, 190,
192*
Mustela erminea, **48**, 49
vison, 187
Myotis myotis, 191
Myrmecocystus, **110**

N

Nelumbo nucifera, 188
Neoceratodus forsteri, 190
Neolithic revolution,
46–7
Nephoecetes niger, 187
Neptune, 16, *186*
Nerve impulses, 53
Nervous system,
autonomic, 11, 132–3,
132
Nettlerash, 141
Neuroterus, 87
Newt, 21, 85
Nicholson, A.J., **89**
Nicotiana tabacum, 45
Nigella damascena, 188
Nightingale, *187*
Nocturnal animals, 21, **113**
Nocturnal (timepiece),
175, **175**
North American
Indians, 75

O

Oak, sessile, *188*
Oat, 47, **87**
Octopus, 127
Onchocerciasis, 92
Oncorhynchus spp, 78, **79**
Ondatra zibethicus, 187
Opossum, 67, **115**, *191*
Virginia, *187*
Oregano, *188*
Orgasm, **69**
Origanum marjorana, 188
*Ornithorhyncus anatinus,
187, 192*
*Oryctolagus cuniculus,
187, 190, 192*
Osprey, *187*

Ostrich, **124**, *190, 192*
Ovis sp, *187, 190*
Ovulation, **66**, **67**, 69
Owl, **113**, *191*
snowy, 89
Oxygen, 17, 27, 28, 29,
100, **101**, 104–5, **106**,
107, 108, **109**, **138**

P

Palmer, Arnold, 145
Pandian haliaetus, 187
Pangaea, 28
Panthera tigris, 69
Pan troglodytes, 187, 192
Papio sp, *187*
Paramecium, **122**
Parasite, **84**, 85–6, 87,
89, **91**, **92**, 111
*Passer domesticus, 190,
191*
Peach, 43
Pear, 43
Pelican, *191*
brown, **90**
Penguin, 72
King, **72**
Performance levels,
human, **113**, **134**,
135, 136
Permocarboniferous, 28
Peromyscus leucopus, 190
Petastites hybridus, 188
*Petrochelidon albifrons,
187*
Petromyzon marinus, 187
Phalaeonoptilus nuttalli,
50, **51**
Pheromone, 68
*Phoca vitulina, 187, 190,
192*
Phocoena phocoena, 187
Photoperiodism, 40–1,
44, 45, **70**, *188*
Photosphere, 19, 24
Photosynthesis, 17, 20,
36, 42, 46, 49, **59**,
100, **101**, **104**, 105
Phoxinus phoxinus, 72
*Phyllostachys
bambusoides, 188*
*Phyllostomus hastatus,
191*
Pig, 67, *187, 190, 192*
Pigeon, **118**, **119**, *191*
Pike, *192*
Pine, bristlecone, 27, **99**
Pineal, 63, 71
Pinworm, 85
*Pipistrellus pipistrellus,
190*
subflarus, 192
Piranga ludoviciana, 187
olivacea, 69
Pituitary gland, **132**,
133, **134**, **135**, **136**,
137

Plague, 93
Plaice, **73**, *187*
Planets, 15, **16**, 17, 18,
149, *186, 192, 193*
Plankton, 112–3
Plants, alternation of
generations, **86**, 87
carbon cycles, 100–101
cells, **96–7**, **100**, 104
cultivated, **42**, **43**, **46**,
47, *189*
daily cycles, **54**, 56–7
flowering, **37**, **43**, **44–5**,
83, *188*
germination, **42**
leaf-fall **48–9**
seasonal growth cycle,
9, **23**, 35, **36–49**, 83
seed distribution, 46
transpiration, **104**
see also Photosynthesis
Plantus alle, 187
Plasmodium malariae, **92**
ovale, 92
vivax, 92
Platynereis, 60
Platypus, *187, 192*
Pleiades, **175**
Pleistocene, 25, 26, **27**,
28, 29, 35
Plesiosaur, **129**
Plethodon cinereus, **112**
Pleuronectes sp, *187*
Plover, crab, **58**
Pluto, 16, 17, *186*
Podiceps cristatus, 74–5
Poinsettia, **44**
Polaris, 21, 22, 30, **175**
Polar years, 39–40
Polar zones, 36, 37, **40**
Poles, geographic, **20**,
21, 22
magnetic, **19**, 20
Pole star *see* Polaris
Poliomyelitis, 93
Pollination, 44, 46
Polyp, 127
Poor-will nightjar, 50,
51, *189*
Populations, 83–93
Populus tremula, 188
Porcupine fish, 126
Porpoise, *187*
Povilla adusta, 81
Precambrian, 28
Premenstrual tension, 67
Primrose, 43, *188*
Primula vulgaris, 188
denticulata, 188
Procyon lotor, 187
Progesterone, **11**, **66**, 67
Protonaria citrea, **70**
Protopterus, **51**
Protozoa, 84, **92**
Psychedelic drugs, 183
Puffinus tenuirostris, 70
Pulse rate, **11**, 108, **109**,
131, **133**, 134, **147**
Pyramid of the Sun,
Mexico City, **177**

Q

Quail, *187*
Quaternary, 25, 27
Quelea, 73
Quercus petraea, 188

R

Rabbit, 66, **67**, 76, 123,
124, 187, 190, 191, 192
Rabies, 83, 93
Raccoon, 70, *187*
Radiocarbon dating, 27
Rainfall, 23, 24, **25**, 37,
38, **39**, 45, 51, *186*
effects on breeding, 73
see also Monsoon
Rain forest, 39
Ramadan, **155**
Rana pipiens, **84**, *190,
192*
Rangifer tarandus, 76, 77
Rat, 67, 71, 111, *187,
190, 191, 192*
kangaroo, **120**
white, *190*
Rattlesnake, **127**, *187*
*Rattus norvegicus, 187,
190, 192*
Ray, manta, **129**
Religious rituals, 162–3
African, **162**
Islamic, **162**
Reproduction, 64–81,
83–93
in amphibians 73
in birds, 69, 70, 71
in fish, 73, 76–81
in mammals, 65, **66**,
67, **68**, 70, 71, 73
see also Breeding
seasons, Migration
Respiration, 10, 103–9,
105–7, 114–5, **138–9**,
190
Rheumatoid arthritis,
136, 138
Rhinoceros unicornis, 192
*Rhinolophus hipposideros,
191*
Rhizome, 43
Rhopalosiphum padi, 87
Ribonucleic acid (RNA),
98
Rice, **46**, 47
*Richmondena cardinalis,
190*
River blindness, 92
RNA *see* Ribonucleic
acid
Robin, American, *190*
European, 73
Rogers, Ginger, 161
Root crops, 83
Rosaceae, 43

Roundworm, 92
Rowan, *188*
Rubicolor minor, 187
Rubus fruticosus, 188
Running, **122–3**, **124–5**,
144
Rye, 47
Rye grass, **44**

S

Sage, *188*
Salamander, 85
red-backed, **112**
Salisbury cathedral,
turret clock, 178, **179**
Salmon, 78, **79**, *187,
192*
Salmonella, 93
Salmo salar, 79, 187, 192
trutta, 192
Salvia officinalis, 188
Sandgrouse, 76
Saturn, 16, *186*
Savanna, 24, **38**, **39**
Scaphiopus bombifrons,
73
Schistosoma, 85, 87
Schwabe, Heinrich, 150
Scomber scombrus, 192
Sea gooseberry, **122**
Seahorse, **126–7**
Seal, *187, 192*
elephant, **75**, 95
Seashore, **30**, **31**, 57,
58–9
Seasnake, 126
Sea urchin, 80, **81**
Seaweed, 30, **59**
Sedum spectabile, 188
Serinus canarius, 190
Sexual cycles *see*
Breeding seasons
Sexual display, **68–9**
Shadow clock *see*
Sundial
Shearwater, 70
Sheep, 67, 70, 71, *187*
Shelduck, 70, *187*
*Shepheard's Kalendar,
The,* **36**
Shepherd's purse, *188*
Shift work, 136, **139**,
166–7
Shortt, William H., 181
Shrew, 70, *187, 190,
192*
Sidereal day, 21, 176
Sidewinder, **127**
Silkmoth, **62**
Silver birch, *188*
Sirius, **176**
Sleep, animal, 114, **115**
human, 9, 11, **12**, 13,
55, 113, **114–5**, **134**,
135, 136, **139**, 155, *191*
Snake, 113, **127**
moccasin, *187*

Snapdragon, *188*
Snowdrop, **37**, **43**, *188*
Social rhythms, 12–13, 113, 134–5, 144–5, 154–69, 171–83
Solar day, 21, 171–7, 179
see also Circadian rhythms
Solar flare, **151**
Solar system, 15–33
Solar year, 171, 176–7, **183**
Sole, **57**
Somateria mollissima, *187*
Sorbus acuparia, *188*
Sorex aranaeus, *187*
 cinereus, *190*, *192*
 palustris, *192*
Sosigenes, 177
Sparrow, *190*, *191*
 Californian white-crowned, 70
Spawning, 76, 78, **80–1**
Speech rhythms, 159, 160
Spider, **122**, *124*
Spinach, **44**
Springhaas, 121–2
Springtail, 120
Squirrel, 70, *191*, *192*
 flying **56**, *190*
 ground, **50**, **66**, **67**, 71
Starling, 74, *190*, *192*
Stars, 16, **21**, 22, 30, **175**, 176
Sterna fuscata, **71**
 paradisaea, **77**
Steroid drugs, 134, 136, **137**
Stickleback, **68**, 69
Stoat, 123
 ermine, **48**, **49**
Stonehenge, **177**
Stork, *191*
Stratosphere, 16, 17
Stress, 115, 134–5, 136–7, 155
Struthio camelus, *190*, *192*
Sturgeon, *192*
Sturnus vulgaris, 74, *190*, *192*
Sufi, **161**
Sugar cane, 47
Summer solstice, 22, **25**, **44**, 176, **177**
Sun, 16, 18, **19**, **22**
 energy from, **16**, 18–19, **20**, 21, 23, 24–5, **28**, **100**, **101**, 103, **104**
Sundial, 174, **175**
Sunflower, 45
Sun sign, **148–9**
Sunspots, **18**, **19**, 150–1, *193*
Sus scrofa, *187*, *190*, *192*

Swallow, barn, *187*
 European, 77, *187*
 Northern cliff, *187*
Swan, mute, *187*
Sweet violet, *188*
Swift, *191*, *192*
 Northern black, *187*
Swimming, **126–7**
 see also Fish
Swoboda, Hermann, 144
Sympterum, **84**

T

Tabanid, 92
Tachyglossus aculeatus, *190*, *192*
Tadorna tadorna, 70, *187*
Tadpole, 73, **85**
Tamias striatus, *187*, *190*, *192*
Tanager, scarlet, **69**
 Western, *187*
Taoists, **173**
Tarsier, 68
Tasmania, introduction of sheep, **88**
Teal, *187*
Teltscher, Alfred, 144
Temperate regions, 20, 21–5, **36–51**, 72–3
Temperature, environmental 20, **21**, **29**, 36, **37**, 38–51, 72–3, 112–3, 180
Termite, 86
Tern, Arctic, 76, **77**
 sooty, **71**
Terrapene carolina, *187*
Testosterone, 73
Testudo hermanni, **98**
Thamnocalamus spathaceus, **45**, *188*
Theropithecus gelada, **68**
Theta waves, 146
Thrip, **119**
Thrush, 70
Thunnus albacares, *187*
Thylacinus cynocephalus, 121
Thylacole carniflex, 121
Thyroid gland, **136**
Tides, 30, 31, 32, **33**, 55, 58–9, **80–1**
 neap, 31, **33**, **59**, 60, 65
 spring, 31, **33**, **59**
Tiger, **69**
Tilia cordata, *188*
Time, 171–83
 cyclical, 172–3
 nautical, **179**
 perception, 182, **183**
Timekeeping, **178–81**
 ancient practices, 174, 175
Tiv people, Nigeria, 154, 158, 159

Toad, 21, 73, 85
 American, *187*
 clawed, *187*
 common, 76
 spadefoot, **73**
Tobacco plant, 45
Tomato, 45
Tortoise, **98**
 giant, *124*
Touch-me-not balsam, *188*
Trade winds, **24**
Trance, 163
Transplant surgery, 138, **141**
Trees, 9
 deciduous, **40**, 43, **48**, 50
 disease, **91**
 growth rings, **99**
 leaf-fall, **48**, 50
 ring-dating, 27, **99**
Trichechus sp, *190*
Troglodytes aëdon, *190*
Trogoderma granarium, 51
Tropical rain forest, 24, 38, 39
Tropic of Cancer, **22**, 38
Tropic of Capricorn, 22, 38
Troposphere, 16, **17**
Trout, **126**
 salmon, *192*
Tsavo National Park, **38**, 39
Tuber, 43
Tulip, 43
Tuna, **126**
 yellow-fin, *187*
Turdus ericetorum, 70
 merula, **74**
 migratorius, *190*
Turkey, *190*
Tursiops truncatus, *190*
Turtle, 78, **115**, *128*
 Eastern box, *187*
 green, **79**
 loggerhead, **129**
Tuttle comet, *186*
Tympanuchus cupidio, **74**
Tyrannus tyrannus, *187*

U

Uca sp, **58**
Ulex europaeus, *188*
Uranus, 16, *186*
Urine flow, **10**, **11**, **13**, *189*
Ursus arctos, *192*
Urticaria *see* Nettlerash
Utah Valley, **49**

V

Vancouver, **37**
Venus, 16, 17, *186*

Vine, **47**
Viola odorata, *188*
Visconti, Palace of the, 178
Viscum album, *188*
Vitis vinifera, **47**
Vole, **89**, *190*
 short-tailed, **111**

W

Walking, **123**, 124–5
Wallaroo, *187*
Wapiti, 60
Warbler, black-polled, *187*
 European garden, **61**
 prothonotary, 70
 willow, **61**
Wasp, 87
Watch, 180, **181**
Water boatman, 85, **128**
Water clock, 174, **175**
Waxwing, 76
Weasel, 70
Weather, 24–7, 30, **39**, 40, 151
 effects on reproductive cycles, 70, 72, **73**
Whale, 68, **78**, **107**, *190*
 blue, *190*
 fin, **106**, *187*, *192*
 humpback, 78
 sperm, **106**
Whooping cough, **93**
Wigeon, *187*
Wildebeest, 76
Wind, **24**, **25**
 solar, 151
 trade, **39**
Winter solstice, 22, **25**, **44**, 174
Wisteria sinensis, *188*
Wolf, Tasmanian, 121
Woodchuck, 50
Woodcock, American, *187*
Working rhythms, 113, **134–5**, 144–5, **164**, **165**
Worm, 33, 80
 palolo, 59, 60, 80
Wren, house, *190*
Wucheria bancrofti, 93, **111**

X

Xenopus laevis, *187*

Y

Yeasts, 84
Yin and yang, **146**, **147**, **173**, *193*

Yoga, 146, 162–3
Yosemite National Park, **27**

Z

Zea mays, **42**
Zebra, 76
Zech, Jacob, **179**
Zeiraphera griseana, 90, **91**
Zenaidura macroura, *190*
Zen Buddhism, **146**
Zodiac, signs of **148**, 149, *192*
Zonotrichia leucophrys, **70**

ACKNOWLEDGEMENTS

The authors contributed text as follows:
Dr Philip Whitfield   8–13, 14–33, 34–51, 94–101, 102–115, 185
Dr John Brady   52–63
Dr D.M. Stoddart   64–81, 116–129
Dr Bryan Turner   82–93
Dr Martin Hetzel   130–141
Kendrick Frazier   142–151
Paul Bohannan   152–169
Chris Morgan   170–183

The Publishers received invaluable help during the preparation of **The Rhythms of Life** from:
Mary Corcoran, picture researcher; Libby Wilson, editorial assistant; Ann Kramer, who compiled the index; Robert De Filipps of the Office of Biological Conservation, Smithsonian Institution; Barbara Anderson; Candy Lee; Nigel O'Gorman

Typesetting by Servis Filmsetting Limited, Manchester
Origination by Adroit Photo Litho Limited, Birmingham

Throughout this book a billion = 1,000,000,000

The Publishers wish to thank the photographers and agencies listed below for their help in providing material for this book. The following abbreviations have been used: *t top; tr top right; tl top left; c centre; l left; r right; br bottom right; bl bottom left; b bottom*.
Cover:
1 Babout/Rapho; 2/3 Tony Stone Associates; 4/5 Courtesy of National Film Archive/Stills Library; 6/7 Stephen Dalton/Bruce Coleman (owls), A. Wetzel/ZEFA (moon), Spectrum Colour Library (cloud); 8/9 Camera Press; 10/11 John Garrett; 12 John Bigg; 13 Lorna Minton; 14/15 Bill Brooks/Masterfile; 16 Space Frontiers Ltd; 17 t Space Frontiers Ltd; 17 b Jet Propulsion Laboratory; 18/19 Space Frontiers Ltd; 20 l Space Frontiers Ltd; 20 r Dr C.T. Scrutton, University of Newcastle upon Tyne/*New Scientist*; 20/21 Lick Observatory Photograph; 22 Michael Boys/Susan Griggs Agency; 24 Space Frontiers Ltd; 25 Santosh Basak/Frank Spooner Pictures; 26 Aerofilms Ltd; 27 l Oxford Scientific Films; 27 r John Cleare; 30/31 Horst Munzig/Susan Griggs Agency; 32/33 Lick Observatory Photographs; 34/35 Denis Waugh; 36 Ann Ronan Picture Library; 37 Adam Woolfitt/Susan Griggs Agency; 38 Heather Angel; 39 P.H. & S.L. Ward/Natural Science Photos; 41 Ardea London; 45 The Photographic Library of Australia; 46 Mireille Vautier; 47 H.W. Silvester/Rapho; 48 Roland & Sabrina Michaud/John Hillelson Agency; 49 Dr J.A.L. Cooke/Oxford Scientific Films; 52/53 David Thompson/Oxford Scientific Films; 56 Stouffer Productions/Bruce Coleman; 57 Wayne Lankinen/Bruce Coleman; 60 Charlie Ott/Bruce Coleman; 62 The Mansell Collection; 64/65 R. Kruschel/Okapia; 68 Varin-Visage/Jacana; 69 Leonard Lee Rue III/Bruce Coleman; 71 Valerie Taylor/Ardea London; 72 Francisco Erize/Bruce Coleman; 73 l Carol Hughes/Bruce Coleman; 73 r Joe McDonald/Oxford Scientific Films; 74 P. Jones-Griffiths/John Hillelson Agency; 75 Massart/Jacana; 76 l P. Bading/ZEFA; 76 r David & Katie Urry/Ardea London; 77 Bryan & Cherry Alexander; 78 Jane Burton/Bruce Coleman; 80 Ted Spiegel/John Hillelson Agency; 82/83 Tadanori Saito/Rex Features; 85 Alan Root/Bruce Coleman; 86 Nigel O'Gorman; 90 Marion Morrison; 91 Swiss National Tourist Office; 93 Mike Abrahams/Network; 94/95 David Thompson/Oxford Scientific Films; 96 Carolina Biological Supply Co/Oxford Scientific Films; 97 E.H. Cook/Science Photo Library; 98 Zig Leszczynski/Oxford Scientific Films; 102/103 Adam Woolfitt/Susan Griggs Agency; 104 F.G. Bass/Natural Science Photos; 106 Ken Balcomb/Bruce Coleman; 107 David Redfern; 109 Oxford Scientific Films; 110 Tibor Hirsch/Susan Griggs Agency; 111 Wayne Lankinen/Bruce Coleman; 113 Spectrum Colour Library; 114 Edward Hausner/NYT Pictures/John Hillelson Agency; 116/117 Pete Turner/The Image Bank; 119 Stephen Dalton/NHPA; 120 Stephen Dalton/Oxford Scientific Films; 121 Jean-Paul Ferrero/Ardea London; 123 The Royal Photographic Society; 124 l Philippe Varin/Jacana; 124 r Root/Okapia; 125 Leo Mason; 127 David Hughes/Bruce Coleman; 128 l Dr Giuseppe Mazza; 128 tr Heather Angel; 128 br Colorsport; 129 Dr Giuseppe Mazza; 130/131 Howard Sochurek/John Hillelson Agency; 133 John Watney; 141 L.L.T. Rhodes/Daily Telegraph Colour Library; 142/143 Michael Friedel/Woodfin Camp & Associates/Susan Griggs Agency; 144 r Kobal Collection; 145 Sporting Pictures (UK) Ltd; 146 Alan Hutchison Library; 147 l Bruno Barbey/John Hillelson Agency; 147 r Roland & Sabrina Michaud/John Hillelson Agency; 150 Bruno Barbey/John Hillelson Agency; 151 Spectrum Colour Library; 152/153 Anthony Crickmay; 154 Phelps/Rapho; 155 l Marc Tulane/Rapho; 150 r Sergio Larrain/John Hillelson Agency; 156 Patrick Thurston; 157 l Clive Barda; 157 r Abbas/Frank Spooner Pictures; 158 l John Bulmer; 158 r Richard & Sally Greenhill; 159 Mike Abrahams/Network; 160/161 Peter Carmichael/Aspect Picture Library; 161 Kobal Collection; 162 l Tony Carr/Colorific!; 162 r Rabout/Rapho; 163 Marc Riboud/John Hillelson Agency; 164/165 Tadanori Saito/*The Sunday Times*, London; 166 John Sturrock/Report, London; 167 Associated Press; 168 l Juliet Highet/Alan Hutchison Library; 168 r Spectrum Colour Library; 169 George Hall/Susan Griggs Agency; 170/171 The Mansell Collection; 172 l Victoria & Albert Museum; 172 r Michael Holford Library/Musée Guimet, Paris; 173 Scala/Vision International; 174 Octopus/California Institute of Technology & Carnegie Institution of Washington; 175 l Ann Ronan Picture Library; 175 r Angelo Hornak; 176 Michael Holford/Courtesy of the Trustees of the British Museum; 177 tl Werner Forman Archive/National Museum of Anthropology, Mexico; 177 bl Giraudon; 177 r Aerofilms Ltd; 178 The Mansell Collection; 179 l *Salisbury Times & Journal* Co Ltd; 179 c Victoria & Albert Museum; 179 r Ann Ronan Picture Library; 180 Victoria & Albert Museum; 181 l Victoria & Albert Museum; 181 c Victoria & Albert Museum; 181 r Paul Brierley; 182 Alain Keler/Sygma/John Hillelson Agency; 183 Peter Newark's Western Americana; 184/185 Science Photo Library